Doing Calculus

with

Scientific Notebook™

Doing Calculus

with

Scientific Notebook™

Darel W. Hardy
Colorado State University

Carol L. Walker
New Mexico State University

TCI Software Research
Las Cruces, New Mexico

Brooks/Cole Publishing Company
Pacific Grove, California

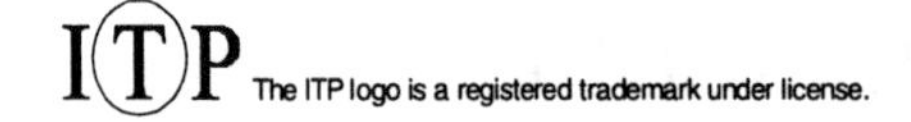

Brooks/Cole Publishing Company
A Division of International Thomson Publishing Inc.

Printed in the United States of America
10 9 8 7 6 5 4 3 2 1

This document was produced with *Scientific WorkPlace*.

Sponsoring Editor: *Robert Evans*
Editorial Associate: *Tami McBroom*
Production Coordinator: *Marlene Thom*
Manuscript Editor: *Carol Dondrea*
Cover Design: *Vernon T. Boes*
Compositor: *TCI Software Research*
Printing and Binding: *Malloy Lithographing, Inc.*

Dedicated to
Chanel, Michelle,
Sarah, Kyle,
Shane, Dylan,
Brendan
and their future siblings
and cousins

To the Instructor

Scientific Notebook provides a ready laboratory in which students can experiment with mathematics to develop new insights and solve interesting problems, and it provides a vehicle for students to produce clear, well-written homework. The essential components of the *Scientific Notebook* interface are free-form editing and natural mathematical notation. The system makes sense of as many different forms as possible, not requiring the user to adhere to a rigid syntax or to just one way of writing an expression. By providing an interface with little or no learning cost, *Scientific Notebook* makes symbolic computation as accessible as any word processor. *Scientific Notebook* is designed to fit the needs of a wide range of users, from the beginning student trying to solve a linear equation to the professional scientist who wants to produce typeset-quality documents with embedded advanced mathematical calculations. It provides a free-form interface to a computer algebra system that is integrated with a scientific word processor. It accepts mathematical formulas and equations entered in natural notation, and its symbolic computation system produces mathematical output inside the document that is formatted in natural notation, can be edited, and can be used directly as input to subsequent mathematical calculations.

Purpose of These Activities

Traditional problems often lose their impact in a technological environment. A new type of exercise is necessary that emphasizes conceptual thinking and general problem-solving strategies rather than computational algorithms, and that emphasizes general techniques for testing validity of answers. The activities *are* designed as a computer-laboratory supplement for calculus, providing problems that encourage experimentation, solution, verification, and communication. Many of the problems are stated in a format that requires a four-step process, and examples are given that illustrate this process.

How to Use the Activities

Each activity includes an introductory problem with a complete solution. These introductory activities probe understanding of basic concepts and illustrate the use of *Scientific Notebook* to solve these problems. Solutions are written as a model for student solutions, providing examples of the form they should use. Hints for using *Scientific Notebook* are interspersed throughout the solution, describing the mechanics of how to use *Scientific Notebook* to solve the introductory activity.

Following the introductory activities are one or more activities consisting of open-ended problems that encourage students to experiment, to generate plausible solutions, to verify that their plausible solutions are consistent with their experiments and known partial results, and to describe and justify their solutions coherently. In support of the emphasis on general techniques for self-testing for validity of answers, formal solutions are not included for these activities, although hints and clues are provided for many of them.

You can select activities that enhance your syllabus for student assignments, as many or as few as you wish and in whatever order you choose. We suggest, however, that you include all of the activities in Chapter 0 near the beginning of your course to get the students started with *Scientific Notebook* and to familiarize them with our notational conventions for *Scientific Notebook* hints. It is assumed that students will have a textbook that they can refer to for basic definitions and examples. A few definitions are provided for topics that may not be included in some textbooks.

The activities in Chapter 0 introduce basic procedures for using the system with examples from precalculus. In later chapters, we introduce additional procedures for using the system as they are needed, and provide examples and exercises to encourage users to practice the ideas presented and to explore possibilities beyond those covered in this manual.

References

Further guidance on using the mathematical features or the document-editing features of *Scientific Notebook*—entering and manipulating mathematical formulas, exploring with interactive graphics, and doing advanced operations with functions—is available on line under the Help menu.

Scientific WorkPlace

Doing Calculus with Scientific Notebook can also be used successfully with *Scientific WorkPlace*. Most of the hints and techniques are exactly the same. Because *Scientific WorkPlace* 2.5 is a front end to Maple V Release 3, whereas *Scientific Notebook* 3.0 is a front end to Maple V Release 4, there are some minor differences in the computational aspects of the two products. One difference you may notice from what is described in this book is that piecewise functions you define will not have full functionality in *Scientific WorkPlace* 2.5. The students can work around this by working with the separate pieces of the functions when differentiating or integrating.

Acknowledgments

Scientific Notebook grew out of the farsighted vision of Roger Hunter, who insisted that there had to be an easier way and actually found one. We thank him for the inspiration he has provided. We appreciate the work of all the referees who read the manuscript and made many helpful comments: David Arnold, College of the Redwoods; Maurino P. Bautista, Rochester Institute of Technology; Elias Y. Deeba, University of Houston–Downtown; John Gosselin, University of Georgia; Jonathan Lewin, Kennesaw State University; Aaron Meyerowitz, Florida Atlantic University; and Donna Molinek, Davidson College. We thank all the people at TCI Software Research who have patiently answered our many questions. We thank the staff at Brooks/Cole for all their help and advice, particularly Bob Evans, Marlene Thom, and Carol Dondrea. We thank David Walker for helpful comments on the manuscript. And we thank Linda and Elbert for their patience, support, and encouragement throughout this project.

Darel W. Hardy
Carol L. Walker

To the Student

Traditionally, mathematicians, engineers, and scientists do mathematics in several steps. They begin by performing algebraic, geometrical, numerical, or physical experiments to observe patterns. They use their results to solve related problems and compare the results with known solutions or verifiable special cases. They describe their methodology in words that justify and explain their solutions. *Scientific Notebook* adds the utility of a computer algebra system and a word-processing system to support this general approach to mathematics. The mathematical skills you learn with the help of *Scientific Notebook* will carry over to mathematical problem solving in future settings, whether or not you have access at that time to *Scientific Notebook*, because you will be learning to do mathematics in natural mathematical notation using general problem-solving methods.

Using Scientific Notebook as a Problem-Solving Tool

When you have access to *Scientific Notebook* or any other advanced computational tool, your instructor will expect more than simple answers. Experimentation, solution, verification, and communication are all crucial components in mathematical problem solving. You will find mathematics to be useful only to the extent that you understand the mathematics. If steps leading to your solution seem mysterious and magical, your solution is flawed and your answer may or may not be correct. Expect to understand what you are doing; read, experiment, and ask questions when there is too much magic involved. A computer algebra system can perform the algebraic and numerical calculations involved in a solution, but the logic behind the solution is in your hands.

Scientific Notebook does not change the way you do mathematics. It acts as an amplifier, letting you do more mathematics and helping you to enjoy mathematics; but it can also amplify mistakes. Take responsibility for your answers. We provide tools to help you do this—plotting tools and other techniques that allow you to test your own work and to validate your answers. The habit of always testing your own work and validating your answers will help you develop confidence in your ability to do mathematics and will prove useful when you work in real-life situations that do not come with built-in answer keys.

Standard Mathematical Notation

Modern mathematical notation has been developed specifically to simplify the communication of complicated ideas. As you become more familiar with the symbolism in mathematics, you will become comfortable with it and appreciate its ability to express ideas succinctly. It is important to be able to express these ideas in words; however, you will find that complicated ideas expressed entirely in words can be much more difficult to comprehend than the same ideas presented symbolically. With *Scientific Notebook*, you have the advantage of using standard mathematical notation directly, rather than being concerned with a specialized syntax specific to the system.

Learning Scientific Notebook

We believe you will adapt quickly to the use of *Scientific Notebook*, even though the environment may be somewhat different from that of your prior experience with technology. *Scientific Notebook* provides a logical and efficient method for entering mathematical expressions and extensive menus that minimize the number of things you need to learn and remember in order to be a productive user. As with any new system, you will need to invest some time and energy at the beginning to become a comfortable user of *Scientific Notebook*. Doing calculus with *Scientific Notebook* through these activities will help you learn the technicalities of *Scientific Notebook* as the need arises. Technical features of *Scientific Notebook* are explained in the on-line Help for convenient reference.

About the Activities
Each activity is divided into the following parts.

Introductory Activity

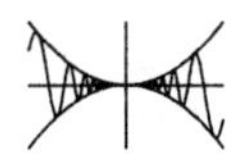

This activity introduces the mathematical concepts developed in the section.

Solution (with Hints for using Scientific Notebook)

These solutions are more than a series of formulas—they include words of explanation and follow general rules of good writing. The solutions are intended as preparation for the solutions of the ensuing activities, as well as models for the write-up of your solutions of the activities.

The hints step through the Scientific Notebook techniques required for a solution of the introductory activity as well as the ensuing activities.

Activities

The activities are open-ended problems that extend the mathematical concepts of the introductory activity and introduce a variety of applications. These are your laboratory/homework assignments. In support of the emphasis on general techniques for self-testing of validity of answers, formal solutions are not given for the open-ended activities, although hints and clues are provided for many of them. You are encouraged to work through all the details of the solution to the introductory activity before attempting the open-ended activity, both as an introduction to the mathematics and to the *Scientific Notebook* techniques. The solutions also provide a variety of models for writing solutions.

We encourage you to experiment, to generate plausible solutions, to verify that your plausible solutions are consistent with your experiments and known partial results, and to describe and justify your solutions coherently. The solutions to the introductory activities provide models in each of these sections to illustrate this process.

Contents

Chapter 0: Introduction

0.1 Introduction to Scientific Notebook 1
Entering text and mathematics: Scientific Notebook conventions.

0.2 Functions and Their Graphs 5
Entering and plotting polynomials and rational functions; defining functions and expressions.
★ In Search of a Solution
Solving equations and systems of equations; finding zeros of functions on specified subintervals.
★ Is Algebra the Clue?
Finding equations for polynomial functions of degree n, given $n+1$ points on the graph.
★ Going in Circles
Finding equations for circles through three specified points.

0.3 Types of Functions: Shifting and Scaling 21
★ Shifting and Scaling
Visualizing simple transformations and translations of parabolas and cubics.

0.4 Graphing Calculators and Computers 25
Comments on floating-point arithmetic, round-off errors, truncation errors, and extended precision.

0.5 Principles of Mathematical Writing 26
Dos and don'ts for writing mathematics.

Chapter 1: Limits and Rates of Change

1.1 The Tangent and Velocity Problems 29
★ The Slope of the Slope
Finding equations for tangent lines; using piecewise-defined functions; using tangent lines as local approximations.

1.2 The Limit of a Function 32
★ That's the Limit (Not!)
Exploring limits through graphs.

1.3 Calculating Limits Using the Limit Laws 36
★ Caught in the Middle
Calculating limits using the limit laws and the squeeze theorem.

1.4 The Precise Definition of a Limit 38
★ Viewing Rectangles for the Epsilon-Delta Game
A visual interpretation of the limit definition.

1.5 Continuity 41

★ Through the Looking Glass

A mental exploration in the world of discontinuities.

★ Dan Took a Bath on This One

Plotting a discontinuous function; examining the intermediate value theorem.

1.6 Limits at Infinity; Horizontal Asymptotes 45

★ Rabbit Island

A logistic growth model, limits at infinity, and horizontal and vertical asymptotes.

1.7 Tangents, Velocities, and Other Rates of Change 47

★ Give Me a Brake

Mathematical models for stopping distance and for the motion of a piston; estimating and computing rates of change.

Chapter 2: Derivatives

2.1 Derivatives 51

★ The Shape of the Derivative

Finding derivatives with Scientific Notebook; relating the shapes of polynomial functions and the shapes of their derivatives.

★ A Double Touch

Finding equations for tangent lines through specified points; plotting functions and tangent lines with special properties.

2.2 Differentiation Formulas 57

★ Basic Differentiation Rules

Defining and differentiating generic functions in Scientific Notebook.

2.3 Rates of Change in the Natural and Social Sciences 58

★ Making the Grade

Finding equations for polynomial functions, given specified function values and slopes of tangent lines at particular points.

2.4 Derivatives of Trigonometric Functions 62

★ Intriguing Derivatives

Graphical approach to differentiating trigonometric functions.

2.5 The Chain Rule 64

★ The Power of the Rules

Using basic rules to differentiate arbitrary elementary functions.

2.6 Implicit Differentiation 66

★ The Plot Is Implicit

Working with implicit functions, their graphs, and their derivatives.

2.7 Higher Derivatives 70
★ Raising Derivatives to New Heights
★ Splines Are Smooth
Finding second, third, and higher order derivatives with Scientific Notebook, and an application of derivatives.

2.8 Related Rates 74
★ The Fastest Ladder in the West
Exploring related rates of change.

2.9 Differentials; Linear and Quadratic Approximations 77
★ What a Difference a Differential Makes
Finding linear and quadratic approximations to functions.

2.10 Newton's Method 80
★ Newton Had a Method
★ Getting Off to a Good Start
A look at Newton's method for estimating zeros of functions.

Chapter 3: Inverse Functions

3.1 Exponential Functions and Their Derivatives 85
★ Getting into Exponential Shape
A First Look at Exponential Functions.

3.2 Inverse Functions 87
★ Speaking Inverse Does Not a Poet Make
Exploring Inverse Functions.

3.3 Logarithmic Functions 90
★ Getting Into Logarithmic Shape
Comparing graphs of logarithmic functions with different bases.

3.4 Derivatives of Logarithmic Functions 91
★ Filling in the Gap
Calculating the derivative of the natural logarithm.

3.5 Exponential Growth and Decay 94
★ Revisiting Rabbit Island
Solving logarithmic differential equations.

3.6 Inverse Trigonometric Functions 97
★ Inverse Sine and Cosine Functions
Using parametric plots to graph inverse trigonometric functions.

3.7 Hyperbolic Functions 99
★ Not Just Hyperbole
A look at hyperbolic functions, their inverses and derivatives.

3.8 Indeterminate Forms and l'Hospital's Rule 104
★ Lining Up l'Hospital's Rule
A graphical interpretation of l'Hospital's Rule.

Chapter 4: The Mean Value Theorem and Curve Sketching

4.1 Maximum and Minimum Values 109
★ Dealing with Extreme Conditions
Finding polynomials with given properties by solving systems of equations.

4.2 The Mean Value Theorem 111
★ The Mean Value Theorem Game
★ The Mean Value Theorem Speed Trap
Applications of the Mean Value Theorem.

4.3 Monotonic Functions and the First Derivative Test 114
★ Controlling the Ups and Downs
Finding polynomials with given properties by solving systems of equations.

4.4 Concavity and Points of Inflection 116
★ Understanding the Shape of a Function
★ More Control over the Shape of a Function
Finding polynomials with given properties; locating inflection points and intervals of concavity.

4.5 Curve Sketching 120
★ Local and Global Shapes
Finding polynomials with given properties by solving systems of equations.

4.6 Graphing with Calculus and Calculators 122
★ Getting Into Polynomial Shape
★ Getting Into Rational Shape
Finding graphs of polynomials and marking extreme values, inflection points, and other features.

4.7 Applied Maximum and Minimum Problems 130
★ Making a Big Box
★ More Braking Distance Models
Solving maximum and minimum problems.

4.8 Applications to Economics 136
★ Compound Interest Is Not Simple
An exponential model for compound interest.

4.9 Antiderivatives 139
★ The Sky Is the Limit
Exploring the flight of a baseball thrown vertically into the air.
★ All Antiderivatives Have the Same Shape
Properties of antiderivatives and their graphs.

Chapter 5: Integrals

5.1 Sigma Notation 143
★ Sums of Squares Are Cubic
Calculating sums using sigma notation.

5.2 Area 146
★ Area Equals Length Times Width is a Good Place to Start
Using rectangles to estimate the area under a curve.

5.3 The Definite Integral 148
★ Areas Are the Limit
Calculating areas as limits of sums.

5.4 The Fundamental Theorem of Calculus 151
★ It Is Fundamental
Making connections between differentiation and integration.

5.5 The Substitution Rule 154
★ When a Variable Goes Flat, Change It
Changing variables to evaluate integrals.

5.6 The Logarithm Defined as an Integral 158
★ No Exact Integral? Just Give It a Name
An indefinite integral defined to be the logarithm function.

Chapter 6: Applications of Integration

6.1 Areas Between Curves 161
★ Caught in the Middle, Again
Calculating areas of regions bounded by curves.

6.2 Volume 163
★ A Revolutionary Volume
Calculating volumes of solids using the disk method.

6.3 Volumes by Cylindrical Shells 166
★ Sometimes Peeling Works Better than Slicing
Calculating volumes of solids using the shell method.

6.4 Work 169
★ Exercise Is Work
Looking at work as an application of integration to physics.

6.5 Average Value of a Function 171
★ It All Averages Out
Calculating the average value of a function on an interval.

Chapter 7: Techniques of Integration

7.1 Integration by Parts 175
★ Integration by Differentiation
Using integration by parts to evaluate integrals.

7.2 Trigonometric Integrals 177
★ From Trig to Algebra
Using algebra to simplify integrals involving trig functions.

7.3 Trigonometric Substitution 180
★ From Algebra to Trig
Using trigonometry to rewrite integrals of ordinary functions.

7.4 Integration of Rational Functions by Partial Fractions 181
★ Writing Rational Functions as Sums
Working with partial fraction decompositions of rational functions.

7.5 Rationalizing Substitutions 185
★ Eliminate the Radicals!
Rationalizing substitutions to simplify integrals.

7.6 Strategy for Integration 188
The importance of memorizing basic formulas.

7.7 Using Tables of Integrals and Computer Algebra Systems 189
Strategies for solving integrals not automatically returned in reasonable form.

7.8 Approximate Integration 190
★ Streams and Virtual Streams
★ You Call This Normal?
Using numerical approximations to estimate definite integrals.

7.9 Improper Integrals 195
★ A Very Wild Function
Understanding convergence and divergence of improper integrals.

Chapter 8: Further Applications of Integration

8.1 Differential Equations 199
★ The Faulty Furnace
Applying differential equations to mixing problems.

8.2 Arc Length 201
★ The Length of a Hanging Cable
Calculating lengths of curves

8.3 Area of a Surface of Revolution 203
★ The Area of a Parabolic Reflector
Calculating areas of surfaces.

8.4 Moments and Centers of Mass 205
★ How to Balance a Sculpture
Calculating the center of mass of a solid object.

8.5 Hydrostatic Pressure and Force 207
★ How to Measure the Dam Force
Calculating the hydrostatic pressure on a dam.

8.6 Applications to Economics and Biology 209
★ The Pressure Builds
Looking at a mathematical model for fluid flow.

Chapter 9: Parametric Equations and Polar Coordinates

9.1 Curves Defined by Parametric Equations 213
★ The Spin Is In
Modeling the flight of a golf ball.

9.2 Tangents and Areas 216
★ The Bezier Controls the Shape
Using Bezier curves to create smooth shapes.

9.3 Arc Length and Surface Area 219
★ How Long is the Rope?
Using area to estimate arc length.

9.4 Polar Coordinates 221
★ You Too Can Be an Artist
Creating interesting and attractive shapes with polar coordinates.

9.5 Areas and Lengths in Polar Coordinates 223
★ Looking at the Rope from Another Angle
Using area to estimate arc length, using polar coordinates.

9.6 Conic Sections 226
★ Generating Random Conic Sections
Studying curves determined by quadratic equations.

9.7 Conic Sections in Polar Coordinates 229
★ Do Parabolas Exist in Nature?
Using polar coordinates to study conic sections.

Chapter 10: Infinite Sequences and Series

10.1 Sequences 233
★ Visualizing Sequences
Exploring the convergence of a sequence visually.

★ Finding a Better Way to Approximate e
Using continued fractions to approximate e.
★ Chaotic Sequences
Looking closely at some special sequences.

10.2 Series 241
★The Cereal Box Problem
Using an infinite series to calculate an expectation.

10.3 The Integral Test and Estimates of Sums 243
★ Connecting Improper Integrals With Series
Using integrals to determine series convergence.

10.4 The Comparison Tests 245
★ Prototype Series
Testing for convergence by comparing against known series.

10.5 Alternating Series 248
★ An Improper Alternative
Using an alternating series to show the convergence of an improper integral.

10.6 Absolute Convergence and the Ratio and Root Tests 250
★ You Control the Sum
Rearranging Conditionally Convergent Series.

10.7 Strategy for Testing Series 252
The importance of understanding strategies for testing series.

10.8 Power Series 252
★Will the Real Interval of Convergence Please Stand Up?
Determining the interval of convergence of a power series.

10.9 Representations of Functions as Power Series 255
★ Series Solutions: One Coefficient at a Time
Finding a power series representation of a function.

10.10 Taylor and Maclaurin Series 257
★ Taylor Made Polynomials
★ Visualizing Taylor Polynomials
Finding and viewing power series representations of functions.

10.11 The Binomial Series 261
★ The Binomial Expansion that Would Not End
Studying the case where a binomial expansion produces an infinite series.

10.12 Applications of Taylor Polynomials 264
★ Keeping a Lid on the Error
★ Approximations by Rational Functions
A close look at errors and approximations.

Index 269

0 INTRODUCTION

In this chapter we introduce *Scientific Notebook* with several sets of precalculus activities and talk about how to write mathematics.

0.1 Introduction to Scientific Notebook

To get started using *Scientific Notebook*, you can proceed in a straightforward fashion. Text entry is natural, with the emphasis on the content itself rather than on its placement. You use familiar mathematical notation to enter and display mathematics. You can think and write at the computer, in much the same way as you have used pencil and paper in the past. But you have the power of a computer algebra system at your fingertips to help you explore and experiment.

The Main Window

The first thing you see when you open the notebook is the *main window*. The largest part of the screen is the *entry area*, which contains the *insertion point*, indicating where *Scientific Notebook* will insert the next character you type. The top bar in the *Scientific Notebook* screen displays the name of the program and the current file. Below that is the Menu bar, listing menu titles:

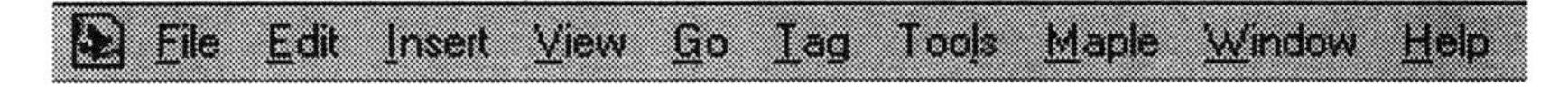

When you choose one of these titles, *Scientific Notebook* displays a pull-down menu listing the available commands.

We will indicate a sequence of choices, such as

- Choose the File menu; then from the File menu, choose the item New

by

- Choose File + New

or

- File + New

Scientific Notebook Toolbars

Beneath the Menu bar, or at the bottom or sides of your screen, may be several icon bars called *toolbars*. To see the names of the buttons that appear on the toolbars, let your mouse pointer pause over a button, and a *tooltip* containing the name of the button will appear on the screen. Two important toolbars for doing mathematics are the Math and Symbol toolbars.

The Math toolbar contains buttons for entering mathematical objects and buttons representing special tools.

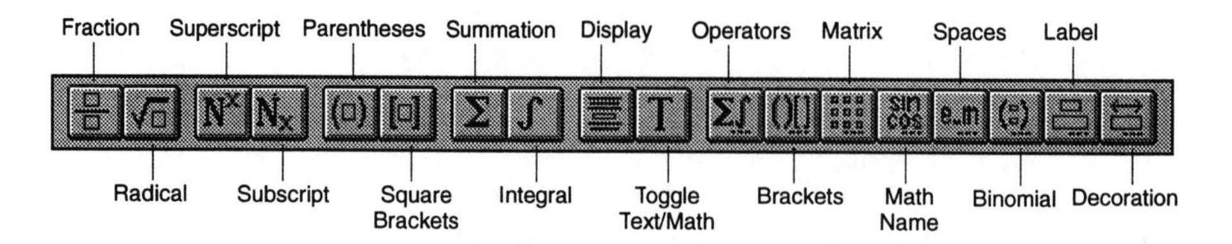

The Symbol toolbar contains buttons for entering symbols and Greek characters.

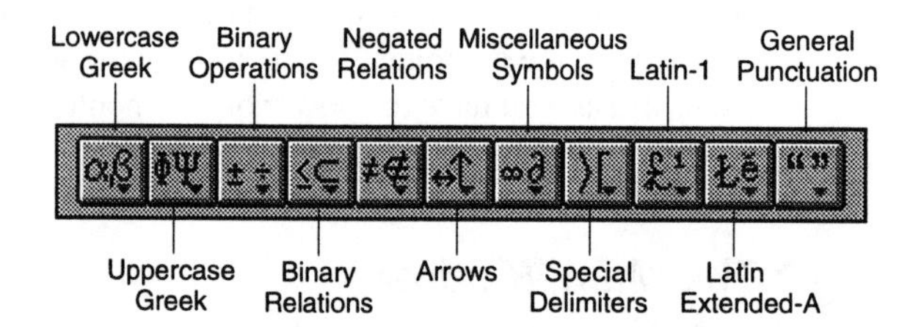

If these two toolbars do not appear on your screen, from the View menu, choose Toolbars and check Math and Symbol to bring them to your screen.

- The Compute toolbar is useful for doing mathematics, as it contains shortcuts for several of the most common mathematical operations on the Maple menu. The Stop toolbar is convenient also.

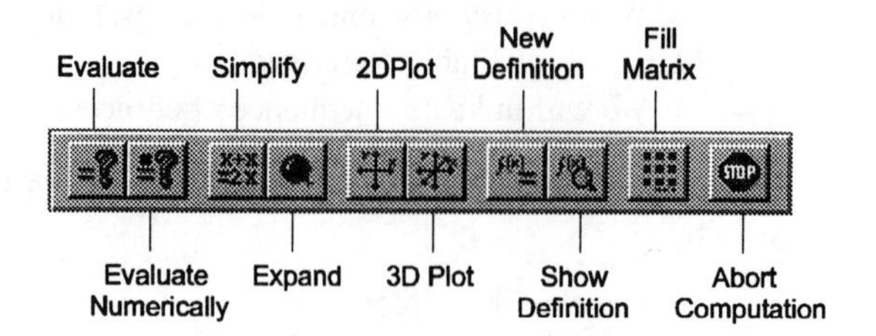

- The toolbars Common Symbols, Standard, and Field contain shortcuts for editing. You will find these very convenient if there is room for them on your screen. All of these commands are also available from other toolbars or from drop-down menus. The toolbars Tag and Fragments will help you add structure to your document as you become more familiar with *Scientific Notebook*.
- The Navigate toolbar helps you move around in a document. This toolbar allows you to jump to section headings and markers by choosing from drop-down lists, and to jump to the previous or next section.

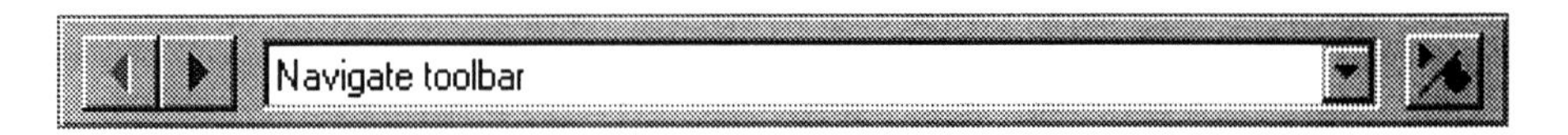

If this toolbar is not on your screen, look at Toolbars on the View menu, and check Navigate to bring it to your screen.

- The Links toolbar helps you move between linked documents, and to retrace your movements with History Back and History Forward.

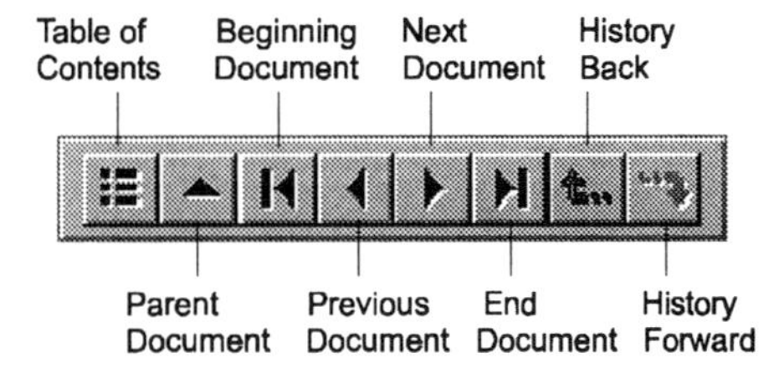

All of the navigating options (except the list of section headings) are also available on the Go menu.

Scientific Notebook Conventions

The *Scientific Notebook* screen presents commands in menus and toolbars. Many of these have associated dialog boxes, as indicated by the ellipsis (. . .) that follows certain menu names and buttons. Like other programs, *Scientific Notebook* responds to commands that you enter using the mouse or the keyboard. Also, *Scientific Notebook* adheres to standard keyboard conventions.

> **Note** We will generally give instructions in terms of only one option—most frequently choosing commands from menus because the configuration of toolbars on your screen may vary. You will find a list of keyboard shortcuts for editing commands under Keyboard Accelerators on the Help menu.

Mathematics and Text

Mathematics and text behave differently in *Scientific Notebook*, both in appearance and function. For example '2+2' and '$2+2$,' entered in text and mathematics, respectively, have a different spacing between characters. They also have a different screen color—in most cases text is black and mathematics is red—to help you interpret objects on the screen. The same structure that gives mathematics its screen color helps *Scientific Notebook* to identify mathematics for computation.

The Toggle Math/Text button [M] or [T] at the top of the screen displays the black [T] when the insertion point is in text and the red [M] when the insertion point is in mathematics. You can toggle between these two modes by placing the mouse pointer on the Toggle Math/Text button and clicking with the left mouse button. You can also toggle from the keyboard with CTRL + M or CTRL + T, or from the Insert menu by choosing Math or Text (or by typing ALT + I, M or ALT + I, T).

Terminology Conventions

We follow terminology conventions established in the *Scientific Notebook* documentation.

- *Choose* means to click on a selection listed in a menu or a toolbar, or shown on a button in a dialog box, to tell *Scientific Notebook* to carry out a command.
- *Check* means to turn on a check box option in a dialog box.
- *Click* means to position the mouse cursor, then press and immediately release the left mouse button without moving the mouse. For example, "Click the numerator."
- *Double-click* means to position the mouse cursor, and then click the left mouse button twice in rapid succession without moving the mouse. For example, "Double-click the Hint Box to see how to use *Scientific Notebook* to solve this problem."
- *Drag* means to position the mouse cursor, press and hold down the left mouse button while you move the mouse, and then release the button. For example, "Drag from the numerator to the denominator to *select* the fraction." (The "selection" will appear on the screen with light and dark reversed.) We may also refer to this operation as selecting "with the mouse."
- *Select* means to select with a drag the part of the document that you want your next action to affect, or to choose a specific option from a dialog box. Within mathematics, the word *select* also means to position the insertion point inside or to the right of a mathematics expression.

Copying, Moving, and Deleting Text and Mathematics

You can move, copy, or delete selections directly with a mouse technique called "drag and drop." This has the same net effect as the Windows or Macintosh "cut and paste" feature—the selection is moved (or copied) from its former position to the point where you released the mouse button.

► To move text with "drag and drop"

1. Select a portion of text or mathematics or both. You can do this with drag, as described above, or by holding down SHIFT and pressing an arrow key.
2. Without pressing any of the mouse buttons, position the mouse pointer within the selection.
3. Press and hold down the (left) mouse button (a scissors icon appears on the screen).
4. While holding the mouse button down, drag the mouse pointer to the position where you want to place the selection. The insertion point follows the mouse cursor as you do this, providing an accurate guide to where your selection will appear.

5. Release the mouse button.

▶ **To copy text with "drag and drop"**

- Follow the same steps used for moving text, except
  - (PC) press either the CTRL key before beginning step 3 or the right mouse button
  - (Macintosh) press the COMMAND key before beginning step 3

  and hold it down until the operation is completed (a copy icon appears on the screen). The net effect is that the selection remains in the original position, and a copy of it is inserted into the new position as well.

▶ **To delete text with "drag and drop"**

- Follow the same steps used for moving text, dragging the mouse pointer to the side of the *Scientific Notebook* window, outside the entry area, such as to the scroll bar to the right of the window. When you release the mouse button, *Scientific Notebook* deletes the text you selected.

▶ **To replace mathematics "in-place"**

- Select a mathematics expression with the mouse, press the CTRL/COMMAND key and, while holding it down, choose Evaluate (or another appropriate command). The expression you selected will be replaced by the result of the command.

The Help Menu

When you choose the Help menu, you will find an on-line help system that describes both the editing and computational features of *Scientific Notebook*. Throughout the activities, we will introduce new *Scientific Notebook* techniques as you need them. In the Help system, you will find alternative approaches and more detail. These are read-only documents. However, you can select paragraphs, copy to the clipboard, and paste into an active document to experiment with examples.

0.2 Functions and Their Graphs

These paragraphs offer experience in techniques for doing mathematics with *Scientific Notebook* that you will use in most of the activities. Additional hints will be given within each activity, as needed.

> Note We will describe *Scientific Notebook* commands in terms of the menu choices.

For keyboard and toolbar shortcuts for choosing commands, choose Help and select Index.

Entering Mathematics

▶ **To enter the polynomial** $3x^2 + x - 5$

1. Choose Insert + Math to toggle to mathematics mode. (On the Math toolbar, you will see the button M when in mathematics mode.)
2. Type $3x$ (This should appear in red on your screen.)
3. Choose Insert + Superscript. Type 2 in the input box.
4. After entering the superscript, press the SPACEBAR or use an arrow key to get out of the superscript and return to the line, and then type the rest of the expression: $+\ x - 5$.

> Note When you enter mathematics, *Scientific Notebook* handles the spacing for you. Pressing the spacebar while in mathematics may change the level of the insertion point, but it will not insert a space.

▶ **To enter the rational expression** $\dfrac{3x^2 + x - 5}{2x^2 - 7}$

1. Choose Insert + Fraction (This automatically places you in mathematics mode.)
2. Type the polynomial $3x^2 + x - 5$ in the upper input box.
3. Press TAB or an arrow key to move to the lower input box.
4. Type the polynomial $2x^2 - 7$ in the lower input box.
5. Press SPACEBAR to exit the fraction.

▶ **To enter the expression** $\sqrt{\dfrac{3x^2 + x - 5}{2x^2 - 7}}$

1. Select the expression $\dfrac{3x^2 + x - 5}{2x^2 - 7}$ with the mouse.
2. Choose Insert + Radical.

Plotting Expressions

▶ **To plot a polynomial**

1. Enter a polynomial expression (say, $3x^2 - 5x + 4$).

2. Leave the insertion point inside or to the right of the polynomial expression (or select the expression with the mouse).

3. Choose Maple + Plot 2D + Rectangular (or click the 2D Plot button on the compute toolbar).

The result should be: $3x^2 - 5x + 4$

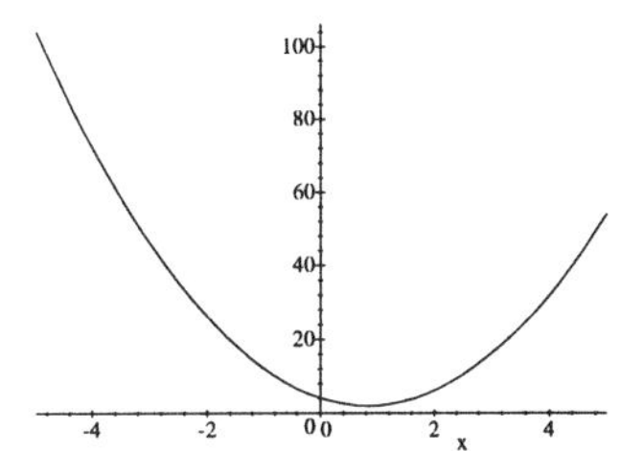

Reminder To apply a command to a mathematical expression, place the insertion point to the right of, or inside of, the expression (or select the expression with the mouse) and choose the command from the Maple menu or from a toolbar.

From now on we will often assume this to be understood, abbreviating the entire list of instructions in the previous example as

- Plot 2D + Rectangular $3x^2 - 5x + 4$

To be recognized as a mathematical expression, the contents of the expression must all be in mathematics (i.e., appear in red or gray on the screen) and not be separated by text spaces.

► To change settings in a plot

- Select the plot and choose Edit + Properties (or click the miniature dialog box icon that appears in the lower right-hand corner of the plot) to obtain a tabbed dialog. Make changes in the pages of the dialog and choose OK.

► To change plot settings in a 2D plot with the plotting tools

- Double-click the plot. Three plotting tools (*Pan*, *Zoom in*, and *Zoom out*) appear in the upper right corner of the plot. These allow you to change the view without changing the size of the frame.
    - *Pan*: With the hand selected, place the mouse cursor over the plot, and press and hold the mouse button. Move the mouse to pan left or right. This will change the domain coordinates of the plot setting.
    - *Zoom in*: With the large mountain selected, place the mouse cursor over a point in the plot, press and hold the mouse button and drag across and down. This will create a box and the contents of the box will expand to fill the frame.
    - *Zoom out*: With the small mountain selected, place the mouse cursor over a point

in the plot, press and hold the mouse button and drag across and down. This will create a box and the contents of the frame will shrink to fill the box, bringing in additional parts of the graph to fill the frame.

The number of degrees of freedom for the zoom boxes depends on the plot settings and the type of plot. For a rectangular plot, the default is one degree of freedom (the domain interval) with the range interval determined automatically by the underlying system. You can change this default on the Axis & View page of the Plot Properties dialog.

► **To plot the expression** $\dfrac{3x^2+x-5}{2x^2-7}$

- With the insertion point in the expression $\dfrac{3x^2+x-5}{2x^2-7}$, choose Maple + Plot 2D + Rectangular.

The result should be: $\dfrac{3x^2+x-5}{2x^2-7}$

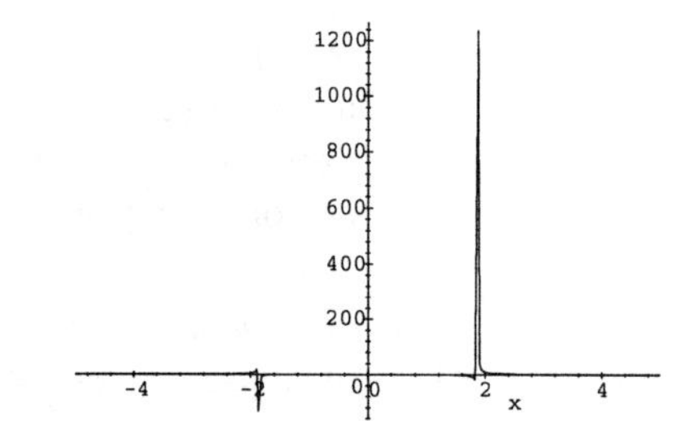

► **To get a better view of this plot**

1. Leave the insertion point to the right of the plot, and choose Edit + Properties (or click the frame and then click the dialog box tool that appears in the lower right corner of the frame).

2. Choose theView page, click the View Intervals Default box to remove the check, and change the settings on the y-interval to $-15 < y < 15$.

This gives the following view:

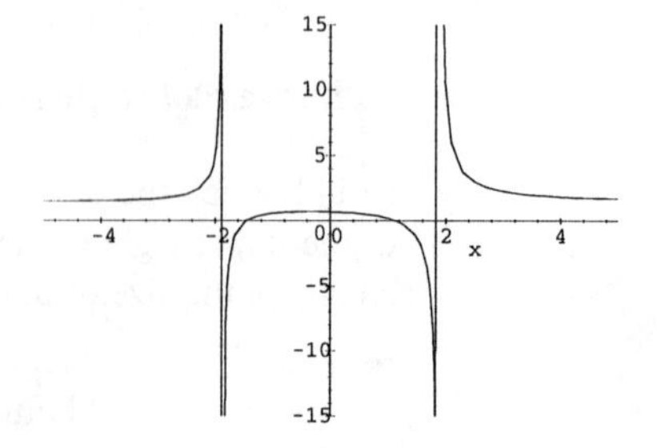

► **To add an expression to a plot**

- From the Plot Components page of the Plot Properties dialog, choose Add Item and type the expression in the Expressions and Relations box.

or

- Select the expression and drag it to the plot with the mouse.

▶ **To plot several expressions at once**

- Start with the expressions in a mathematical list. The expressions in the list must be separated by red commas, with no text spaces. Place the insertion point in the mathematics expression $x^2, 2x + 1, x^3 - 2x$ and choose Plot 2D + Rectangular.

The result should be: $x^2, 2x{+}1, x^3{-}2x$

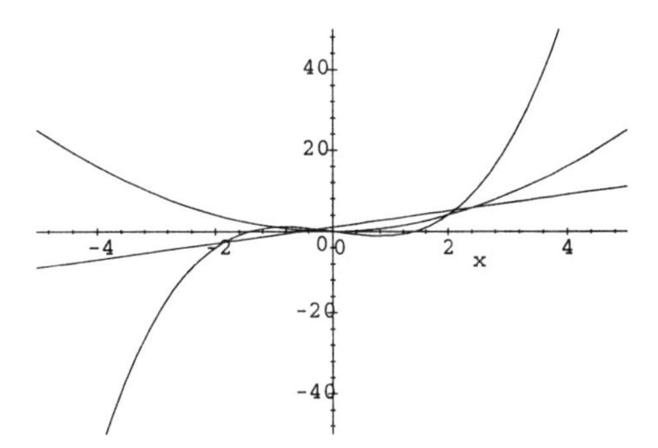

Defining Expressions and Functions

Some functions, such as $\sin x$, $\ln x$, and e^x are automatically recognized by *Scientific Notebook* when entered in mathematics mode. You can also create temporary user-defined expressions and functions.

▶ **To give the expression $ax + b$ the name z**

1. Place the insertion point in the equation $z = ax + b$ (or select the equation with the mouse).

2. From the Maple menu, choose Define + New Definition (or choose the New Definition button from the Compute toolbar).

▶ **To create the user-defined function $p(x) = ax + b$**

1. Place the insertion point in the equation $p(x) = ax + b$ (or select the equation with the mouse).

2. From the Maple menu, choose Define + New Definition.

▶ To evaluate a defined function

1. Place the insertion point in an expression of the form $p(3)$.
2. From the Maple menu, choose Evaluate.
3. If you have defined $p(x) = ax + b$, then the result should be $p(3) = 3a + b$.

▶ To define a generic function

1. Place the insertion point in the expression $f(x)$.
2. From the Maple menu, choose Define + New Definition.

The expression $f(x)$ will then be recognized as a mathematical function rather than simply a sequence of four characters. For example, evaluating $p(f(x))$ and $f(p(x))$ with p and f defined as above gives the results

$$p(f(x)) = af(x) + b \qquad \text{and} \qquad f(p(x)) = f(ax + b)$$

▶ To view the list of active user-defined expressions and functions

- Choose Maple + Define + Show Definitions (or click Show Definitions on the Compute toolbar).

▶ To remove all active user-defined definitions

- Choose Maple + Define + Clear Definitions.

▶ To remove one user-defined definition

- Select the function or expression and choose Maple + Define + Undefine.

or

- Make a new definition using the same function or expression name.

▶ To make a table of values

1. Select the function or expression and choose Maple + Define + New Definition.
2. Choose Insert + Matrix and specify 6 rows (or whatever number of values you want) and 1 column. Fill it in with x-values near 0. For this example, we use 0.1, 0.01,

0.001, −0.001, −0.01, −0.1. (As you type these in the matrix, you can press TAB to move between the entries.)

3. Select the matrix with the mouse and choose Insert + Brackets (or click a brackets button); then place the function name (in mathematics mode) just to the left of the brackets.

4. With the insertion point in the expression $f\begin{pmatrix} 0.1 \\ 0.01 \\ 0.001 \\ -0.001 \\ -0.01 \\ -0.1 \end{pmatrix}$, choose Evaluate.

▶ **To plot a defined function**

- Define + New Definition $f(x) = 3x^2 - 5x - 1$
- Plot 2D + Rectangular $f(x)$

or

- Plot 2D + Rectangular f

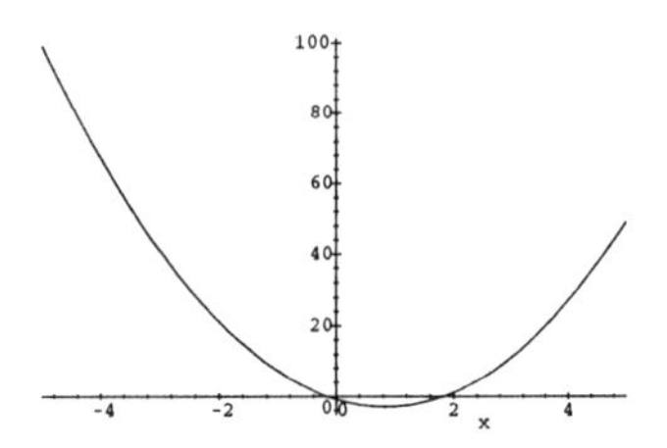

Note You can use either an expression, such as $f(x)$ or $3x^2 - 5x - 1$, or the function name f to obtain a plot of the function $f(x) = 3x^2 - 5x - 1$.

If you place the insertion point in an equation, such as $f(x) = 3x^2 - 5x - 1$ or $y = 3x^2 - 5x - 1$, and ask for a 2D rectangular plot, *Scientific Notebook* will select the expression that is on the same side of the equals sign as the insertion point, not the entire equation. *Scientific Notebook* selects equations only for special types of plots, such as implicit plots.

- Plot 2D + Rectangular (insertion point on the left-hand side)

$y = 3x^2 - 5x - 1$

- Plot 2D + Rectangular (insertion point on the right-hand side)

$y = 3x^2 - 5x - 1$

Searching for Solutions

There are several choices on the Solve submenu. Most situations call for Solve + Exact. Some special situations that we address below call for Solve + Numeric. The other choices will rarely be used in calculus.

Solving Equations Exactly

▶ **To solve an equation**

1. Place the insertion point in an equation, such as $2x - 3 = 7x$ (or select the equation with the mouse).

2. Choose Maple + Solve + Exact. (You will get the response "Solution is : $\{x = -\frac{3}{5}\}$.")

> **Hint** To make definitions, it is often useful to select the equation with the mouse ("highlight" it) before making the definition. To define $x = -\frac{3}{5}$ from the solution above, you can simply place the insertion point in the equation and choose Define + New Definition. However, in the following example, placing the insertion point in the solution would select both equations, and this would not allow you to make a definition. By selecting each equation with the mouse, you can make definitions without any editing required.

▶ **To solve the system of equations $2x + 3y = 41$ and $x + y = 146$**

1. Choose Insert + Display (or click the Display button  on the Math toolbar) to open a display.

2. Enter the equation $2x + 3y = 41$ in the input box.

3. With the insertion point on the right of the equation, press ENTER to create a new line in the display.

4. Enter the equation $x + y = 146$ in the input box in the second line of the display.

5. With the insertion point in the display,
$$2x + 3y = 41$$
$$x + y = 146$$
choose Solve + Exact. (*Scientific Notebook* will return the result, "Solution is : $\{y = -251, x = 397\}$.")

Note You can enter a system of equations in any one-column "matrix" with the number of rows equal to the number of equations. The display feature is a convenient way to do this, but you can also begin with Insert + Matrix and specify one column and two rows. This matrix would appear "in-line" rather than displayed.

Solving Equations Numerically

In some of the activities, you will want to search for solutions to equations on specified intervals. For this purpose, use the command Solve + Numeric and the notation

$$x \in (a, b)$$

to indicate that x is in the interval (a, b)—that is, that $a < x < b$. (You will find the symbol $\in$ in the Binary Relations panel that drops down when you click the Binary Relations button on the Math toolbar.) The general procedure for specifying an interval for a solution is as follows.

► **To locate a solution to an equation $c = f(x)$ by specifying a search interval (a, b)**

1. Choose Insert + Display.
2. Enter $f(x) = c$ into the input box and press ENTER.
3. Enter $x \in (a, b)$ into the second input box.
4. Choose Solve + Numeric.

In Search of a Solution

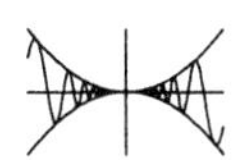

The graph of $y = \sin(x + x\cos x)$ has many x-intercepts. Find the x-intercept(s) given by Solve + Exact and Solve + Numeric and compare the results. Specify a search interval and use Solve + Numeric to find the x-intercept between 6 and 6.4. Find a solution to $\sin(x + x\cos x) = 0.5$ with x between 2 and 3. Explain how you would find all the x-intercepts in an interval $a \leq x \leq b$. Include a graph for the interval $a = -2$, $b = 8$ in your discussion.

Sally's Solution

Applying Solve + Exact to the equation $\sin(x + x\cos x) = 0$ gives the x-intercepts $\{x = 0\}, \{x = \pi\}$. Applying Solve + Numeric to the equation $\sin(x + x\cos x) = 0$ gives the x-intercept $\{x = 3.1416\}$. In this example, Solve + Exact gave two exact solutions, and Solve + Numeric found an approximation to one of the same solutions.

With the insertion point in the equation $\sin(x + x\cos x) = 0$, choose Solve + Exact. With the insertion point in the equation $\sin(x + x\cos x) = 0$, choose Solve + Numeric.

Plotting the function makes it obvious that there are many x-intercepts for the function $y = \sin(x + x\cos x)$ between -2 and 8, and shows approximately where they are.

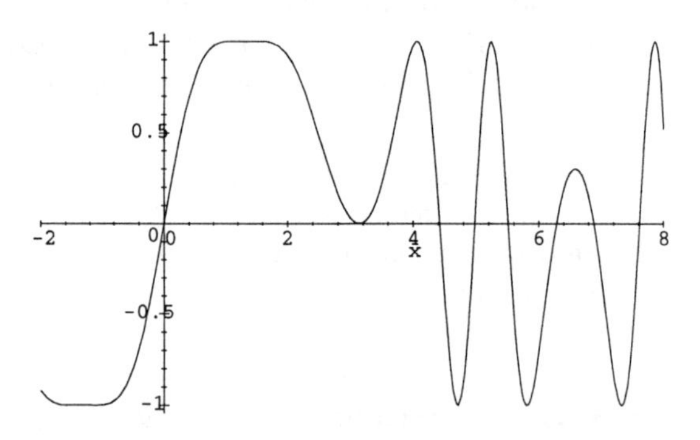

$y = \sin(x + x\cos x)$

With the insertion point in the expression $\sin(x + x\cos x)$, choose Maple + Plot 2D + Rectangular. Click the plot and choose Edit + Properties. Click the Labeling tab and type $y = \sin(x + x\cos x)$ in the Caption Text box. Click the Plot Components tab and change the Domain Interval to $-2 < x < 8$. Choose OK.

To find an x-intercept between 6 and 6.4, we apply Solve + Numeric to the display

$$\begin{aligned} \sin(x + x\cos x) &= 0 \\ x &\in (6, 6.4) \end{aligned}$$

This gives the solution $\{x = 6.2832\}$.

Choose Insert + Display. Type $\sin(x + x\cos x) = 0$ in the input box and press ENTER. Type $x \in (6, 6.4)$ in the input box. With the insertion point in the display, choose Solve + Numeric.

To find a solution to the equation $\sin(x + x\cos x) = 0.5$ between 2 and 3, we apply Solve + Numeric to the display

$$\begin{aligned} \sin(x + x\cos x) &= 0.5 \\ x &\in (2, 3) \end{aligned}$$

This gives the solution $\{x = 2.4797\}$.

Choose Insert + Display. Type $\sin(x + x\cos x) = 0.5$ in the input box and press ENTER. Type $x \in (2, 3)$ in the input box and choose Solve + Numeric.

If there are multiple solutions, Solve + Exact may not find them all. In this case it found only two solutions when there are obviously infinitely many. For the equations we are looking at, Solve + Numeric finds only one solution at a time, but it allows you to find different solutions by specifying an interval the solution lies in.

To find all the solutions in an interval $a \leq x \leq b$, I would first plot the function on that interval. Then from the graph I could locate x-intercepts approximately, and by specifying an interval about each x-intercept that contains only one crossing, I would find a numerical approximation to each x-intercept in the interval. In the graph above, I used the interval $-2 < x < 8$. For this interval, I would then specify eight different subintervals to find all of the x-intercepts between -2 and 8.

> **Note** You can enter an equation and specify an interval for its solution in a one-column, two-row matrix instead of in a two-line display, if you prefer. Remember, you must use Solve + Numeric for this technique (not Solve + Exact).

Activity

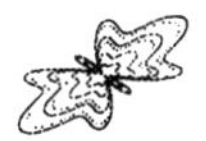

Find all the solutions to $\sin(x + x \cos x) = 0$ between 22 and 23. First use the zoom-in tool to estimate the solutions to as many decimal places as you want; then specify an interval for the solution and use Solve + Numeric. Which method do you prefer and why?

> **Reminder** To create the graph, place the insertion point in the expression $\sin(x + x \cos x)$ and choose Plot 2D + Rectangular. To find the zoom-in tool, double-click the graph. Three plotting tools (*Pan*, *Zoom in*, and *Zoom out*) appear in the upper right corner of the frame. Select the large mountain (zoom-in tool), place the mouse cursor over a point in the graph, press and hold the mouse button, drag to the right, and release the mouse button. This will create a box and the contents of the box will expand to fill the frame.

Is Algebra the Clue?

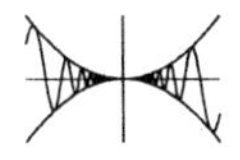

Algebra can be used for keeping (and telling) secrets. Company ALG needs to protect a secret file, the key to which is a hidden number. They are asking for a scheme so that any two employees working together could learn the hidden number but no one working alone could find it. You can develop such schemes using *polynomial functions*—that is, functions of the form

$$f(x) = a_n x^n + \cdots + a_2 x^2 + a_1 x + a_0$$

These include, in particular,

- *Lines*: $f(x) = ax + b$
- *Parabolas*: $f(x) = ax^2 + bx + c$
- *Cubics*: $f(x) = ax^3 + bx^2 + cx + d$

The schemes explored in these activities are based on geometric properties.

- Two points determine a line, and there are many lines passing through any one given point.
- Three points, not on a line, determine a parabola, and there are many parabolas passing through any two given points.
- Four points on a cubic curve determine the cubic, and there are many cubics passing through any three such points.

Because these lines and curves are determined by a polynomial of degree 1, 2 or 3, you can use algebra to implement the schemes. The scheme you are asked to explore is based upon finding the y-intercept of a line.

At ALG, Inc., each employee is given the coordinates of a different point on a line. The following table lists the points for eight employees:

x	1	2	3	4	5	6	7	8
y	−196	41	278	515	752	989	1226	1463

Suppose the employees with x-coordinates 2 and 8 put their clues together.

1. Find the y-intercept determined by these two employees' coordinates.
2. Choose another set of two employees. Show that they determine the same y-intercept.
3. Choose one employee. Find and graph several different solutions that use this employee's coordinates. Verify that they cross the y-axis in different places.

Reminder For each command on the Maple menu, place the insertion point in the mathematical expression and then choose the command.

Carol's and Darel's Solution

The system for revealing a hidden number to any two employees working together, but hiding the number from anyone working alone, is based upon finding the y-intercept of a line. Each employee is given the coordinates of a different point on the line. The following table lists coordinates for a company of eight employees.

x	1	2	3	4	5	6	7	8
y	−196	41	278	515	752	989	1226	1463

To make this table, choose Insert + Display. Choose Insert + Matrix. Set Rows to 2, Columns to 8, and choose OK. Type numbers in the input boxes.

When the employees with x-coordinates 2 and 8 put their clues together, their coordinates determine a unique line passing through the two points $(2, 41)$ and $(8, 1463)$. To find the y-intercept determined by these coordinates, we define the function $p(x) = ax + b$ and solve the system of equations $p(2) = 41$ and $p(8) = 1463$. Solving this system gives the solution $\{a = 237, b = -433\}$ and hence the y-intercept is -433.

As a precaution, first choose Maple + Define + Clear Definitions. With the insertion point in the equation $p(x) = ax + b$, choose Maple + Define + New Definition. Choose Insert + Display and, in the input box, type $p(2) = 41$. Press ENTER to add a line to the display, and type $p(8) = 1463$ in the input box. With the insertion point in the display

$$\begin{aligned} p(2) &= 41 \\ p(8) &= 1463 \end{aligned}$$

choose Maple + Solve + Exact.

When the employees with x-coordinates 1 and 5 put their clues together, their coordinates determine a unique line passing through the two points $(1, -196)$ and $(5, 752)$. Solving the system $p(1) = -196$ and $p(5) = 752$ gives the same solution as before, $\{a = 237,\ b = -433\}$, and hence the same y-intercept of -433.

Choose Insert + Display. In the input box, type $p(1) = -196$. Press ENTER, and type $p(5) = 752$. With the insertion point in the display

$$\begin{aligned} p(1) &= -196 \\ p(5) &= 752 \end{aligned}$$

choose Maple + Solve + Exact

The employee with x-coordinate 3 determines a family of lines passing through the point $(3, 278)$. Solving $p(3) = 278$ for b gives the solution $\{b = -3a + 278\}$.

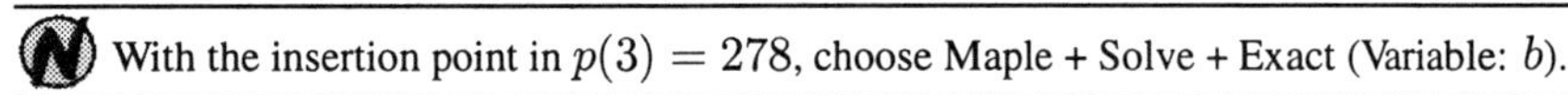

With the insertion point in $p(3) = 278$, choose Maple + Solve + Exact (Variable: b).

Taking $a = 1, 2, 3$ gives the lines $y = x + 275$, $y = 2x + 272$, and $y = 3x + 269$ with three different y-intercepts 275, 272, and 269 shown in the following graph:

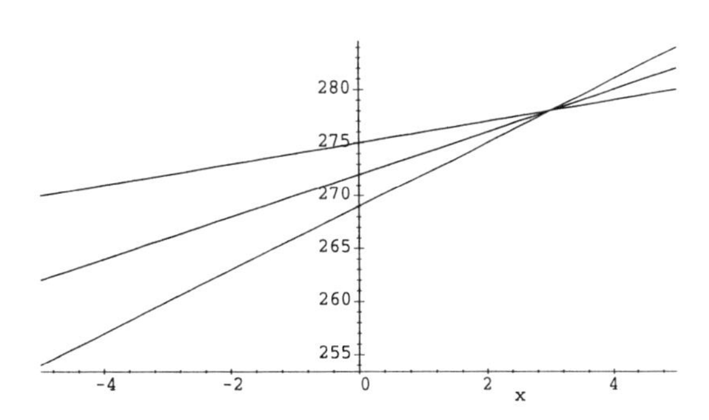

Choose Insert + Math. Type the list $x + 275, 2x + 272, 3x + 269$ with the three expressions separated by red commas. With the insertion point in the list, choose Maple + Plot 2D + Rectangular.

An individual working alone has no clue to the hidden number, because the y-intercept for a line passing through a single point off the y-axis could be any number.

Remark The line $y = 237x - 433$ is shown in the following plot, together with the eight employees' coordinates:

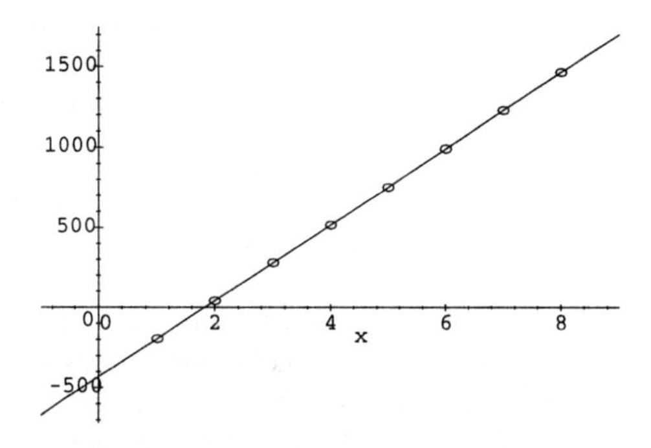

To obtain this result, plot the expression for the line: $237x - 433$. Select and drag to the frame a list of the employees' points, enclosed in parentheses:

$$(1, -196, 2, 41, 3, 278, 4, 515, 5, 752, 6, 989, 7, 1226, 8, 1463)$$

Choose Edit + Properties, and select the Plot Components tab. Choose Item Number 2 and change the graph of employees' points to a "point" plot with symbol "circle." Change Domain Interval to $-1 \leq x \leq 9$. Choose OK.

Activities

1. Company ABC is asking for a system that would permit any three employees working together to learn a secret number, but that would hide the secret number from any one or two employees.

   a. Describe a scheme based on finding the y-intercept of a polynomial $p(x)$ of degree 2 and giving each employee the coordinates of a different point that lies on the graph of this polynomial. Use the following table of points:

x	1	2	3	4	5	6	7	8
y	6758	6397	6110	5897	5758	5693	5702	5785

b. Suppose three employees with x-coordinates 2, 4, and 8 get together to authorize access. Find the y-intercept determined by these three employees' coordinates.
c. Choose another set of three employees. Show that they determine the same y-intercept.
d. Choose a set of two employees. Find and graph several different solutions. V erify that they cross the y-axis in different places.

2. Company XYZ is asking for a system that would permit any four employees working together to learn a secret number, but that would hide the secret number from any group of three or fewer employees.

   a. Describe a scheme, based upon finding the y-intercept of a polynomial $p(x)$ of degree 3, in which you give each employee the coordinates of a different point that lies on the graph of this polynomial. Use the following table of points:

x	1	2	3	4	5	6	7	8
y	27807	32919	36919	39993	42327	44107	45519	46749

   b. Suppose four employees with x-coordinates 2, 4, 7, and 8 get together to authorize access. Find the y-intercept determined by these four employees' coordinates.
   c. Choose another set of four employees. Show that they determine the same y-intercept.
   d. Choose a set of three employees. Find and graph several different solutions. V erify that they cross the y-axis in different places.

3. Develop a scheme for a company with ten employees plus four supervisors that allows any three employees working together, or any two supervisors working together, to find the secret number 52293.

Reminder Don't forget to check your answers by looking at the problem several ways, including *algebraically, numerically*, and *geometrically.*

Going in Circles

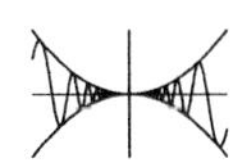

Every circle in the plane satisfies an equation of the form $x^2 + y^2 + Ax + By + C = 0$ for some constants A, B, and C. Through any three points in the plane that do not lie on a straight line, there is exactly one circle that passes through these three points. Use these two facts to find an equation for the circle that passes through the points $(57, -59)$, $(48, -8)$, and $(-93, 92)$ by defining $f(x, y) = x^2 + y^2 + Ax + By + C$ and solving the system

$$\begin{aligned} f(57,-59) &= 0 \\ f(48,-8) &= 0 \\ f(-93,92) &= 0 \end{aligned}$$

of equations. Plot the graph and visually verify that the three points lie on the circle.

> **Hint** To graph a circle, you can create an "implicit" plot that depends on the equation $f(x, y) = 0$. To graph a circle with a "rectangular" plot, you would need to solve the equation for y and graph both the function that describes the upper half of the circle and the function that describes the lower half of the circle. Note that an implicit plot still uses the rectangular coordinate system.

Sue's Solution

Equations for circles can be written in the form $x^2 + y^2 + Ax + By + C = 0$. To find an equation for the circle through the three points $(57, -59)$, $(48, -8)$, and $(-93, 92)$, we define the function $f(x, y) = x^2 + y^2 + Ax + By + C$ and solve the system of three equations

$$\begin{aligned} f(57, -59) &= 0 \\ f(48, -8) &= 0 \\ f(-93, 92) &= 0 \end{aligned}$$

The equation for the circle certainly must satisfy these three equations. The function $f(x, y)$ has only three unknown coefficients, so these three equations will completely determine the circle. Solving this system of equations gives the three coefficients $A = \frac{396065}{2097}$, $B = \frac{83083}{699}$, and $C = -\frac{2442536}{233}$, and consequently the equation

$$f(x, y) = x^2 + y^2 + \frac{396065}{2097}x + \frac{83083}{699}y - \frac{2442536}{233}$$

(N) Choose (Maple +) Define + Clear Definitions. With the insertion point in the equation $f(x, y) = x^2 + y^2 + Ax + By + C$, choose Define + New Definition. Choose Insert + Display. Type $f(57, -59) = 0$ in the input box. Press ENTER. Type $f(48, -8) = 0$ in the input box. Press ENTER. Type $f(-93, 92) = 0$ in the input box. With the insertion point in the system of equations, choose (Maple +) Solve + Exact.

Following is a graph of the circle determined by the equation $f(x, y) = 0$, with each of the three given points marked with a small circle:

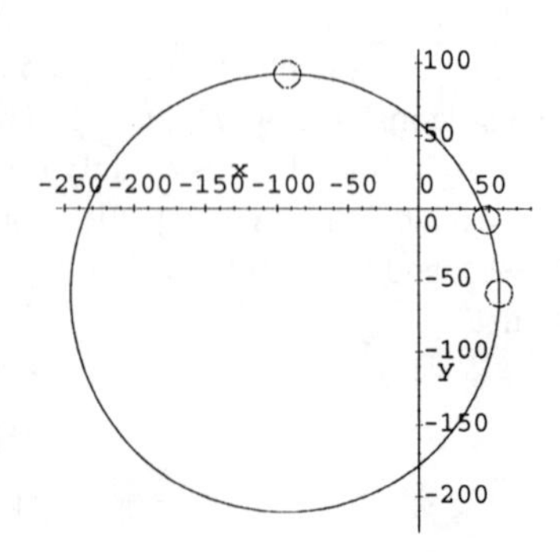

With the insertion point in the equation $x^2+y^2+\frac{396065}{2097}x+\frac{83083}{699}y-\frac{2442536}{233}=0$, choose Maple + Plot 2D + Implicit. Choose Edit + Properties (or click the frame and click the dialog tool that appears in the lower right corner of the frame).The default domain intervals lie entirely inside this circle, so you will see no part of the graph until you increase the domain intervals. Click the Plot Components tab; set Domain Intervals to $-260 \leq x \leq 70$ and $-220 \leq y \leq 100$. To make the circle appear like a circle rather than an ellipse, click the Axes tab and check Equal Scaling Along Each Axis. Choose OK. To add the small circles about the three points, write the three equations

$$\begin{aligned}(x-57)^2+(y+59)^2 &= 10^2\\ (x-48)^2+(y+8)^2 &= 10^2\\ (x+93)^2+(y-92)^2 &= 10^2\end{aligned}$$

Select and drag each of these equations to the plot (or copy them to the clipboard and paste them on the plot). To improve the appearance of the small circles, on the Plot Components page change Sample Size to 50 for each axis. This would be a good graph for practicing with the zoom-in and zoom-out tools (see page 7) to find appropriate domain intervals.

0.3 Types of Functions: Shifting and Scaling

The following activities will help you become more comfortable with *Scientific Notebook* while helping you understand the effects of shifting and scaling.

Shifting and Scaling

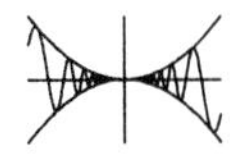

Consider the function $f(x) = x^3 - x$. Experiment with the graphs of $y = f(x)$, $y = f(x-a)$, $y = f(ax)$, $y = af(x)$, and $y = a + f(x)$ for different values of a to study the effect of shifting and scaling. Describe in words the effect of each. Plot the graphs of $y = f(x)$ and $y = 3f(2x-1) - 4$ together and describe the role of each of the numbers 1, 2, 3, and 4.

Dylan's Solution

The following graphs show the effect on the graph of $f(x) = x^3 - x$ of several choices of shifting and scaling. In each figure the starting graph is drawn with a fat pen and the transformed graph is drawn with a thin pen. Replacing x by $x - 2$ shifts the graph to the right 2 units, as illustrated in the following figure:

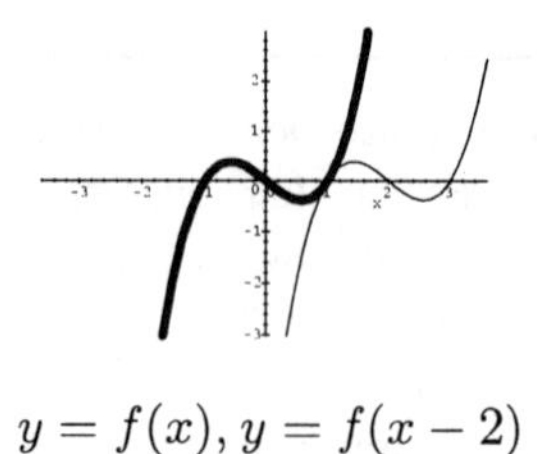

$y = f(x), y = f(x-2)$

With the insertion point in the equation $f(x) = x^3 - x$, choose Define + New Definition. With the insertion point in the expression $x^3 - x$, choose Plot 2D + Rectangular. Select and drag to the plot $f(x-2)$. Choose Edit + Properties. Click Plot Components tab; change Thickness to Medium for Item Number 1; set Domain Interval to $-3.5 < x < 3.5$. Choose Axis & View page; click Default to turn it off; change View Intervals to $-3.5 < x < 3.5$ and $-3 < y < 3$. Choose Frame page; change Placement Option to Displayed. Choose Labeling page; type names of functions in Caption Text box; choose OK.

Replacing x by $2x$ compresses the graph horizontally by a factor of 2.

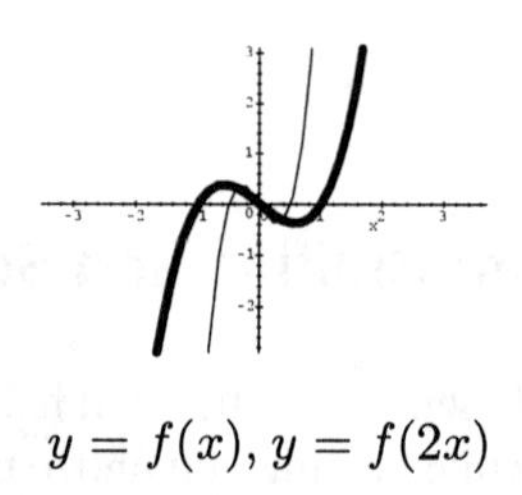

$y = f(x), y = f(2x)$

With the insertion point in the expression $f(x)$, choose Plot 2D + Rectangular. Select and drag to the plot $f(2x)$. Choose Edit + Properties. Choose Plot Components page; change Thickness to Medium for Item Number 1; set Domain Interval to $-3.5 < x < 3.5$. Choose Axis & View page; check Default to turn it off; change View Intervals to $-3.5 < x < 3.5$ and $-3 < y < 3$. Choose Frame page; change Placement Option to Displayed. Choose Labeling page; type name of function in Caption Text box; choose OK.

Adding 2 to the function value shifts the graph up 2 units.

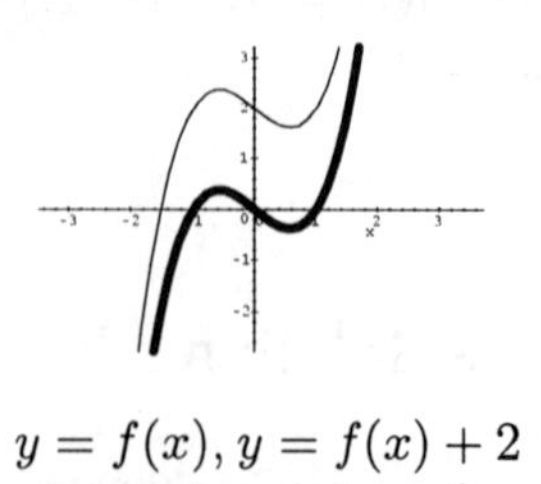

$y = f(x), y = f(x) + 2$

With the insertion point in the expression $f(x)$, choose Plot 2D + Rectangular. Select and drag to the plot $f(x) + 2$. Choose Edit + Properties. Choose Plot Components page; change

Thickness to Medium for Item Number 1; set Domain Interval to $-3.5 < x < 3.5$. Choose Axis & View page; check Default to turn it off; change View Intervals to $-3.5 < x < 3.5$ and $-3 < y < 3$. Choose Frame page; change Placement Option to Displayed. Choose Labeling page; type name of function in Caption Text box; choose OK.

Multiplying the function values by 2 stretches the graph vertically from the x-axis by a factor of 2.

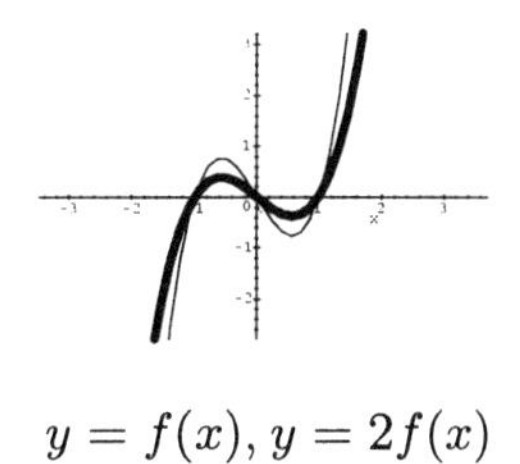

$y = f(x), y = 2f(x)$

With the insertion point in the expression $f(x)$, choose Plot 2D + Rectangular. Select and drag to the plot $2f(x)$. Choose Edit + Properties. Choose Plot Components page; change Thickness to Medium for Item Number 1; set Domain Interval to $-3.5 < x < 3.5$. Choose Axis & View page; check Default to turn it off; change View Intervals to $-3.5 < x < 3.5$ and $-3 < y < 3$. Choose Frame page; change Placement Option to Displayed. Choose Labeling page; type name of function in Caption Text box; choose OK.

Replacing $f(x)$ by $3f(2x-1)-4$ combines all of these effects. Note that $3f(2x-1) - 4 = 3f(2(x-\frac{1}{2}))-4$. The 2 compresses the graph horizontally, the $\frac{1}{2}$ shifts the graph to the right, the 3 stretches the graph vertically, and then the -4 shifts the graph down.

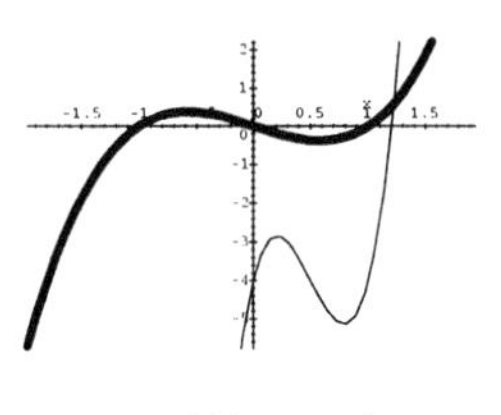

$y = 3f(2x - 1) - 4$

With the insertion point in the expression $f(x)$, choose Plot 2D + Rectangular. Select and drag to the plot $3f(2x - 1) - 4$. Choose Edit + Properties. Choose Plot Components page; change Thickness to Medium for Item Number 1; set Domain Interval to $-3.5 < x < 3.5$. Choose Axis & View page; click Default to turn it off; change View Intervals to $-3.5 < x < 3.5$ and $-3 < y < 3$. Choose Frame page; change Placement Option to Displayed. Choose Labeling page; type name of function in Caption Text box; choose OK.

To understand the precise shifting and scaling, it is necessary to analyze the transformed expression in terms of a sequence of simple transformations. Rewriting the expression $3f(2x - 1) - 4$ in the equivalent form $3f\left(2(x - \frac{1}{2})\right) - 4$, note that we first translate the graph $\frac{1}{2}$ unit to the right, then compress the graph horizontally by a factor of 2, then stretch the graph vertically by a factor of 3, then shift down 4 units. The following figures illustrate this sequence of transformations. As before, in each figure the starting graph is drawn with a fat pen and the transformed graph is drawn with a thin pen.

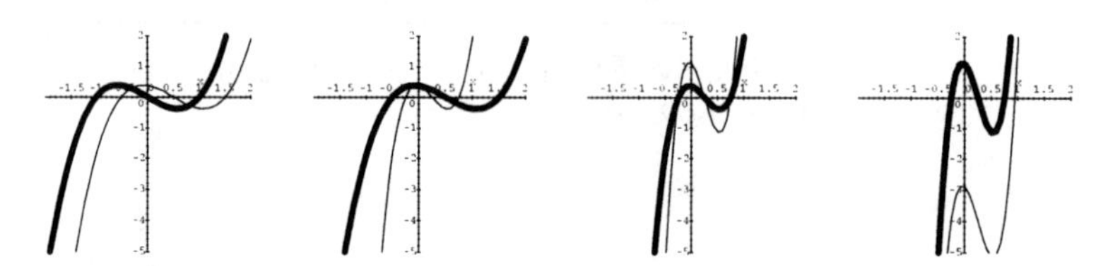

To center a group of several frames, select each frame and choose Edit + Properties. Choose Frame page; change Placement Option to In Line; choose OK. With the insertion point in the group of frames, from the Tag bar, choose Body Center.

Activities (Do All Parabolas Have the Same Shape?)

1. Let $f(x) = x^2$ and consider the parabola $y = 2x^2 - 4x + 1$. Plot the graph of $y = f(x)$ using the viewing window $-1 \leq x \leq 1$ by $-1 \leq y \leq 1$. In a separate frame, plot a graph of $y = 2x^2 - 4x + 1$. Adjust the second viewing window so that the curve is visible in the extreme upper left-hand corner, the extreme upper right-hand corner, and the lower middle. Compare and contrast the shapes of the two graphs. Rewrite the expression $2x^2 - 4x + 1$ in the form $a(x-h)^2 + k$. Explain the geometric significance of the numbers a, h, and k.

2. Let $f(x) = \sin x$. Experiment to determine the roles of the numbers a, b, h, and k in the expression $af(b(x-h)) + k$. Plot a sequence of four graphs that illustrate these roles for a particular (nonzero) choice of the numbers a, b, h, and k. Plot a graph of $y = f(x)$ that looks similar to your graph of $y = af(b(x-h)) + k$ by selecting an appropriate viewing window. Describe the connection between the numbers a, b, h, and k and your choice of viewing window for the graph of $y = f(x)$.

3. Let $f(x) = 6x^2 + x - 2$. Describe the connection between the graph of $y = f(x)$ and the graph of $y = |f(x)|$, where

$$|a| = \sqrt{a^2} = \begin{cases} -a & \text{if} \quad a < 0 \\ a & \text{if} \quad a \geq 0 \end{cases}$$

denotes absolute value.

Describe the connection between the graph of $y = f(x)$ and the graph of $y = \frac{1}{f(x)}$. Include graphs of $y = f(x)$, $y = \frac{1}{f(x)}$, $y = 1$, and $y = -1$ in your discussion. In particular, explain the significance of the lines $y = 1$ and $y = -1$. Use a viewing window such as $-3 \leq x \leq 3$ by $-3 \leq y \leq 3$ to display graphs of $y = f(x)$ and $y =$

$\frac{1}{f(x)}$. (Recall that for a rectangular plot, the default is one degree of freedom—the domain interval—with the range interval determined automatically by the underlying system. You can change this default on the Axis & View page of the Plot Properties dialog.)

4. The volume of a box in the shape of a rectangular solid is given by

$$\textit{Volume} = (\textit{Base Area}) \times \textit{Height}$$

Regulations of the United States Postal Service state that a box can be mailed only if the length plus the girth is at most 108 inches. (The *girth* means the distance around. My girth is my belt size. For a rectangular solid, the girth is just another name for the perimeter of the base.) Let y be the height of a box and assume the base is an $x \times x$ square. The postal regulation states that $4x + y \leq 108$. To find the largest volume possible, we use equality: $4x + y = 108$, or $y = 108 - 4x$. Thus the volume of the largest with base x^2 that can be shipped is given by $V = x^2 y = x^2(108 - 4x) = 108x^2 - 4x^3$. Give a reasonable domain for this volume function. Discuss whether or not the shape of the volume function $V(x) = 108x^2 - 4x^3$ is the same as the shape of the cubic $y = x^3 - x$. Use suitable graphs in your discussion. Discuss how your answer uses the ideas of shifting and scaling.

0.4 Graphing Calculators and Computers

Numerical analysis may be described as the study of errors introduced by using *floating-point* arithmetic (*round-off errors*) and by using a finite number of terms when an infinite number of terms are required for exactness (*truncation errors*). Floating point is a data type that is machine-dependent. It is important to understand that a floating-point number is neither rational nor irrational; indeed, each floating-point number represents an infinite number of possible rational numbers and an infinite number of possible irrational numbers. (An *irrational number* is a real number such as $\sqrt{2}$ and π that cannot be expressed as a quotient of integers.)

Computer algebra systems use what is called *infinite precision* or *extended precision* to represent integers and rationals exactly. (Because computers are necessarily finite, there is really no such thing as "infinite precision." The term *extended precision* is probably a better term. Sooner or later, any computer attempting to perform sufficiently complicated extended-precision computations will run out of memory.)

Numbers such as $\sin 1$, $\sqrt{2}$, and π are additional examples of numbers that are represented exactly. These numbers cannot be further simplified. Numerical evaluation leads to the following approximations.

$$\begin{aligned} \sin 1 &= .84147\,09848\,07896\,50665\,25023 \\ \sqrt{2} &= 1.41421\,35623\,73095\,04880\,1689 \\ \pi &= 3.14159\,26535\,89793\,23846\,2643 \end{aligned}$$

N Choose Maple + Settings, Engine Parameters page. Set Digits Used in Computations and Digits Used in Display both to 25. Choose OK. Choose Maple + Evaluate Numerically.

Notice that the approximations are broken into blocks, each 5 decimal digits in length, in order to make them more readable. The numbers on the left are exact, whereas the numbers on the right are merely approximations. In particular, $\left(\sqrt{2}\right)^2 = 2$ exactly, and

$$1.41421\,35623\,73095\,04880\,1689^2 = 2.00000\,00000\,00000\,00000\,0001$$

The remarkable fact is that modern applications of mathematics in fields such as cryptology require absolute precision. A single missing bit would completely scramble a message and make it unrecoverable (even to users with authorized passwords).

The following example illustrates typical products in extended precision. If a requires n BYTES and b requires m BYTES of memory, then the product requires $n+m$ BYTES, roughly proportional to the length of the decimal representation.

$$93478934923678895 \times 37893267207 = 354\,22222\,59288\,72851\,97014\,96265$$

With the insertion point in the expression $93478934923678895 \times 37893267207$, choose Evaluate.

Another place where the finite nature of computers is evident is with plotting. The following two pictures are versions of the same graph, with the plot style on the left being Point and on the right Line. Line plots are created by connecting the points with a smooth curve. You can see that the system concentrates the plotted points in the regions where the graph has the greatest curvature.

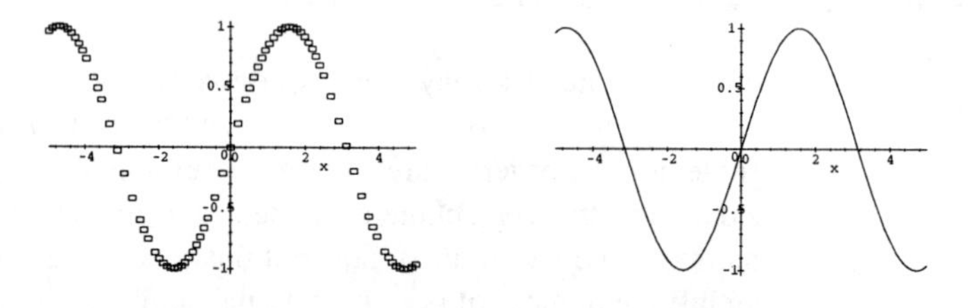

Graphing calculators utilize similar principles for computing and graphing, and if you own a graphing calculator, then you can use it to do many of the activities in this book. Other activities require symbol manipulation and extended-precision arithmetic and may be beyond the reach of your calculator.

Because *Scientific Notebook* encourages you to do mathematics using ordinary words and symbols, it will help you understand the underlying mathematical principles. When you go back to paper and pencil or graphing calculators, your approach to mathematical problem solving should be the same. Think of *Scientific Notebook* as a faithful assistant that knows how to spell words and perform messy calculations, but that understands nothing about setting up problems and interpreting the solution. Working together, you will become a mathematical problem-solving monster team.

0.5 Principles of Mathematical Writing

Writing mathematics well requires thought and effort. Over many years, mathematicians have developed several principles that allow them to communicate effectively and precisely. The following list of hints provides a solid foundation for mathematical writing. Your instructor may suggest additional hints as you master these.

★ **Write in complete sentences. Avoid disjointed series of formulas.**

- *BAD:* $3x - 4 = 5x + 6$, $3x - 5x = 6 + 4$, $-2x = 10$, $x = -5$
- *GOOD:* To solve the equation $3x - 4 = 5x + 6$, add $4 - 5x$ to each side to get $3x - 5x = 6 + 4$. Combine like terms to get $-2x = 10$, and divide both sides by -2 to get $x = -5$.

★ **Never begin a sentence with a symbol.**

- *BAD:* x is bigger than 10.
- *GOOD:* The number x is bigger than 10.

★ **Separate formulas by words.**

- *BAD:* If $x < 3$, $2x < 6$.
- *GOOD:* If $x < 3$, then $2x < 6$.

★ **Set off important or complicated formulas in a display.**

- *BAD:* The solution to $ax^2 + bx + c = 0$ is given by the formula $x = \frac{-b \pm \sqrt{b^2-4ac}}{2a}$.
- *GOOD:* The solution to the equation
$$ax^2 + bx + c = 0$$
is given by the formula
$$x = \frac{-b \pm \sqrt{b^2 - 4ac}}{2a}$$

★ **Avoid unnecessary symbolism when words will provide a clear statement.**

- *BAD:* If $\varepsilon > 0 \Longrightarrow \exists \delta > 0 \ni 0 < |x - a| < \delta \Longrightarrow |f(x) - f(a)| < \varepsilon$, f is continuous at a, $\lim_{x \to a} f(x) = f(a)$.
- *GOOD:* The function f is continuous at a if for every real number $\varepsilon > 0$ there exists a real number $\delta > 0$ such that whenever $0 < |x - a| < \delta$ it follows that $|f(x) - f(a)| < \varepsilon$. In this case we write
$$\lim_{x \to a} f(x) = f(a)$$

★ **Do not use the same notation to represent two different things in the same expression.**

- *BAD:* Define h by the equation $h(x) = \dfrac{f(x+h) - f(x)}{h}$.
- *GOOD:* Define g by the equation $g(x) = \dfrac{f(x+h) - f(x)}{h}$.

★ **Commas and periods should be placed inside quotation marks. Colons, semicolons, question marks, and exclamation points stay outside quotation marks unless they are part of the quotation.**

- *BAD:* She said, "I like mathematics".
- *BAD:* Why did you say, "I like mathematics?"
- *GOOD:* Why did you say, "I like mathematics"?
- *GOOD:* She said, "I like mathematics."
- *GOOD:* He said, "Why do you like mathematics?"

★ **Parentheses and brackets should be strictly logical. Sentences may be entirely enclosed inside parentheses. When parentheses enclose a partial sentence, ending punctuation should appear outside the parentheses.**

- *BAD:* I can read, (but I cannot write.)
- *GOOD:* I can listen (but I cannot sing).
- *GOOD:* I can do mathematics. (However, I cannot write about it.)

★ **Use the spelling checker.**

- *BAD:* If I find a mispelt werd, I shud fix it.
- *GOOD:* Misspelled words are a distraction to the reader.

With the spelling checker available on the Tools submenu, there are no excuses for misspelled words on homework assignments! You can apply the spelling checker to selected parts of your document or to your entire document.

★ **Be consistent.**

- Many rules are still open to debate. For example, many authors use "x-axis" and "x-coordinate," whereas others prefer "x axis" and "x coordinate" to describe axes and coordinates. Some authors include ending punctuation for sentences that end in displayed mathematics, whereas others insist that a display is itself a form of punctuation and that an additional period is not necessary. Adopt a set of standards that are generally acceptable and stick with these standards.

Those who read your writing will love you for helping them to read and understand what you have written. Good writing can also improve your course grade.

Activity

Produce your own good and bad examples for each of the points described above. Find three more guidelines for good mathematical writing, and give good and bad examples for each.

1 LIMITS AND RATES OF CHANGE

The study of motion is the central idea of calculus and is what distinguishes calculus from earlier mathematics.

1.1 The Tangent and Velocity Problems

The slope of a line is a measure of rate of change. The slope measures how fast a path is rising compared with the horizontal distance traversed. A level path has slope 0. A path rising at an angle of 45° has slope 1, because it rises at the same rate it goes forward. A path rising at an angle near to 90° is changing altitude very rapidly, and can have slope as high as you like.

The Slope of the Slope

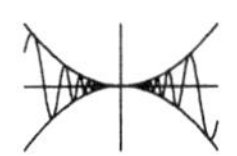

A Colorado ski slope starts off gently, but gets steeper and steeper as a skier starts down the mountain. Give the top of the mountain the coordinates $(0, 5000)$ and suppose the coordinates of the ski slope are given by (x, y), where $y = 5000 - \frac{x^2}{1000}$ on the interval $0 \leq x \leq 500$; then follow the tangent line to the bottom of the mountain. What is the slope of the slope at $x = 500$? What is the equation of the mountain for $x > 500$? Plot the ski slope and discuss whether or not you would like to ski this slope.

Brian's Solution

To view the top of the mountain we plot $y = 5000 - \frac{1}{1000}x^2$ and set the Domain Interval to $[0, 500]$.

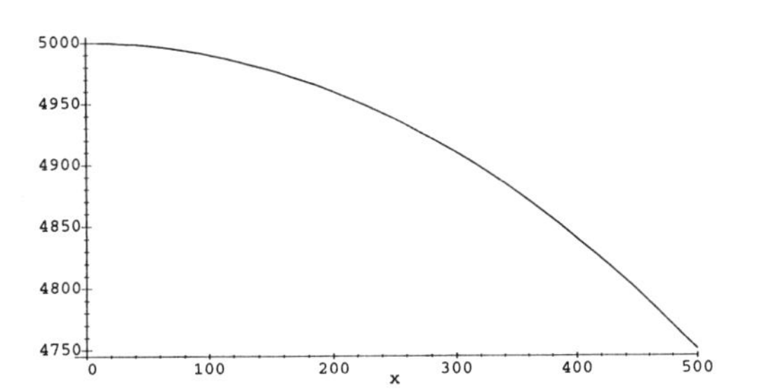

With the insertion point in the expression $5000 - \frac{1}{1000}x^2$, choose Plot 2D + Rectangular. Choose Edit + Properties, Plot Components page. Change the Domain Interval to $0 < x < 500$. Choose OK.

The tangent line can be approximated by zooming in on the point $(500, 4750)$ until the curve appears to be a straight line.

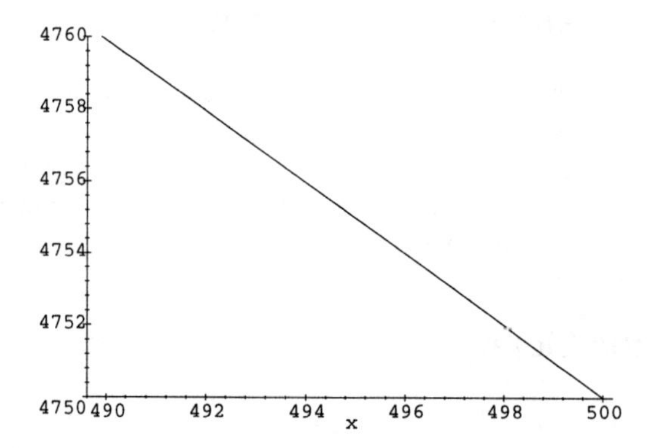

To zoom in on the point $(500, 4750)$, double-click the frame. Choose the middle plotting tool that appears in the upper right corner of the frame (the large mountain). Place the insertion point to the left of the point $x = 500$, and, with the mouse button pressed, drag the dashed-line rectangle to the point $x = 500$. Release the mouse button. Repeat this procedure until the graph looks like a straight line.

The point $(490, 4760)$ appears to lie approximately on the curve. Thus the slope of the tangent line is given approximately by $\frac{4760-4750}{490-500} = -1$, and an approximate equation of the tangent line is $y - 4750 = -1(x - 500)$, which can be rewritten in the form $y = 5250 - x$.

Determine the slope of what appears to be a straight line by locating a second point such as $(490, 4760)$ that appears to be on the curve and evaluating $\frac{4760-4750}{490-500}$. Write the equation $y - 4750 = -1(x - 500)$ and choose Solve + Exact (Variable to Solve For: y) to get an approximate equation for the tangent line.

We can visualize the ski slope by plotting the parabola $y = 5000 - \frac{x^2}{1000}$ together with the line $y = 5250 - x$.

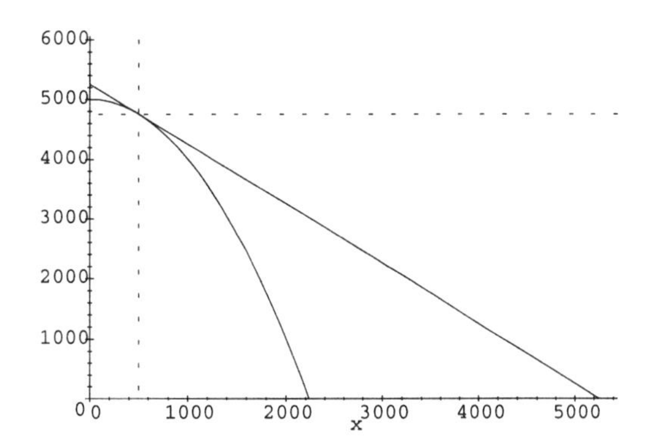

Select $5250 - x$ and drag it to the frame. To zoom out to view the curved and straight portions of the ski slope, either (1) Choose Edit + Properties, Plot Components page, and set Domain Interval to $0 < x < 5000$; or (2) use the small-mountain zoom tool.

Note that the ski slope follows the curve for $0 \leq x \leq 500$ and follows the straight line for $x > 500$.

Activities

1. Define $f(x) = \dfrac{x+1}{x-1}$. Plot a graph of f. From the View page of the Plot Properties tabbed dialog, turn off Default for the View Intervals. Find an equation for the tangent line at $(0, -1)$ by zooming in on the graph of f and estimating the slope of the tangent line. Add the tangent line to your plot. Zoom out to view the curve and line. Explain how you know that your equation is correct.

   Evaluate $\dfrac{f(h) - f(0)}{h - 0}$ for several values of h near 0. Explain the connection between these values and the slope of the tangent line at 0.

   Simplify the expression $\dfrac{f(h) - f(0)}{h - 0}$. Explain the connection between this simplified expression and the slope of the tangent line at 0.

2. Let $f(x) = x \ln x$. Find an equation for the tangent line at $(1, 0)$ by zooming in on the graph of f and estimating the slope of the tangent line. Add the tangent line to your plot. Zoom out to view the curve and line. Explain how you know that your equation is correct.

   Remark When you type the letters "ln" in math mode the first letter should appear in red; then when "n" is typed, the whole expression should turn to gray. (This shows that *Scientific Notebook* interprets ln as a single object rather than a product of two variables.) If the letters "ln" remain red, choose Tools + Automatic Substitution and turn off Disable Automatic Substitution and try typing "ln" again.

3. (Slope of the Slope Reprise) The ski slope starts at the top of the mountain with a shape determined by a parabola; then partway down the mountain the ski slope follows a straight line. Such composite shapes can be viewed by using *piecewise-defined functions*. Use the following hint to define f; then plot it using the viewing rectangle $0 \le x \le 6000$ by $0 \le y \le 6000$.

Hint To define a piecewise function, put the insertion point in math mode and type

$$f(x) =$$

Choose Insert + Brackets, select left brace and right null delimiter 

and [⋮], and choose OK. Choose Insert + Matrix, select 2 Rows, 3 Columns, and choose OK. Enter expressions in the left column, the word "if" in the middle column (in either math or text), and the interval of definition in the right column in the form $a \le x$ or $x \le b$ or $x \ge a$ or $a < x \le b$. With the insertion point in the expression

$$f(x) = \begin{cases} 5000 - \frac{x^2}{1000} & \text{if} \quad x \le 500 \\ 5250 - x & \text{if} \quad x > 500 \end{cases}$$

choose Define + New Definition.

To graph a piecewise-defined function, select the function name f or the expression $f(x)$ with the mouse and choose Plot 2D + Rectangular. You can also graph a piecewise-defined function by selecting the

matrix $\begin{cases} 5000 - \frac{x^2}{1000} & \text{if} \quad x \le 500 \\ 5250 - x & \text{if} \quad x > 500 \end{cases}$, including the expanding brackets.

1.2 The Limit of a Function

The concept of a limit is fundamental to the study of calculus. The notion, which encompasses subtle concepts such as instantaneous velocity, can be fully understood only through experience and experimentation.

That's the Limit (Not!)

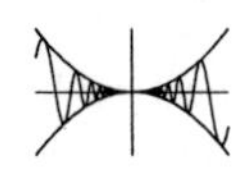

Explore $\lim_{x \to 0} \sin \frac{1}{x}$ by zooming in on the graph of $y = \sin \frac{1}{x}$. Does this limit exist? How do you know? Verify that Evaluate produces $\lim_{x \to 0} \sin \frac{1}{x} = -1..1$, which is

Maple syntax for the interval $-1 \leq y \leq 1$. Does this output make sense? Why?

Chanel's Solution

A graph of $y = \sin \frac{1}{x}$ exhibits wild behavior for x near zero. In fact, the smaller the viewing rectangle, the wilder the behavior. The curve apparently bounces back and forth between -1 and 1 an infinite number of times in any interval containing zero.

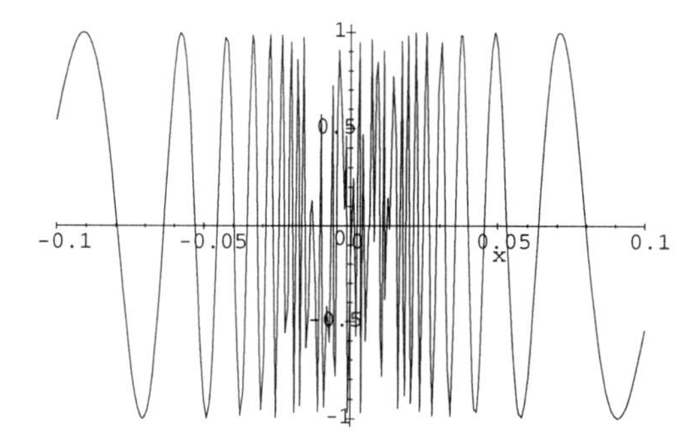

With the insertion point in the expression $\sin \frac{1}{x}$, choose Plot 2D + Rectangular.

Evaluation of $\lim_{x \to 0} \sin \frac{1}{x}$ produces the output $-1..1$, which is notation for the interval $[-1, 1]$. This is consistent with the observation that the curve takes on every value in the interval $[-1, 1]$ infinitely often in any interval containing zero. However, this does not mean that the limit exists. It merely indicates the way in which the limit fails to exist.

To enter and evaluate the expression $\lim_{x \to 0} \sin \frac{1}{x}$, type lim in mathematics mode. (It should turn gray when you type the 'm'. If not, choose Insert + Math Name, and select lim from the list.) Choose Insert + Subscript and enter $x \to 0$ in the input box. (Click the arrow on the Common Symbols toolbar, or click the Arrows button on the Symbols toolbar and click a right-arrow button.) Press Spacebar to get out of the subscript box. Type sin in mathematics (it should turn gray), or choose sin from the Insert + Math Name list. Choose Insert + Fraction and enter the 1 and x in the input boxes. With the insertion point in the expression, choose Evaluate.

Activities (Taking it to the Limit)

1. Define $f(x)$ and $g(x)$ to be generic functions (see page 10) and use **Evaluate** to calculate each of the following limits:

- $\lim_{x \to a} [f(x) + g(x)]$
- $\lim_{x \to a} [f(x) - g(x)]$
- $\lim_{x \to a} [cf(x)]$
- $\lim_{x \to a} [f(x)g(x)]$
- $\lim_{x \to a} \frac{f(x)}{g(x)}$

What are the subtle assumptions that were made by the underlying computational system?

2. Explore the limit

$$\lim_{t \to 0} \frac{\sqrt{t^2+4}-2}{t^2}$$

by doing the following. Define

$$f(t) = \frac{\sqrt{t^2+4}-2}{t^2}$$

and create a table of values for t close to 0.

On the basis of your table, guess a value for the limit. Explain your reasoning.

Plot a graph of f. Zoom in on $t = 0$ and make a guess about the limit based on the graph. Zoom in more until something strange happens. Is what you are seeing real, or is it related to numerical problems with differences?

Duplicate the following by using an in-place replacement. (Select the expression $\left(\sqrt{t^2+4}-2\right)\left(\sqrt{t^2+4}+2\right)$ in the numerator and press the CTRL (PC) or COMMAND (Mac) key and choose Evaluate.)

$$\frac{\sqrt{t^2+4}-2}{t^2} \times \frac{\sqrt{t^2+4}+2}{\sqrt{t^2+4}+2} = \frac{\left(\sqrt{t^2+4}-2\right)\left(\sqrt{t^2+4}+2\right)}{t^2\left(\sqrt{t^2+4}+2\right)}$$

$$= \frac{t^2}{t^2\left(\sqrt{t^2+4}+2\right)}$$

$$= \frac{1}{\sqrt{(t^2+4)}+2}$$

Use these calculations to evaluate the limit. Verify your calculations by applying Evaluate to $\lim_{t \to 0} \frac{\sqrt{t^2+4}-2}{t^2}$.

> **Hint** To make a table of values, choose Insert + Matrix and specify 6 rows (or whatever number of values you want) and 1 column.
> Fill it in with t-values near 0. For this example, we use 0.1, 0.01, 0.001, −0.001, −0.01, −0.1. (As you type these in the matrix, you can press TAB to move between the entries.)
> Select the matrix with the mouse and choose Insert + Brackets (or click a brackets button). Then place an f (in mathematics mode) just to the

left of the brackets. With the insertion point in the expression

$$f\begin{pmatrix} 0.1 \\ 0.01 \\ 0.001 \\ -0.001 \\ -0.01 \\ -0.1 \end{pmatrix}$$

choose Evaluate.

3. Investigate

$$\lim_{x\to 0} \frac{1}{x^2}$$

by looking closely at the graph of $y = \dfrac{1}{x^2}$ for x close to zero. Does this limit exist? How do you know?

4. Explore

$$\lim_{x\to 3} \frac{x^2+x-12}{(x-3)(x-4)}$$

and

$$\lim_{x\to 4} \frac{x^2+x-12}{(x-3)(x-4)}$$

By looking at the form of these limits, predict their values. Plot the curve

$$y = \frac{x^2+x-12}{(x-3)(x-4)}$$

Does your plot indicate a vertical asymptote $x = 3$? What about $x = 4$? Evaluate each of the following limits and explain connections between these limits and your graph:

$$\lim_{x\to 3^+} \frac{x^2+x-12}{(x-3)(x-4)}$$

$$\lim_{x\to 3^-} \frac{x^2+x-12}{(x-3)(x-4)}$$

$$\lim_{x\to 4^+} \frac{x^2+x-12}{(x-3)(x-4)}$$

$$\lim_{x\to 4^-} \frac{x^2+x-12}{(x-3)(x-4)}$$

Use in-place replacement to Factor the numerator. How does this relate to what you observed earlier?

Hint To enter the one-sided limit $\lim_{x\to 3^+} \frac{x^2+x-12}{(x-3)(x-4)}$, put the insertion point in mathematics mode and type lim. Choose Insert + Subscript and type $x \to 3$ in the input box. Choose Insert + Superscript and type $+$ in the input box. Press SPACEBAR two times to return to the line. Enter the rational function $\frac{x^2+x-12}{(x-3)(x-4)}$.

5. Investigate the limit
$$\lim_{x\to 0} \frac{\sin x}{x}$$
by looking closely at the graph of
$$y = \frac{\sin x}{x}$$
for x close to zero. What is the limit and how do you know this is the limit?

1.3 Calculating Limits Using the Limit Laws

The limit laws express properties of limits. You investigated several of these limit laws in the previous activity "Taking it to the Limit." For example, when the limits exist separately, the limit of a sum is the sum of the limits, and the limit of a product is the product of the limits. Here, we investigate another law, variously referred to as the "squeeze theorem," the "sandwich theorem," or the "pinching theorem," which says that if values of a function lie between the values of two other functions, both having the same limit at some point a, then the middle function must have the same limit at that point a.

Theorem (The Squeeze Theorem) *If $f(x) \le g(x) \le h(x)$ for all x in an open interval that contains a (except possibly at a) and*
$$\lim_{x\to a} f(x) = \lim_{x\to a} h(x) = L$$
then
$$\lim_{x\to a} g(x) = L$$

Caught in the Middle

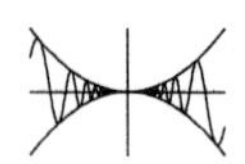

Study the limit $\lim_{x\to 0} x^2 \sin \frac{1}{x}$ by doing the following. Plot a graph of $g(x) = x^2 \sin \frac{1}{x}$ for $-0.1 \le x \le 0.1$. Find appropriate functions $f(x)$ and $h(x)$ that satisfy the hypotheses of the Squeeze Theorem, as applied to $g(x)$ at 0. Explain how you know that the required inequalities are satisfied. Add plots of f and h to your graph of g. Use your plot to predict a value for the limit. Calculate the limits $\lim_{x\to 0} f(x)$ and $\lim_{x\to 0} h(x)$ and explain the answers. Apply the Squeeze Theorem to find $\lim_{x\to 0} x^2 \sin \frac{1}{x}$, and verify your answer by evaluating $\lim_{x\to 0} x^2 \sin \frac{1}{x}$.

Lynn's Solution

We are interested in the behavior of the graph of the function $g(x) = x^2 \sin \frac{1}{x}$ near $x = 0$. This graph appears to bounce back and forth a lot.

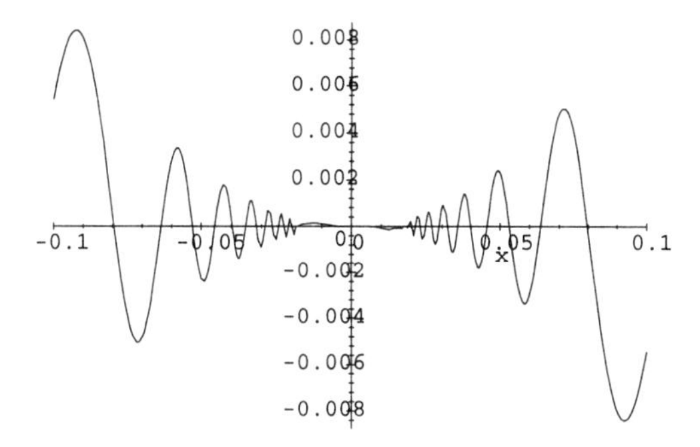

With the insertion point in the expression $x^2 \sin \frac{1}{x}$, choose Plot 2D + Rectangular. With the plot selected, choose Edit + Properties; choose the Plot Components page, set the Domain Interval to $-0.1 \leq x \leq 0.1$, and choose OK.

Because $-1 \leq \sin \frac{1}{x} \leq 1$ for all $x \neq 0$ and $x^2 > 0$ for $x \neq 0$, it follows that $-x^2 \leq x^2 \sin \frac{1}{x} \leq x^2$ for all $x \neq 0$. Let $f(x) = -x^2$ and $h(x) = x^2$, so that $f(x) \leq g(x) \leq h(x)$ for all $x \neq 0$. The function $h(x) = x^2$ is a continuous function, so $\lim_{x \to 0} h(x) = h(0) = 0$. Similarly, $\lim_{x \to 0} f(x) = f(0) = 0$. Then it follows from the Squeeze Theorem that $\lim_{x \to 0} g(x) = 0$.

To enter $\lim_{x \to 0} x^2$, choose Insert + Math and type lim. (It should automatically turn gray when you type the m.) Choose Insert + Subscript and enter $x \to 0$ in the input box. Press SPACEBAR. Type x^2. With the insertion point in the expression $\lim_{x \to 0} x^2$, choose Evaluate. With the insertion point in the expression $\lim_{x \to 0} \left(-x^2\right)$, choose Evaluate. With the insertion point in the expression $\lim_{x \to 0} \left(x^2 \sin \frac{1}{x}\right)$, choose Evaluate.

This is illustrated graphically in the following figure:

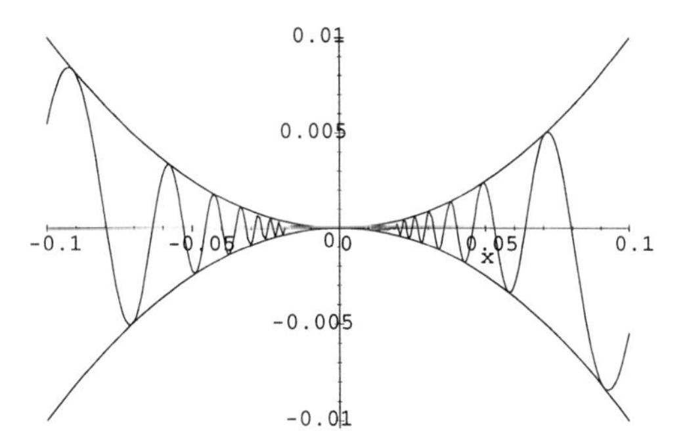

Select and drag $-x^2, x^2$ to the previous graph.

Activities

1. Explain why $\lim_{x\to 1}\left(\frac{x^2}{x-1}-\frac{1}{x-1}\right) \neq \lim_{x\to 1}\left(\frac{x^2}{x-1}\right)-\lim_{x\to 1}\left(\frac{1}{x-1}\right)$. Use graphs of $y=\frac{x^2}{x-1}-\frac{1}{x-1}$, $y=\frac{x^2}{x-1}$, and $y=\frac{1}{x-1}$ in your discussion. Describe how this inequality is related to the statement, "If the limits $\lim_{x\to a} f(x)$ and $\lim_{x\to a} g(x)$ exist, then $\lim_{x\to a}(f(x)-g(x)) = \lim_{x\to a}(f(x)) - \lim_{x\to a}(g(x))$."

2. Explain why $\lim_{x\to 0}\left(\frac{1}{x}\cdot\sin x\right) \neq \lim_{x\to 0}\left(\frac{1}{x}\right)\cdot\lim_{x\to 0}(\sin x)$. Include appropriate graphs in your discussion. Describe how this inequality is related to the statement, "If the limits $\lim_{x\to a} f(x)$ and $\lim_{x\to a} g(x)$ exist, then $\lim_{x\to a}(f(x)\cdot g(x)) = \lim_{x\to a}(f(x))\cdot\lim_{x\to a}(g(x))$."

3. Use the limit laws and rules of algebra to calculate each of the following limits. V erify your results visually by creating appropriate graphs.

   a. $\lim_{x\to 1}\frac{x^2+x-2}{x^2-1}$

   b. $\lim_{h\to 0}\frac{(3+h)^3-27}{h}$

   c. $\lim_{x\to 0}\frac{x}{\sqrt{4+3x}-2}$

   d. $\lim_{x\to 2^-}\lfloor x\rfloor$ and $\lim_{x\to 2^+}\lfloor x\rfloor$ (where $\lfloor x\rfloor$ denotes the greatest integer $\leq x$).

1.4 The Precise Definition of a Limit

Let f be a function defined on an open interval that contains the number a, although not necessarily defined at a itself. Then we say that **the limit of $f(x)$ as x approaches a** is L, and we write

$$\lim_{x\to a} f(x) = L$$

if given any number $\varepsilon > 0$ there exists a number $\delta > 0$ such that

$$0 < |x-a| < \delta \quad \text{implies} \quad |f(x)-L| < \varepsilon$$

Viewing Rectangles for the Epsilon-Delta Game

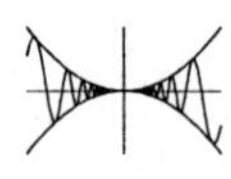

Define $f(x) = \frac{\sin x}{x}$ and plot the graph. From the graph, it will appear that $f(0) = 1$. However, $\lim_{x\to 0}\sin x = 0$ and $\lim_{x\to 0} x = 0$ so $\lim_{x\to 0}\frac{\sin x}{x}$ is indeterminate of type $\frac{0}{0}$.

Thus the formula $\frac{\sin x}{x}$ does not define $f(0)$. To understand this limit, take a closer look at the graph inside of appropriate viewing rectangles.

Make four graphs of $\frac{\sin x}{x}$ that vertically center viewing rectangles about $y = 1$. Taking ε (epsilon) to have the values $1, 0.1, 0.01, 0.001$, visually estimate the δ (delta) needed for each of these ε's (epsilons) to capture the graph inside the rectangle, and reset the limits on x accordingly. Explain how this relates to the definition of a limit for $\frac{\sin x}{x}$ at $x = 0$.

Dorothy's Solution

The graph of $f(x) = \frac{\sin x}{x}$ for $-5 < x < 5$ is shown on the left, and the portion of the same graph that lies inside the viewing rectangle $-5 < x < 5$ and $0 < y < 2$ is shown on the right.

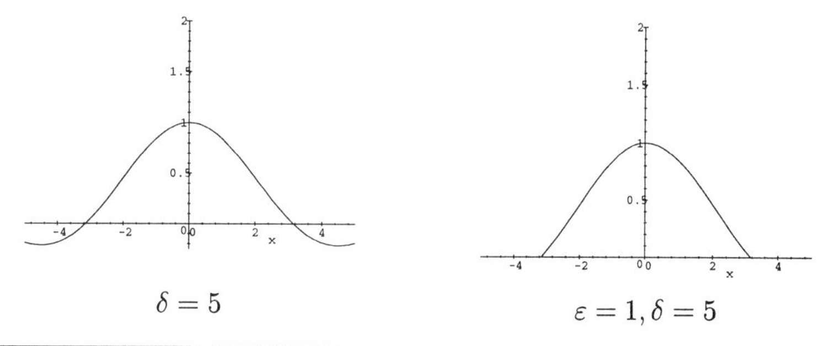

$\delta = 5$ $\qquad\qquad$ $\varepsilon = 1, \delta = 5$

With the insertion point in the expression $\frac{\sin x}{x}$, choose Plot 2D + Rectangular. Choose Edit + Properties, View page. Click Default to turn it off and set the y-interval to $0 < y < 2$. Choose OK.

Comparing these two pictures, you can see that part of the graph lies below the $\varepsilon = 1, \delta = 5$ viewing window, so we need to make the viewing window narrower. To get the graph inside the viewing window for $\varepsilon = 1$, we can take $\delta = 3$.

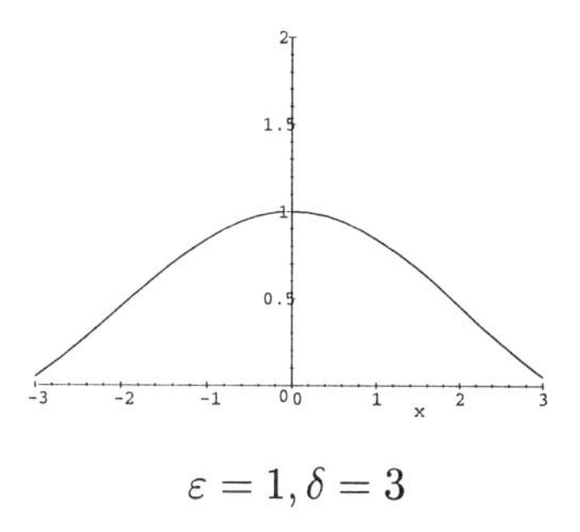

$\varepsilon = 1, \delta = 3$

Looking at graphs for $\varepsilon = .1$ $(0.9 < y < 1.1)$, $\varepsilon = .01$ $(0.99 < y < 1.01)$, and $\varepsilon = .001$ $(0.999 < y < 1.001)$, I found I could use $\delta = .7$, $.2$, and $.06$ to get the graph inside the viewing angle. Smaller deltas would have worked just as well.

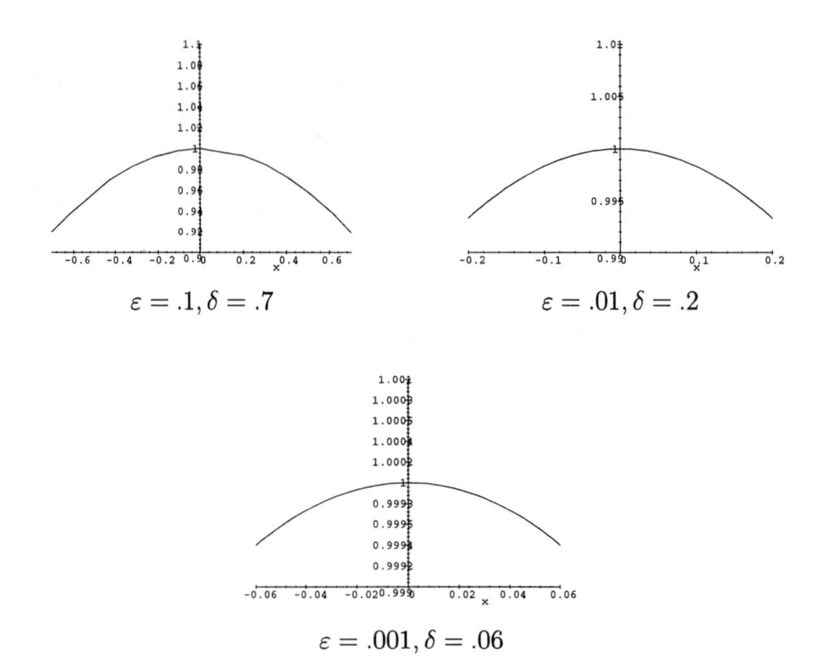

$\varepsilon = .1, \delta = .7$

$\varepsilon = .01, \delta = .2$

$\varepsilon = .001, \delta = .06$

These four graphs support the hypothesis that the function $f(x) = \frac{\sin x}{x}$ has the limit 1 at $x = 0$. The definition of limit says that for each $\varepsilon > 0$ there must be a $\delta > 0$ for which the graph of the function lies entirely inside the viewing window centered at $(0, 1)$ with height 2ε and width 2δ. To create the preceding graphs, we first set the vertical limits at $1 - \varepsilon < y < 1 + \varepsilon$ and then examined the graph to determine horizontal limits $-\delta < x < \delta$ that would capture the graph inside the viewing rectangle. We showed that it is possible to do this for $\varepsilon = 1, 0.1, 0.001, 0.0001$.

> **Remark** While working on this problem, you may notice that, as the viewing rectangle gets very small, the graph gets jagged and does not even appear symmetric. Something is clearly wrong if that happens! To avoid this problem, reset the domain intervals to match the viewing intervals. Otherwise, because the system is computing the graph for the domain interval, it will be computing a lot of points outside the viewing rectangle and not many points inside the very small viewing rectangle. After resetting the domain intervals to match the viewing intervals you will get the graphs pictured here.

Activity (A Winning Strategy)

When Elaine read the definition of a limit, her cunning mind quickly developed a winning strategy for a new game called Epsilon-Delta. Given a function and a point a in its domain, person A picks a number L she guesses to be the limit of the function at the point

a. Then person B picks a positive number ε and person A must find a positive number δ so that the graph of the function is captured inside the view window centered at (a, L) with height 2ε and width 2δ. The game ends when one person gives up. Elaine played the game with the function $f(x) = \frac{\sin x}{x}$, asking her friends to pick an epsilon for an interval about 1 and then cleverly coming up every time with a delta for an interval about 0 that captured the function in the view window centered at $(0, 1)$. Then she turned the tables, using the function $f(x) = \sin \frac{1}{x}$. She offered to pick the epsilons for an interval about 0 and asked her friends to pick a number L and then to find a delta interval about L for each of her epsilons that captured the function in the view window centered at $(0, L)$. Explain how she outwitted her friends.

Which side would you choose to be on in the Epsilon-Delta game for each of the following situations? What numbers L would you pick when you chose to be person A?

$f(x)$	$\frac{\sin x^2}{x}$	$\frac{\sin x}{x^2}$	$\frac{\sin x^2}{x^2}$	$\frac{x^4-4}{x^2-2}$	$\frac{x+2}{x-1}$	$\frac{2t^2-11t+5}{5-t}$	$\frac{\sqrt{x}-1}{x-1}$	$x \sin \frac{1}{x}$
a	0	0	0	$\sqrt{2}$	1	5	1	0
L	?	?	?	?	?	?	?	?

Describe the behavior of each of these functions near the indicated point. Include graphs in your discussion.

1.5 Continuity

A continuous function not only is always headed somewhere (has a limit at every point of its domain), but passes through all of the expected places (the value of the function is equal to the limit). When a function is continuous, it passes through all the intermediate points between values.

Through the Looking Glass

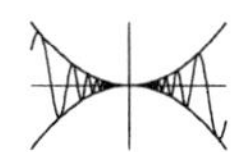

When Alice stepped through the looking glass and found herself in a quite unexpected place, she had experienced a *discontinuity*. Make a graph that shows the living room sofa at the origin $(0, 0)$, shows her heading in a straight path toward the looking glass at the point $(1, 1)$, but finds her in Wonderland at the point $(1, 0)$, walking in a straight line toward her next adventure at the point $(2, 2)$. Imagine the first coordinate depicts time passing and the second coordinate depicts her location. Put this graph into a short story that describes Alice's progress along the path. In your story, make a connection between the continuous and discontinuous parts of her path and the idea of limits discussed in the previous section. Where does the intermediate value theorem fail to hold?

George's Solution

Alice sat in her living room, relaxed and quite content. She saw something in the mirror that roused her curiosity, so she stood up and walked straight to the glass. Leaning forward to look into the glass she suddenly found herself on the other side in a quite unexpected place. Looking back she saw no continuous path back to her living room; in fact, she could not see her living room at all! Feeling most amazed and more than a little nervous, she started walking toward something she could see in the distance. Her feelings of curiosity overcome her nervousness and she was excited by the prospect of adventure.

Her adventure so far is depicted in the following graph:

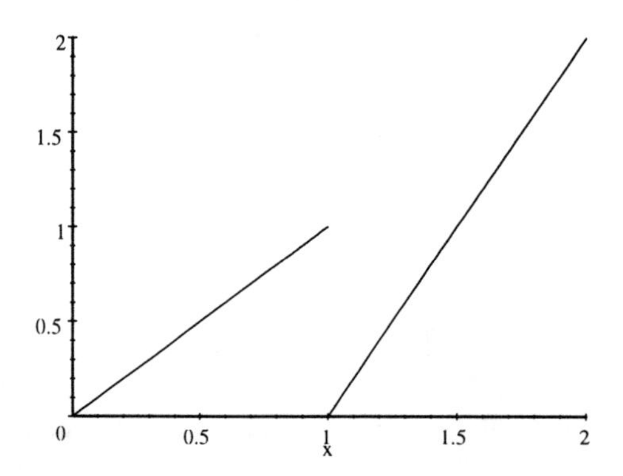

Choose Insert + Math and type $f(x) =$. Choose Insert + Brackets and select Left brace, Right null delimiter (dashed line), and OK. Choose Insert + Matrix. Set at 2 Rows, 3 Columns, and choose OK. Enter expressions in the left column, the word "if" in the middle column (in either math or text), and the interval in the right column:

$$f(x) = \begin{cases} x & \text{if} \quad x < 1 \\ 2x - 2 & \text{if} \quad x \geq 1 \end{cases}$$

With the insertion point in the expression, choose Define + New Definition. To plot this function, select f with the mouse and choose Plot 2D + Rectangular. Choose Edit + Properties. Choose Plot Components page; set Domain Interval to $0 < x < 2$. Choose Frame page; set Placement to Displayed. Choose OK.

On the first leg of her journey, she is traveling according to the rule $y = x$ and at each moment t less than 1, she finds herself at the point $\lim_{x \to t} x = t$, as expected. However, at the moment $t = 1$, she expects to be at $\lim_{x \to 1} x = 1$, but unexpectedly finds herself at $y = 0$. She went from almost $y = 1$ to $y = 0$ without passing through any of the points in between. Undeterred, she takes off twice as fast, in the same direction.

Dan Took a Bath on This One

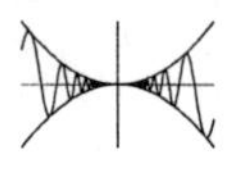

Dan decides to take a bath. He turns on the bath water at 7:00 a.m., and the depth in inches of water in the tub, where t measures the number of minutes after 7:00 a.m., is given approximately by the following piecewise-defined function.

$$h(t) = \begin{cases} t+5-\frac{10}{t+2} & \text{if} \quad t \le 8 \\ 12 & \text{if} \quad 8 < t \le 10 \\ \frac{4}{t-8}+13 & \text{if} \quad 10 < t \le 20 \\ 10 & \text{if} \quad 20 < t \le 21 \\ 10-2(t-21) & \text{if} \quad 21 < t \end{cases}$$

Plot this function on the interval $0 \le t \le 26$ and write a story that describes what is happening on each of the five subintervals. Describe any discontinuities in the function h. Explain these discontinuities in your story. Does the intermediate value theorem hold?

Alicia's Solution

Dan decides to take a bath and turns on the bath water at 7:00 a.m. The depth of water in inches after t minutes is given approximately by the piecewise-defined function

$$h(t) = \begin{cases} t+5-\frac{10}{t+2} & \text{if} \quad t \le 8 \\ 12 & \text{if} \quad 8 < t \le 10 \\ \frac{4}{t-8}+13 & \text{if} \quad 10 < t \le 20 \\ 10 & \text{if} \quad 20 < t \le 21 \\ 10-2(t-21) & \text{if} \quad 21 < t \end{cases}$$

The graph of the function describing the depth of water versus time is given in the following figure:

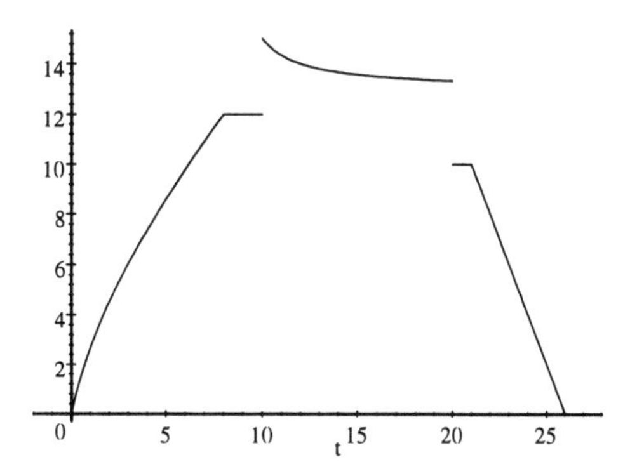

In math, type $h(t) =$. Choose Insert + Brackets; select Left brace and Right null delimiter (dashed line), and choose OK. Choose Insert + Matrix (5 Rows, 3 Columns). Type formulas in left column, "if" in middle column, and intervals in right column. With the insertion point in the expression, choose Define + New Definition. Select the expression $h(x)$ with the mouse, and choose Plot 2D + Rectangular. Change the Domain Interval to $0 < x < 26$.

During the first 8 minutes the bathtub is filling. It fills faster initially because the cross-sectional area of the tub is smallest at the bottom of the tub. At 7:08 Dan turns off the water and the water level is constant for the next two minutes. At 7:10 Dan steps into the tub and water starts to go out the overflow. Dan steps out of the tub at 7:20 and grabs a towel. At 7:21 Dan pulls the plug and the water level drops. The straight line between 7:21 and 7:26 may be explained by the effects of water pressure and surface area balancing each other. There are two discontinuities in the mathematical function h. One of these jumps takes place when Dan steps into the tub and the other when Dan steps out.

Although the water level changes very quickly, there are no actual physical discontinuities, and hence a connected graph is more appropriate. If it takes Dan 3 seconds to get into the tub and 7 seconds to get out, the real depth of water increases over a 3-second interval and decreases over a 7-second interval. The discontinuities in the mathematical model correspond to rapid but continuous changes in the physical situation. The intermediate value theorem does not hold in the model, but in the actual physical situation, although the water level may change very quickly, it does pass through all the intermediate values! This situation is reflected in the following figure, where the discontinuities show up as vertical line segments.

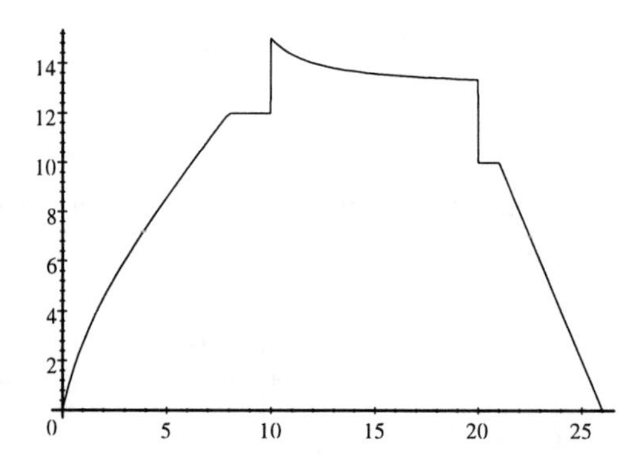

Select the function name h with the mouse, and choose Plot 2D + Rectangular. Change the Domain Interval to $0 < x < 26$.

Activities

1. Find two different choices for the numbers a and b that make the following piecewise-defined function continuous.

$$f(x) = \begin{cases} x^2 & if \quad x \leq 1 \\ ax + b & if \quad x > 1 \end{cases}$$

Plot the variations together. Find a choice of a and b so that f is discontinuous. Plot the graph.

2. Imagine more adventures for Alice and make a piecewise-defined graph to illustrate

the story. Make some abrupt changes that are continuous and some that are not, and discuss the differences.

3. Find a piecewise-defined function corresponding to each of the following graphs. V erify your solution by using a viewing rectangle that displays a graph that looks like the given graph. Describe the techniques you used to obtain your solution.

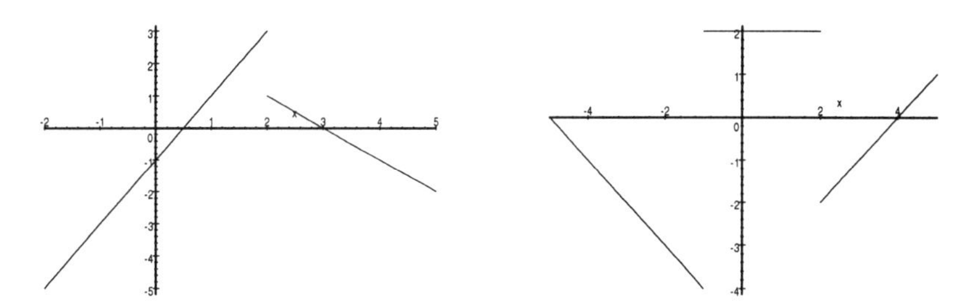

1.6 Limits at Infinity; Horizontal Asymptotes

Sometimes, as x gets larger and larger, $f(x)$ gets closer and closer to some number. We write $\lim_{x\to\infty} f(x) = L$ when we can make the value of $f(x)$ as close to L as we like by taking x sufficiently large. This is similar to the Epsilon-Delta game—given $\varepsilon > 0$ there is always some number N such that $|f(x) - L| < \varepsilon$ whenever $x > N$.

Continuous mathematical models illustrate average behavior rather than the jumps that would be expected in a discrete model. Much of the power of calculus lies in the observation that many problems that are discrete (where, for example, only integer values make sense) can be described in a useful way using continuous models.

Rabbit Island

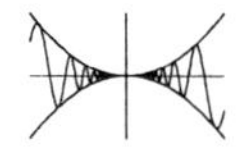

The function $r(t) = 50000 + \dfrac{50000t}{\sqrt{100 + t^2}}$ was used as a model to describe the number of rabbits on Rabbit Island, where t represents the number of years before and after the base year of 1900. Plot the graph for the years between 1860 and 1940. Interpret the significance of the horizontal asymptotes $r = 0$ and $r = 100\,000$.

Cheryl's Solution

The number of rabbits on Rabbit Island between the years of 1860 and 1940 is depicted in the following graph:

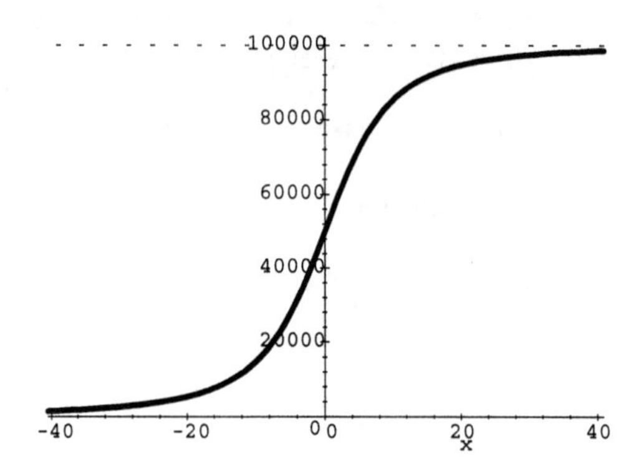

With the insertion point in the expression $50000 + \frac{50000t}{\sqrt{100+t^2}}$, choose Plot 2D + Rectangular. Select and drag to the frame 100000. Select the frame and choose Edit + Properties, Plot Properties page. Set Domain Interval at -40 to 40. Select Item Number 1 and, under Thickness, choose Medium. Select Item Number 2 and, under Line Style, choose Dots.

The horizontal asymptote $r = 100000$ represents the maximal sustainable population on the island. Only a fixed amount of vegetation grows on the island, and this vegetation can support a maximum of 100000 rabbits. The horizontal asymptote $r = 0$ represents the limiting level of the population many years ago. The initial population was certainly at least 2, when caged rabbits were set free just before their ship was sunk.

Activities

1. Find the horizontal and vertical asymptotes of $y = \frac{x}{\sqrt{x^2-1}}$. Draw the graph and indicate asymptotes with dotted lines. Explain any gaps you find in the domain. Calculate $\lim_{x\to\infty} \frac{x}{\sqrt{x^2-1}}$ and $\lim_{x\to-\infty} \frac{x}{\sqrt{x^2-1}}$ by hand, using the technique of multiplying and dividing by $\frac{1}{x}$. Explain why
$$\frac{1}{x}\sqrt{x^2-1} = -\sqrt{1-\frac{1}{x^2}}$$
for $x < 0$.

2. Draw a graph of $y = \frac{x^3+1}{x^3-2x^2-5x+6}$, using a viewing window that clearly displays all of its horizontal and vertical asymptotes. Factor the denominator and explain the connection between this factorization and the location of the vertical asymptotes. Calculate one-sided limits at each of the vertical asymptotes and describe connections between these limits and the graph.

1.7 Tangents, Velocities, and Other Rates of Change

Using limits, you can compute rates of change.

Give Me a Brake

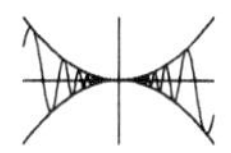

The total stopping distance for a car in a panic stop depends on reaction distance and braking distance. Reaction distance is approximately proportional to the velocity v, as it takes most drivers the same amount of time to react by putting their foot on the brake pedal at one driving speed as another, usually between 0.5 and 1.0 second. Braking distance is proportional to v^2, and hence the total stopping distance is given by $D(v) = av + bv^2$ for some unknown coefficients a and b. Following is a sample of experimental data:

20	25	30	35	40	45	50	55	60	65	70	75	80
42	56	74	92	116	142	173	210	248	292	343	401	464

with the number in the top row indicating velocity in miles per hour and the number immediately below each velocity indicating stopping distance in feet. Use the approximations $D(40) \approx 116$ and $D(60) \approx 248$ and solve the system

$$\begin{aligned} D(40) &= 116 \\ D(60) &= 248 \end{aligned}$$

for the coefficients a and b. Plot the model $D(v)$ together with the data points.

Estimate a and b by using two other data points. Add the graph of $av + bv^2$ to your figure and describe any discrepancies you see.

Jean's Solution

Because the reaction distance is proportional to v and braking distance is proportional to v^2, the total stopping distance should be given by $D(v) = av + bv^2$ for some numbers a and b. We will test this model on the following set of experimental data:

v	20	25	30	35	40	45	50	55	60	65	70	75	80
D	42	56	74	92	116	142	173	210	248	292	343	401	464

We first solve for a and b using the two data points $D(40) \approx 116$ and $D(60) \approx 248$. The solution gives the model $D(v) = \frac{13}{30}v + \frac{37}{600}v^2$.

With the insertion point in the expression $D(v) = av + bv^2$, choose Define + New Definition. Choose Insert + Display and type $D(40) = 116$ in the input box. Press ENTER and type $D(60) = 250$. With the insertion point in the display,

$$\begin{aligned} D(40) &= 116 \\ D(60) &= 248 \end{aligned}$$

choose Solve + Exact.

A graph of this model together with the original data points is given in the following figure:

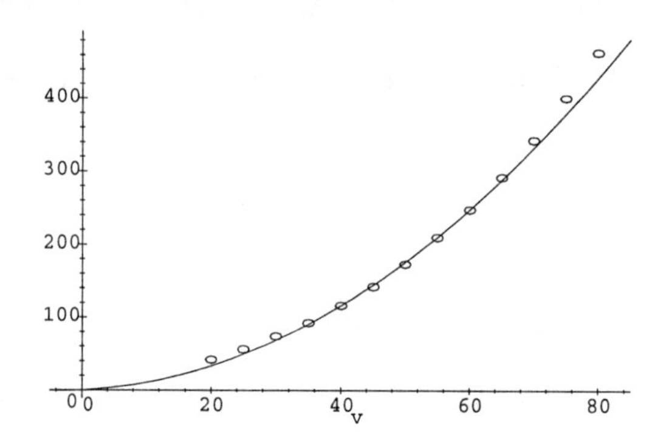

With the insertion point in the expression $\frac{13}{30}v+\frac{37}{600}v^2$, choose Plot 2D + Rectangular. Select and drag to the frame each of the following

$$(20, 42, 25, 56, 30, 74, 35, 92) \quad (40, 116, 45, 142, 50, 173)$$
$$(55, 210, 60, 248, 65, 292) \quad (70, 343, 75, 401, 80, 464)$$

Select the frame and choose Edit + Properties, Plot Components page. For each of Items 2, 3, 4, 5, under Plot Style, choose Point; under Point Symbol, choose Circle. Choose OK.

(Alternate method) To plot the data without retyping the numbers, you can use some matrix operations. With the insertion point in the matrix

$$\begin{matrix} 20 & 25 & 30 & 35 & 40 & 45 & 50 & 55 & 60 & 65 & 70 & 75 & 80 \\ 42 & 56 & 74 & 92 & 116 & 142 & 173 & 210 & 248 & 292 & 343 & 401 & 464 \end{matrix}$$

choose Matrix + Transpose. With the insertion point in the (vertical) matrix, choose Matrix + Reshape. (Specify 26 columns.) With the insertion point in the (horizontal) matrix, choose Plot 2D + Rectangular.

Note that the graph goes through the points $(40, 116)$ and $(60, 248)$ but appears to miss some of the other data points.

We can also solve for a and b using the data points $(20, 42)$ and $(70, 343)$. Here we get $a = \frac{49}{50}$ and $b = \frac{7}{125}$. Adding the graph of $\frac{49}{50}v + \frac{7}{125}v^2$ produces

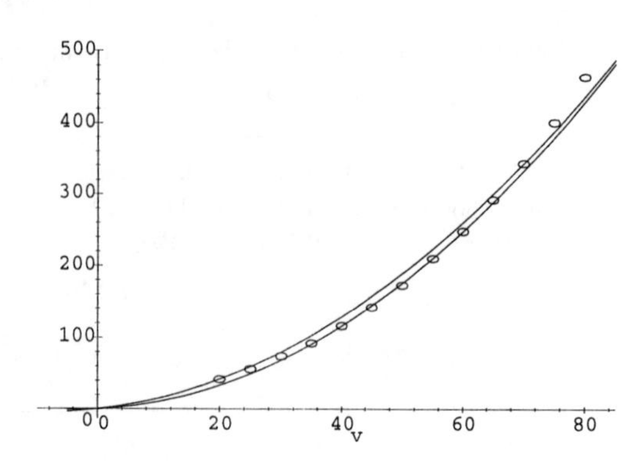

 With the insertion point in the system of equations

$$\begin{aligned} D(20) &= 42 \\ D(70) &= 343 \end{aligned}$$

choose Solve + Exact. Select and drag to the frame $\frac{49}{50}v + \frac{7}{125}v^2$.

This seems to indicate that the model one gets depends rather strongly on which pair of data points are chosen. I suspect there are some small errors in the original data, and that picking just two points to represent the entire data set is questionable at best. There should be a method that depends in some way on the entire data set.

Remark Stay tuned. Such a method will be given in a later chapter.

Activities

1. A crankshaft turns with its center at the origin. A piston is attached to a rod at $(0, y)$.

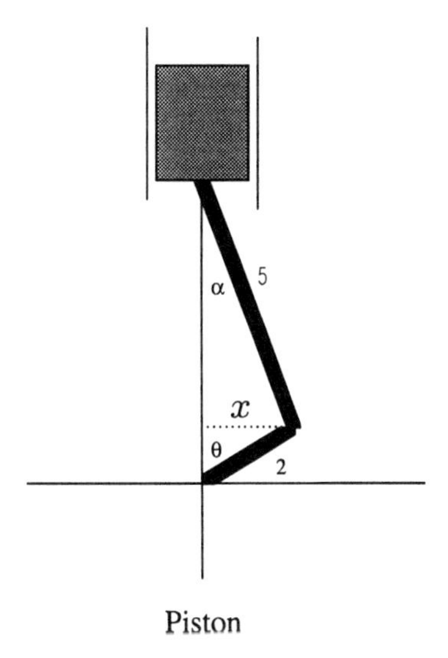

Piston

From the figure, $y = y_1 + y_2$, where $y_1 = 2\cos\theta$ and $y_2 = 5\cos\alpha$. The angles α and θ are related by the fact that $x = 2\sin\theta = 5\sin\alpha$, which implies $\alpha = \arcsin\left(\frac{2}{5}\sin\theta\right)$. Thus $y = 2\cos\theta + 5\cos\left(\arcsin\left(\frac{2}{5}\sin\theta\right)\right)$.

Assume that the crankshaft turns at a constant rate, so that $\frac{d\theta}{dt} = c$ radians/second. Visually study the motion of the piston by graphing (on the same axes) y versus θ, $\frac{dy}{d\theta}$ versus θ, and $\frac{d^2y}{d\theta^2}$ versus θ. At what point is the piston moving the fastest? At what point is the acceleration of the piston the greatest? Over what interval(s) is the acceleration of the piston nearly constant?

2. Find numbers a and b that make the piecewise-defined function

$$f(x) = \begin{cases} x^2 + 1 & \text{if} \quad x < 1 \\ -x^2 + ax + b & \text{if} \quad x \geq 1 \end{cases}$$

differentiable at $x = 1$. Verify your solution by drawing a graph of $y = f(x)$ on the interval $0 \leq x \leq 2$.

3. Consider the piecewise-defined function

$$g(x) = \begin{cases} x^2 & if \quad x < 0 \\ 1 & if \quad x \geq 0 \end{cases}$$

Show that g is not continuous at $x = 0$. Cite a theorem that states that if a function is differentiable at a point, then it is continuous at that point. Evaluate $g'(x)$. Show that g' is (apparently) continuous at $x = 0$. Discuss the apparent contradiction and what caused it.

2 DERIVATIVES

The derivative of a function measures the rate of change of the function. It has a geometric interpretation as the slope of a tangent. The derivative is also a function in its own right.

2.1 Derivatives

The **derivative of a function** f **at a number** a, denoted by $f'(a)$, is given by

$$f'(a) = \lim_{h\to 0} \frac{f(a+h) - f(a)}{h}$$

provided this limit exists.

If $f'(a)$ exists, then an **equation of the tangent line** to the curve $y = f(x)$ at the point $(a, f(a))$ is

$$y - f(a) = f'(a)(x-a)$$

or

$$y = f(a) + f'(a)(x-a)$$

You can use a variety of standard notation for a derivative in *Scientific Notebook*. To take the derivative of the expression $x^2 + \sin x$, evaluate one of the expressions

$$\frac{d}{dx}\left(x^2 + \sin x\right) \qquad \text{or} \qquad D_x\left(x^2 + \sin x\right)$$

(Use the fraction template to enter $\frac{d}{dx}$, even though the expression $\frac{d}{dx}$ is not a fraction in the usual sense.) You can also evaluate the expression $\frac{dy}{dx}$ if you first define y as an expression.

After defining $f(x)$ either as a generic function (see page 10) or in terms of a mathematical expression, you can find its derivative by evaluating one of the following expressions:

$$f'(x) \qquad \frac{d}{dx}f(x) \qquad D_x\left(f(x)\right)$$

To evaluate a derivative at a, use any of the equivalent forms

$$f'(a) \qquad \left[f'(x)\right]_{x=a} \qquad \left[\tfrac{d}{dx}\left(x^2 + \sin x\right)\right]_{x=a}$$

$$\left[\frac{dy}{dx}\right]_{x=a} \qquad \left[\tfrac{d}{dx}f(x)\right]_{x=a} \qquad \left[D_x\left(x^2 + \sin x\right)\right]_{x=a}$$

Note Use standard notation and apply Evaluate to find a derivative. There is no special menu item for general derivatives and none is needed. (The choice Implicit Differentiation on the Calculus submenu is needed only for special situations that will be explained later.)

The Shape of the Derivative

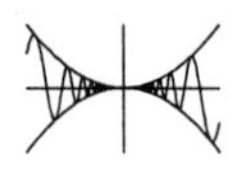

For the following two figures, discuss the properties of the thin and thick graphs relative to one another. Explain why it is reasonable to claim that the thin graph depicts the derivative of the function depicted by the thick graph in each figure.

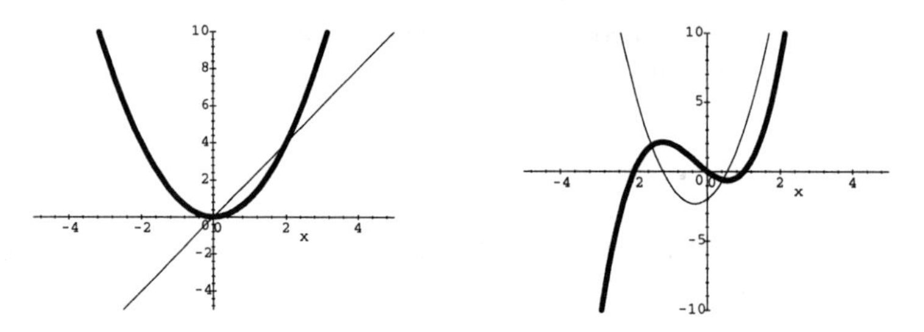

Bernice's Solution

In the figure

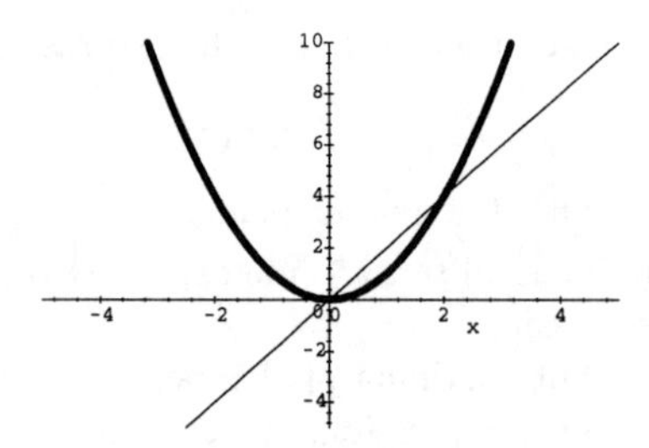

the thick graph has only one point with a horizontal tangent, and at that point the thin graph has the value zero.

With the insertion point in the expression x^2, choose Plot 2D + Rectangular. Select x with the mouse and drag to the plot. With the insertion point to the right of the graph, choose Edit + Properties, Plot Components page. For Item Number 1, change Thickness to Medium. Choose OK. Use Scientific Notebook to write down your observations.

Although the slope of the thick graph is sometimes negative and sometimes positive, the value of the slope is always increasing as you move from left to right along the curve, consistent with the property that the thin graph is always increasing. These observations are consistent with the possibility that the thin graph represents the derivative of the function described by the thick graph.

In the second figure

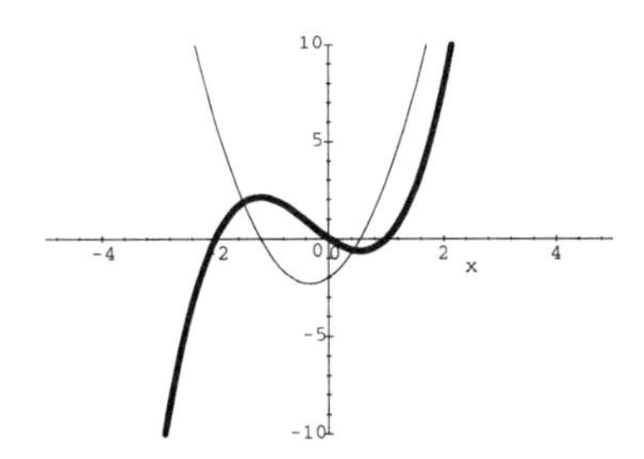

the thick graph has two points with a horizontal tangent, and at those points the thin graph has the value zero.

With the insertion point in the expression $(x-1)(x+2)x$, choose Plot 2D + Rectangular. Select $3x^2+2x-2$ with the mouse and drag it to the plot. With the insertion point to the right of the graph, choose Edit + Properties, Plot Components page. For Item Number 1, change Thickness to Medium. Choose OK. Use Scientific Notebook to write down your observations.

In the left side of the thick graph, the value of the slope is decreasing as you move from left to right along the curve, consistent with the fact that the thin graph is decreasing on that interval. In the right side of the thick graph, the value of the slope is increasing as you move from left to right along the curve, consistent with the fact that the thin graph is increasing on that interval. These observations are consistent with the possibility that the thin graph represents the derivative of the function described by the thick graph.

Activity

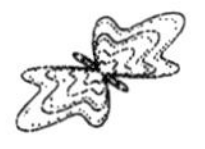

For the following figure, discuss the properties of the thin and thick graphs relative to one another. Explain why it is reasonable to claim that the thin graph depicts the derivative of the function depicted by the thick graph.

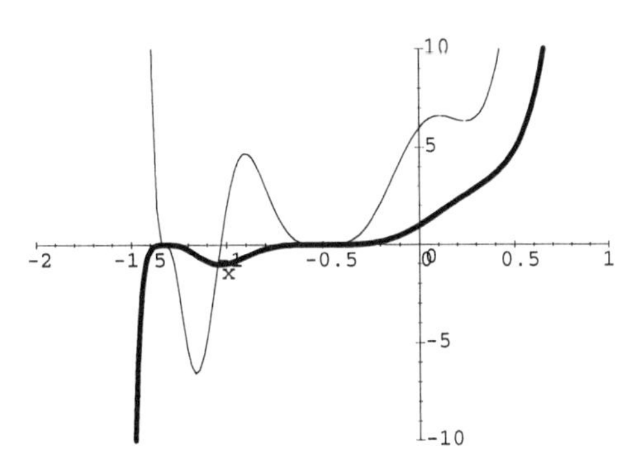

$$y=(2x+1)^5\left(x^3-x+1\right)^4$$

A Double Touch

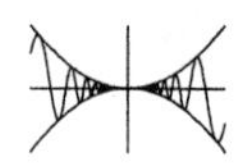

Find an equation of a line that is tangent to the graph of

$$f(x) = x^4 - 10x^3 + 27x^2 - 18x$$

at two different points: $(a, f(a))$ and $(b, f(b))$. Plot a graph of f and estimate a and b by drawing a line by hand that appears to be tangent at two points. Find a and b exactly by defining f as a function and then solving the system

$$\begin{aligned} f'(a) &= m \\ f'(b) &= m \\ \frac{f(b) - f(a)}{b - a} &= m \end{aligned}$$

of equations for a, b, and m. Explain why this methodology works by equating an average rate of change with two instantaneous rates of change. Write the equation of the tangent line in simplified form. Verify your solution by plotting the curve and line together. Zoom in far enough on the points of tangency to observe that the curve and the tangent line are indistinguishable in a very small window about the point.

June's Solution

The graph of $y = x^4 - 10x^3 + 27x^2 - 18x$ appears to rest on a line (shown as a dashed line) that is tangent at two points. The two points are given approximately by $(0.2, -3)$ and $(4.8, -40)$.

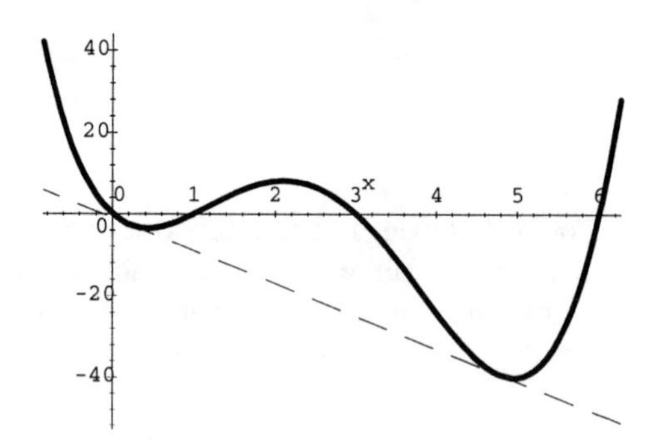

Given the function $f(x) = x^4 - 10x^3 + 27x^2 - 18x$, the slope of the line segment joining $(a, f(a))$ and $(b, f(b))$ is given by

$$m = \frac{f(b) - f(a)}{b - a}$$

This slope must equal the slope of the tangent line at a and the slope of the tangent line at b, hence $m = f'(a) = f'(b)$. Solving the system

$$f'(a) = m$$

$$\begin{aligned} f'(b) &= m \\ \frac{f(b)-f(a)}{b-a} &= m \end{aligned}$$

yields two solutions: $\{a = b, m = 4b^3 - 30b^2 + 54b - 18, b = b\}$ and $\{b = \rho, a = 5 - \rho, m = -8\}$, where ρ is a root of $Z^2 - 5Z + 1$.

With the insertion point in the equation $f(x) = x^4 - 10x^3 + 27x^2 - 18x$, choose Define + New Definition.

The first solution $a = b$ violates the condition that $(a, f(a))$ and $(b, f(b))$ be two distinct points. The polynomial $Z^2 - 5Z + 1$ has the two roots

$$a = \frac{5}{2} - \frac{1}{2}\sqrt{21} \approx .20871$$

and

$$b = \frac{5}{2} + \frac{1}{2}\sqrt{21} \approx 4.7913$$

With the insertion point in the equation $Z^2 - 5Z + 1 = 0$, choose Solve + Exact.

The tangent line is given by $y = f(a) + f'(a)(x - a)$, which simplifies to $y = -8x-1$. To check this solution, we plot the graph of f together with the line $y = -8x-1$.

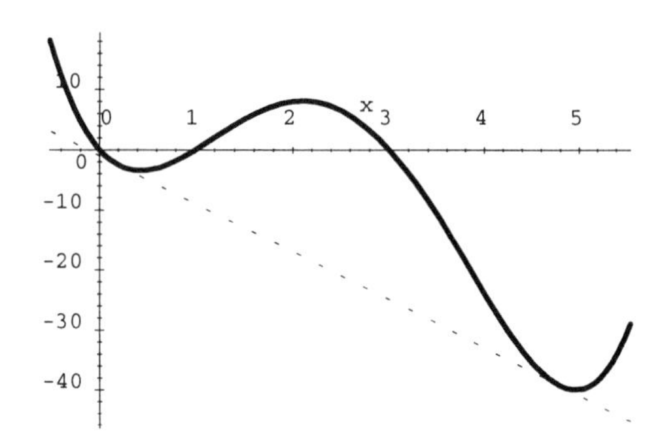

The line indeed appears to be tangent at two points.

With the insertion point in the equation $a = \frac{5}{2} - \frac{1}{2}\sqrt{21}$, choose Define + New Definition. With the insertion point in the equation $b = \frac{5}{2} + \frac{1}{2}\sqrt{21}$, choose Define + New Definition. With the insertion point in the expression $f(x)$, choose Plot 2D + Rectangular. Select and drag to the frame $f(a) + f'(a)(x - a)$. Choose Edit + Properties, Plot Components page; change Thickness to Medium for Item Number 1, and change Line Style to Dots for Item Number 2. Choose OK.

The two points of tangency are approximately $(.20871, -2.6697)$ and $(4.7913, -39.33)$, reasonably close to our original estimate.

Zooming in on the two points of tangency shows that the curve and tangent line almost coincide near the two points of tangency.

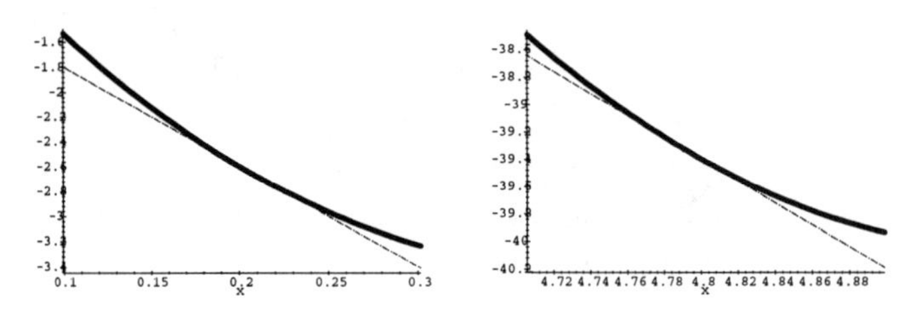

Zooming in further is even more convincing.

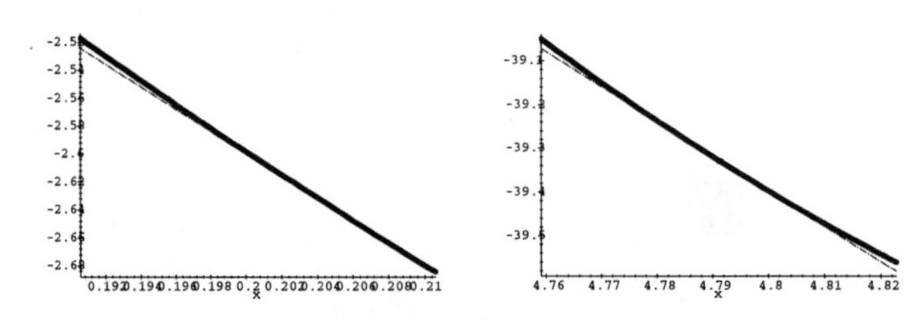

Eventually the two are indistinguishable.

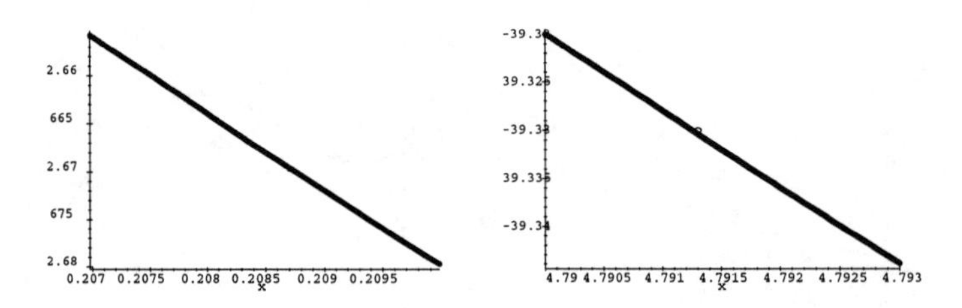

Activities

1. Plot $y = x^2$ and visually find two lines that pass through the point $(0, -4)$ that are tangent to the curve $y = x^2$. Use some calculus to find these lines exactly and verify your equations by plotting the two lines on the same screen as $y = x^2$.

2. Given the function $f(x) = x^5 - 19x^4 + 79x^3 + 171x^2 - 792x$, visually locate three lines, each of which is tangent to f at two different points $(a, f(a))$ and $(b, f(b))$. Use these estimates one line at a time to generate search intervals for a and b; then solve a system of equations numerically. Verify your solutions visually by attaching three lines to a plot of f.

2.2 Differentiation Formulas

With differentiation formulas you can differentiate many functions using only a small number of basic formulas.

Basic Differentiation Rules

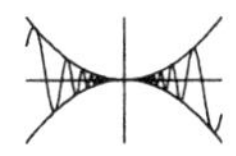

Define $f(x)$ and $g(x)$ as generic functions. Evaluate each of the following expressions with *Scientific Notebook.*

$$\frac{d}{dx}(f(x)+g(x))$$
$$\frac{d}{dx}(f(x)g(x))$$
$$\frac{d}{dx}\left(\frac{f(x)}{g(x)}\right)$$
$$\frac{d}{dx}(x^n)$$
$$\frac{d}{dx}(f\circ g)(x)$$
$$\frac{d}{dx}(f(g(x)))$$

Verify that your results are compatible with the usual differentiation formulas. What are the subtle assumptions made by *Scientific Notebook* about the functions f and g?

Shane's Solution

If f and g are defined as generic functions, then evaluation leads to the formulas

$$\begin{aligned}
\frac{d}{dx}(f(x)+g(x)) &= f'(x)+g'(x) \\
\frac{d}{dx}(f(x)g(x)) &= f'(x)\,g(x)+f(x)\,g'(x) \\
\frac{d}{dx}\left(\frac{f(x)}{g(x)}\right) &= \frac{f'(x)}{g(x)}-\frac{f(x)}{g^2(x)}g'(x) \\
\frac{d}{dx}(x^n) &= x^n\frac{n}{x} \\
\frac{d}{dx}(f(g(x))) &= f'(g(x))\,g'(x)
\end{aligned}$$

Ⓝ Choose Define + Clear Definitions. With the insertion point in the expression $f(x)$, choose Define + New Definition. With the insertion point in the expression $g(x)$, choose Define + New Definition. With the insertion point in each of the expressions stated in the problem, choose Evaluate. To display your results, select the first equation with your mouse and choose Insert + Display.

Press ENTER. Drag and drop the second equation into the input box. Press ENTER again, and continue in similar fashion until all of the equations are displayed.

Most of these results agree exactly with the usual differentiation formulas. Factoring yields

$$\frac{f'(x)}{g(x)} - \frac{f(x)}{g^2(x)} g'(x) = \frac{f'(x) g(x) - f(x) g'(x)}{g^2(x)}$$

which gives the usual expression for the derivative of a quotient. Simplification gives

$$x^n \frac{n}{x} = x^{n-1} n$$

which is the usual expression for the power rule.

With the insertion point in the expression $\frac{f'(x)}{g(x)} - \frac{f(x)}{g^2(x)} g'(x)$, choose Factor. With the insertion point in the expression $x^n \frac{n}{x}$, choose Simplify.

Throughout, *Scientific Notebook* has treated all the generic functions as differentiable. Also, for the quotient rule, it has assumed that $g(x) \neq 0$ and for the power rule that $x \neq 0$ (or in the case n is negative, that x is positive).

Activities

1. Define $f(x)$ as a generic function. Evaluate $\frac{d}{dx}\left(x^2 f(x)\right)$. Explain the result.
2. For what values of x does the graph of $f(x) = 2x^3 - 3x^2 - 6x + 87$ have a horizontal tangent? Estimate these values using a plot, then find them exactly by applying Solve + Exact to $f'(x) = 0$.

2.3 Rates of Change in the Natural and Social Sciences

You saw in Section 2.1, given the graph of a function, how to zoom in on the graph to find the slope of the tangent line at a particular point. In the following activities, you will discover how to find an equation for a polynomial function, given certain details about function values and slopes of tangent lines at particular points.

You already know to do this for linear functions. For example, find the equation of a line given that its slope is 3 and that the graph goes through the point $(-2, 1)$. An appropriate answer is given by the point-slope form of a straight line: $y - 1 = 3(x + 2)$, which may be written as $y = 3x + 7$.

Here is another approach that builds on ideas in calculus. One advantage to this method is that it can be easily extended to polynomials of higher degree, as you will see in the activities.

Making the Grade

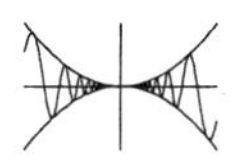

Find an equation of the line with slope 3 whose graph goes through the point $(-2, 1)$. Verify your equation by plotting it together with a small circle at the point $(-2, 1)$.

Marilyn's Solution

To find an equation for the line with slope 3 whose graph goes through the point $(-2, 1)$, we begin by defining the function $f(x) = ax + b$.

Define + Clear Definitions. With the insertion point in the equation $f(x) = ax + b$, choose Define + New Definition.

The slope of the line at x is the derivative of the function f at the point x, so we have the condition $f'(x) = 3$. The points on the graph are of the form $(x, f(x))$, so another condition we have is $f(-2) = 1$. Solving this system of two equations

$$\begin{aligned} f(-2) &= 1 \\ f'(x) &= 3 \end{aligned}$$

gives the solution $\{a = 3, b = 7\}$. Thus the function f is given by $f(x) = 3x + 7$.

Choose Insert + Display and type $f(-2) = 1$ in the input box. Press ENTER and type $f'(x) = 3$ in the input box. With the insertion point in the display

$$\begin{aligned} f(-2) &= 1 \\ f'(x) &= 3 \end{aligned}$$

choose Solve + Exact.

This formula can be verified geometrically. The following plot shows the line $y = 3x + 7$ and the point $(-2, 1)$. It also shows the line $y = 3x$ to check the slope.

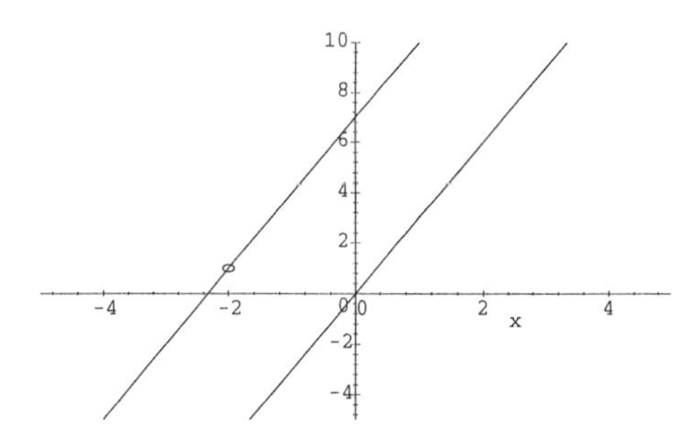

$y = 3x + 7, (-2, 1), y = 3x$

With the insertion point in the expression $3x + 7$, choose Plot 2D + Rectangular. Select the expression $(-2, 1)$ and drag it to the frame. Click the frame to select it. Choose Edit + Properties. Click the Plot Components tab. Choose Item Number 2. Change Plot Style to Point. Change Point Symbol to Circle. Choose OK. Select the expression $3x$ and drag it to the frame.

Activities

Reminder Begin each activity by choosing Define + Clear Definitions.

1. Find a quadratic polynomial function $f(x) = ax^2 + bx + c$ that has the following three properties:

   a. The graph goes through the point $(5, 8)$.
   b. The slope of the tangent line at $x = 3$ is equal to 7.
   c. The slope of the tangent line at $x = 7$ is -2.

   Plot a graph of your function with its tangent lines at $x = 3$ and $x = 7$. Explain how you know that your function f has the stated properties. Include both algebraic and geometric evidence in your discussion.

2. Find a cubic polynomial function $f(x) = ax^3 + bx^2 + cx + d$ that has the following properties:

   a. The graph of f goes through the points $(-2, 10)$ and $(5, 0)$.
   b. The slope of the tangent line at $(-2, 10)$ is 0 and the slope of the tangent line at $(5, 0)$ is 0.

   Explain how you know that your function has all of the stated properties. Include both algebraic and geometric evidence in your discussion. Find equations for a tangent line at $x = 3$ and a tangent line at $x = 7$. Add these two tangent lines to your plot.

3. A portion of a highway is being designed to go over a mountain pass. The grade going up one side is 4%, and the grade down the other side is 7%. (The *grade* of a highway is the same as the *slope* of the tangent line. Thus a 4% grade indicates that the elevation changes by 4 feet for each 100 feet of horizontal change.) A transition curve in the shape of an inverted parabola is needed to connect the two steep stretches of highway. Define a function $f(x) = ax^2 + bx + c$ for the transition curve assuming the parabola highway changes from 4% grade (slope of $4/100$) to parabola $y = f(x)$ at the point $(0, f(0))$ and that the highway changes from parabola to a downhill 7% grade (slope of $-7/100$) at the point $(2000, f(2000))$, where x and

$f(x)$ are measured in feet. For simplicity, assume the elevation at the first transition point is 0—that is, $f(0) = 0$. Write down two equations that will ensure that the transition points look smooth. Find a, b, and c by solving a system of three equations.

Plot the 4% uphill grade, the transition parabola $y = f(x)$, and the 7% downhill grade. What is the elevation of the transition point on the 7% downhill grade? What is the elevation at the highest point on the stretch of highway?

4. The solution to the preceding highway problem may look smooth, but it may not feel smooth because the piecewise-defined function does not have a continuous second derivative. Design a section of a roller coaster by piecing together two parabolas: $f(x) = ax^2 + bx + c$ and $g(x) = rx^2 + sx + t$. Assume the two parabolas share a transition point at $(0, 0)$, and that the slope at the transition point is 100%, so that the angle with the horizontal is 45°. Assume that $f''(0) = \frac{1}{200}$ and $g''(0) = -\frac{1}{200}$. Plot a graph of the two parabolas using the viewing rectangle $-400 \leq x \leq 400$ by $-100 \leq y \leq 100$. Describe what a ride on this roller coaster would feel like when you cross the transition point at $(0, 0)$.

5. Your goal is to design a better highway by using a piecewise-defined function on the interval $0 < x < 2000$ that has a continuous second derivative. Define the quadratic $f(x) = ax^2 + bx + c$ for use on the interval $100 \leq x \leq 1900$; define the cubic $g(x) = dx^3 + jx^2 + kx + l$ for use on the interval $0 \leq x \leq 100$; and define the cubic $h(x) = mx^3 + nx^2 + px + q$ for the interval $1900 \leq x \leq 2000$. Explain why it is necessary and sufficient to solve the following system of 11 equations and 11 unknowns:

$$
\begin{aligned}
g(0) &= 0 \\
g'(0) &= .04 \\
g''(0) &= 0 \\
g(100) &= f(100) \\
g'(100) &= f'(100) \\
g''(100) &= f''(100) \\
f(1900) &= h(1900) \\
h'(1900) &= f'(1900) \\
h''(1900) &= f''(1900) \\
h'(2000) &= -.07 \\
h''(2000) &= 0
\end{aligned}
$$

Solve the system and plot the graphs of f, g, and h together with the linear approaches at 4% and 7% on either side.

2.4 Derivatives of Trigonometric Functions

The derivatives of trigonometric functions are again trigonometric functions. This activity uses a graphical approach to explore this phenomenon.

Intriguing Derivatives

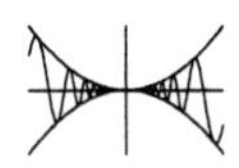

Define f by

$$f(x) = \frac{\sin(x+h) - \sin x}{h}$$

and then define $h = 0.1$. Plot the graph of $y = f(x)$, and then select $\frac{d}{dx} \sin x$ and drag it to the frame. Describe what you see on the screen, and explain why this is reasonable. Explain why it is reasonable that

$$\frac{d}{dx} \sin x = \cos x$$

Diana's Solution

Define f by

$$f(x) = \frac{\sin(x+h) - \sin x}{h}$$

and then define $h = 0.1$.

Choose Define + Clear Definitions. (In other words, start with a clean sheet of paper!) With the insertion point in the equation $f(x) = \frac{\sin(x+h)-\sin x}{h}$, choose Define + New Definition. With the insertion point in the equation $h = 0.1$, choose Define + New Definition

---

The plots of $y = \dfrac{\sin(x+0.1) - \sin x}{0.1}$ and $y = \frac{d}{dx} \sin x$ are given in the following figure:

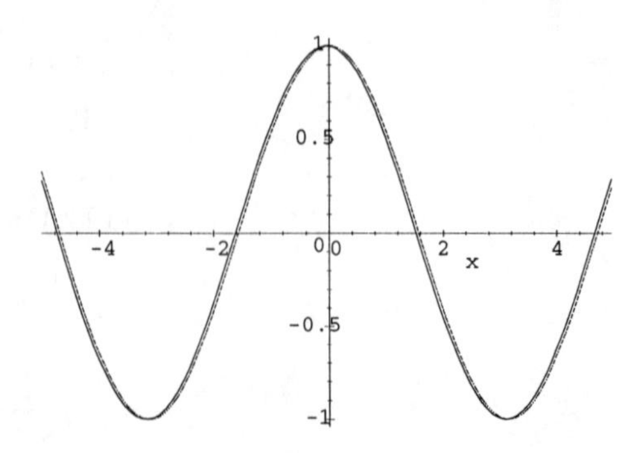

With the insertion point in the list $\frac{\sin(x+h)-\sin x}{h}$, $\frac{d}{dx} \sin x$, choose Plot 2D + Rectangular.

The two curves seem to have similar shapes, although one curve is slightly translated horizontally. It it difficult to distinguish one curve from the other. This is reasonable because

$$\frac{d}{dx}\sin x = \lim_{h\to 0}\frac{\sin(x+h)-\sin x}{h}$$

and therefore the two curves should look similar for h close to zero.

The curves look like the graph of $y = \cos x$. Indeed, $\cos x$ and $\frac{d}{dx}\sin x$ plotted together give the following plot. It appears to be a single curve.

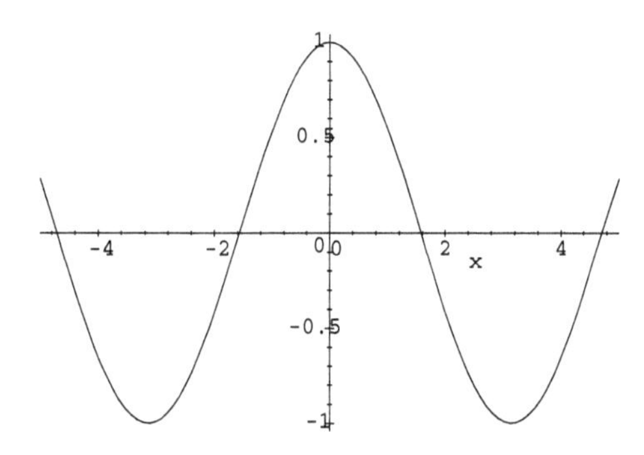

 With the insertion point in the list $\cos x$, $\frac{d}{dx}\sin x$, choose Plot 2D + Rectangular.

Activities

1. Use the definition of derivative to approximate the derivative of $y = \tan x$. Explain why it is reasonable that
$$\frac{d}{dx}\tan x = \sec^2 x$$
How do you know by looking at the graph that the derivative of $\tan x$ is always positive?

2. By creating appropriate plots, tell why you believe that

   a. $\lim_{\theta\to 0}\sin\theta = 0$
   b. $\lim_{\theta\to 0}\cos\theta = 1$
   c. $\lim_{\theta\to 0}\frac{\sin\theta}{\theta} = 1$
   d. $\lim_{\theta\to 0}\frac{1-\cos\theta}{\theta} = 0$

3. Construct a table of derivatives for the six basic trigonometric functions. Plot each function with its derivative and describe the geometric features that indicate that one graph represents the derivative of the other.

4. Show that the two functions $f(x) = \sin(\tan x)$ and $g(x) = \sin x \tan x$ are different by plotting their graphs together. Explain why one of the graphs has vertical asymptotes and the other does not. Describe other essential differences between the two functions.

2.5 The Chain Rule

The chain rule describes the derivative of the composition of functions in terms of the derivatives of the functions. With the chain rule added to the differentiation rules you already know, and knowing only a very small number of simple formulas, you can compute the derivative of all of the functions generally dealt with in calculus: the *elementary functions.* The elementary functions are the polynomials, rational functions, power functions, exponential functions, logarithmic functions, trigonometric and inverse trigonometric functions, hyperbolic and inverse hyperbolic functions, and all functions that can be obtained from these by the five operations of addition, subtraction, multiplication, division, and composition. For example, a function that appears as complicated as this

$$f(x) = \sqrt[3]{\frac{x^5 - 6x^3 + 2x - 1}{3x^{\frac{1}{4}}}} + \sinh\left(\ln x^3\right) - \cos\left(e^{x^4 + \ln x} - \tan x\right)$$

is considered an elementary function. In this activity you will practice combining basic differentiation rules and formulas.

The Power of the Rules

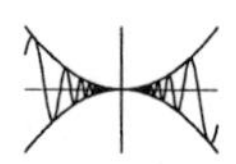

Waldo knew the power rule

$$\frac{d}{dx}x^n = nx^{n-1}$$

and his sister May knew the chain rule

$$\frac{d}{dx}\left(f(g(x))\right) = f'\left(g\left(x\right)\right)g'\left(x\right)$$

They discovered that, working together, they had the generalized power rule

$$\frac{d}{dx}g(x)^n = ng(x)^{n-1}g'(x)$$

ready to use. Explain their reasoning. After their brother Wendell learned the product rule,

$$\frac{d}{dx}\left(f(x)g(x)\right) = f'\left(x\right)g\left(x\right) + f\left(x\right)g'\left(x\right)$$

the three of them, working together, found that they knew the quotient rule

$$\frac{d}{dx}\left(\frac{f(x)}{g(x)}\right)=\frac{f'(x)\,g(x)-f(x)\,g'(x)}{g(x)^2}$$

as well. Explain how they knew this.

Ana's Solution

As soon as they saw that the function $y=g(x)^n$ can be viewed as a composition $y=f(g(x))$, where $f(x)=x^n$, they had the problem solved. With Waldo's power rule, they knew that $f'(x)=nx^{n-1}$, so that $f'(g(x))=ng(x)^{n-1}$

Define + Clear Definitions. With the insertion point in the expression $g(x)$, choose Define + New Definition. With the insertion point in the equation $f(x)=x^n$, choose Define + New Definition. With the insertion point in the expression $f'(g(x))$, choose Evaluate, and then choose Simplify.

Putting this together with May's chain rule they had

$$\frac{d}{dx}g(x)^n=\frac{d}{dx}(f(g(x)))=f'(g(x))\,g'(x)=ng(x)^{n-1}g'(x)$$

After their brother Wendell learned the product rule, the three of them looked at a quotient $\frac{f(x)}{h(x)}$ and observed that it could be rewritten as a product $f(x)\,(h(x))^{-1}$. Now they could put their generalized power rule to good use. Taking $g(x)=h(x)^{-1}$, they knew that $g'(x)=-h(x)^{-2}h'(x)$, so they had

$$\begin{aligned}\frac{d}{dx}\left(\frac{f(x)}{h(x)}\right) &= \frac{d}{dx}\left(f(x)\left(h(x)^{-1}\right)\right)\\ &= \frac{d}{dx}(f(x)g(x))=f'(x)\,g(x)+f(x)\,g'(x)\\ &= f'(x)\,h(x)^{-1}-f(x)\,h(x)^{-2}h'(x)\end{aligned}$$

Use Scientific Notebook to enter these expressions. Because you are "showing all the steps," you cannot simply evaluate. You can use some shortcuts such as Copy and Paste and "in-place replacement." Recall that in-place replacement is done by selecting an expression with the mouse and pressing the CTRL key while choosing Evaluate.

Then, calling on their expertise in algebra, they simplified this expression to get the usual form of the quotient rule.

$$\frac{d}{dx}\left(\frac{f(x)}{h(x)}\right)=\frac{f'(x)\,h(x)-f(x)\,h'(x)}{(h(x))^2}$$

Activity (Together, We Can Do it All)

After Waldo learned the power rule, May learned the chain rule, and Wendell learned the product rule, the rest of the family decided to get into the act. Esther studied trigonometry and learned the derivative

$$\frac{d}{dx}\sin x = \cos x$$

of the sine function. Marjorie studied exponential functions and learned the derivative

$$\frac{d}{dx}e^x = e^x$$

of the exponential function base e, and Oscar learned the sum rule

$$\frac{d}{dx}(f(x) + g(x)) = f'(x) + g'(x)$$

For each of the following functions, explain who has to work together (that is, which rules are needed) in order to find the derivative of the function and how they can combine their rules to find the derivative. (You may assume that everyone in the family has a thorough knowledge of algebra and trigonometry. They know facts like $\cos x = \sin\left(x + \frac{\pi}{2}\right)$ and $\ln e^x = x$, for example.)

$$\begin{aligned} y &= 3x^3 + 2x \\ y &= x\cos x \\ y &= \sin\left(x^2 + x\right) \\ y &= xe^{\tan x} \\ y &= \ln x \\ y &= \sqrt{x + e^x} \\ y &= \frac{\sin x}{x + 1} \end{aligned}$$

What is your opinion of their claim, "Together, we can do it all"?

2.6 Implicit Differentiation

Something implied or understood though not directly expressed is *implicit*. An implicit function is determined by an equation that cannot necessarily be solved in an elementary fashion. The technique of implicit differentiation allows you to differentiate such functions without expressing them directly.

The Plot Is Implicit

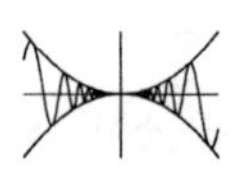

Make an implicit plot of the equation $2y^2 + xy = x^2 + 2$ with viewing rectangle $-5 < x < 5$ and $-5 < y < 5$. Verify algebraically that $(1, 1)$ is a point on the graph of the equation. Solve for y in terms of x and plot the curve that goes through the point $(1, 1)$. Find an equation of the tangent line at $(1, 1)$ by differentiating to find the slope of the tangent line. Differentiate the equation implicitly to find the slope of the tangent line at $(1, 1)$, and compare the results of the explicit and implicit methods. Verify that the equation of the tangent line is reasonable by graphing the equation together with its tangent line.

Carolyn's Solution

The graph of a quadratic equation is a conic section, and an implicit plot of the equation $2y^2 + xy = x^2 + 2$ reveals that it is an hyperbola.

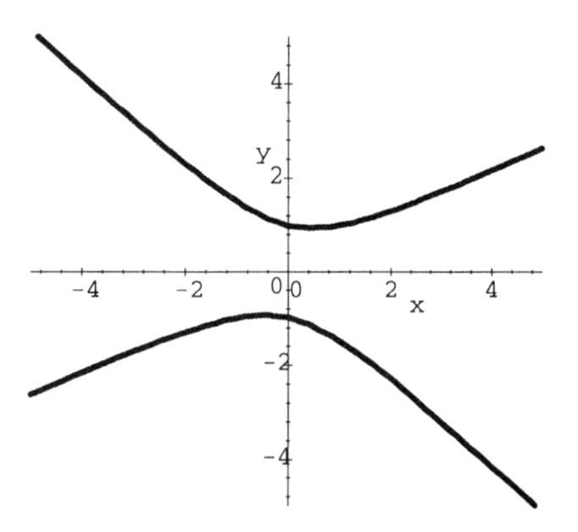

With the insertion point in the equation $2y^2 + xy = x^2 + 2$, choose Plot 2D + Implicit.

Solving the equation $2y^2 + xy = x^2 + 2$ for y, we get the two solutions

$$\left\{y = -\frac{1}{4}x + \frac{1}{4}\sqrt{(9x^2 + 16)}\right\}, \left\{y = -\frac{1}{4}x - \frac{1}{4}\sqrt{(9x^2 + 16)}\right\}$$

Because

$$\left[-\frac{1}{4}x + \frac{1}{4}\sqrt{(9x^2 + 16)}\right]_{x=1} = -\frac{1}{4} + \frac{1}{4}\sqrt{25} = 1$$

it follows that $(1, 1)$ is a point on the graph of $y = -\frac{1}{4}x + \frac{1}{4}\sqrt{(9x^2 + 16)}$.

With the insertion point in the equation $2y^2+xy = x^2+2$, choose Solve + Exact (Variable(s) to Solve For: y). Select the expression $-\frac{1}{4} + \frac{9}{4\sqrt{(9x^2+16)}}x$ with the mouse, choose Insert + Brackets, select Square Brackets, and choose OK. Choose Insert + Subscript and type $x = 1$ in the input box. With the insertion point in the expression $\left[-\frac{1}{4} + \frac{9}{4\sqrt{(9x^2+16)}}x\right]_{x=1}$, choose Evaluate, and then choose Simplify.

The graph of this branch of the curve is given by

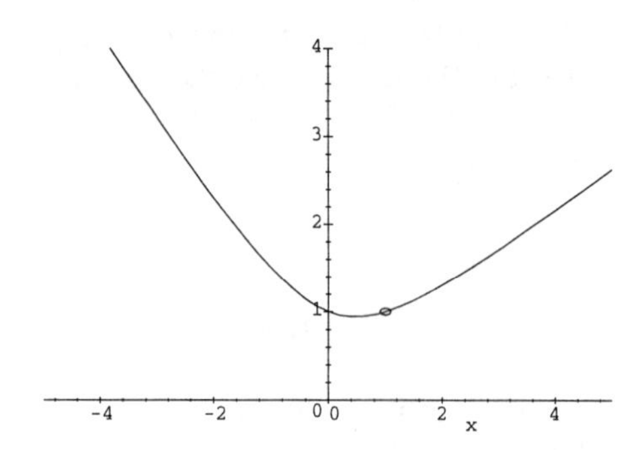

We add the point $(1,1)$ to the plot to verify visually that this point is actually on the curve.

With the insertion point in the expression $-\frac{1}{4}x + \frac{1}{4}\sqrt{(9x^2+16)}$, choose Plot 2D + Rectangular. Select the expression $(1,1)$ and drag it to the plot.

The derivative at $x = 1$ is given by

$$\left[\frac{d}{dx}\left(-\frac{1}{4}x + \frac{1}{4}\sqrt{(9x^2+16)}\right)\right]_{x=1} = \left[-\frac{1}{4} + \frac{9}{4\sqrt{(9x^2+16)}}x\right]_{x=1} = \frac{1}{5}$$

and hence the slope of the graph at $(1,1)$ is $\frac{1}{5}$ and the tangent line is given by $y - 1 = \frac{1}{5}(x-1)$, which simplifies to $y = \frac{4}{5} + \frac{1}{5}x$.

To show one of the steps in evaluating the derivative, use Edit + Copy and Paste to make a copy of $\left[\frac{d}{dx}\left(-\frac{1}{4}x + \frac{1}{4}\sqrt{(9x^2+16)}\right)\right]_{x=1}$ and type $=$ between the two expressions. For an "in-place replacement," select the copy of $\frac{d}{dx}\left(-\frac{1}{4}x + \frac{1}{4}\sqrt{(9x^2+16)}\right)$ with the mouse, and while pressing the CTRL key, choose Evaluate. With the insertion point in the expression $\left[-\frac{1}{4} + \frac{9}{4\sqrt{(9x^2+16)}}x\right]_{x=1}$, choose Evaluate. To simplify the equation for the tangent line, with the insertion point in the equation $y - 1 = \frac{1}{5}(x-1)$, choose Solve + Exact (Variable(s) to Solve For: y).

Alternatively, implicit differentiation applied to the equation $2y^2 + xy = x^2 + 2$ yields $4yy' + y + xy' = 2x$, and solving for y', we get

$$y' = -\frac{y-2x}{4y+x}$$

With the insertion point in the equation $2y^2 + xy = x^2 + 2$, choose Calculus + Implicit Differentiation (Differentiation Variable: x). With the insertion point in the equation $4yy' + y + xy' = 2x$, choose Solve + Exact (Variable(s) to Solve For: y').

Because

$$\left[-\frac{y-2x}{4y+x}\right]_{x=1,y=1}=\frac{1}{5}$$

it follows that an equation of the tangent line is $y-1=\frac{1}{5}(x-1)$, which can be rewritten as $x-5y=-4$.

Select the expression $-\frac{y-2x}{4y+x}$ with the mouse and choose Insert + Brackets (select Square Brackets). Choose Insert + Subscript and type $x=1, y=1$ in the input box. With the insertion point in the expression $\left[-\frac{y-2x}{4y+x}\right]_{x=1,y=1}$, choose Evaluate.

The implicit and explicit techniques lead to equivalent answers, but the two forms for the derivative have quite a different appearance:

$$y' = -\frac{1}{4}+\frac{9}{4\sqrt{(9x^2+16)}}x$$

$$y' = -\frac{y-2x}{4y+x}$$

The implicit form appears simpler, but one must remember that x and y remain related by the original equation.

The following is a graph of the equation $2y^2+xy=x^2+2$ together with the line $x-5y=-4$.

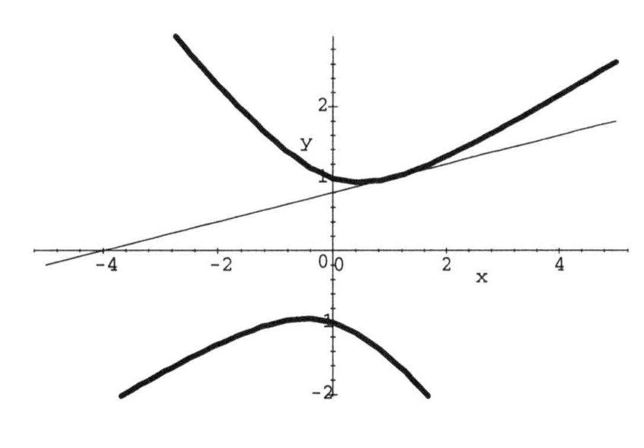

$2y^2+xy=x^2+2$, $x-5y=-4$

With the insertion point in the equation $2y^2+xy=x^2+2$, choose Plot 2D + Implicit. Select the equation $x-5y=4$ and drag it to the frame.

From the appearance of this graph, the equation we used for the tangent line is reasonable.

Activities

1. Consider the equation $x^3 + y^3 = 6xy$. Verify that $(3, 3)$ is a point on the graph of the equation. Find an equation of the tangent line at $(3, 3)$ by solving for y in terms of x and differentiating. Differentiate the equation implicitly to find the slope of the tangent line at $(1, 1)$. Verify that the equation of the tangent line is reasonable by plotting the equation with its tangent line. Discuss the pros and cons of using implicit differentiation versus solving for y in terms of x and differentiating explicitly.

2. Apply Plot 2D + Implicit to get a graph of $x^2y^2 = (y+1)^2(4-y^2)$. Verify that $(0, -1)$ is a point on the graph. Find two lines that are "tangent" to the graph at $(0, -1)$. Plot the curve together with these two tangent lines.

   Hint Solve for x in terms of y and assume that $\dfrac{dy}{dx} = \dfrac{1}{dx/dy}$.

3. Apply Plot 2D + Implicit to get a graph of $xy = 1$. Select and drag $x^2 - y^2 = 4$ to the frame. On the Axes page of the Plot Properties tabbed dialog, check Equal Scaling Along Each Axis. Find the point in the first quadrant where the two curves intersect, and attach tangent lines to each curve at that point. Do the tangent lines appear to be orthogonal? Explain why they are perpendicular or show that they are not.

2.7 Higher Derivatives

Scientific Notebook recognizes standard notation such as $\dfrac{d^2y}{dx^2}, \dfrac{d^2}{dx^2}f(x)$, and $f''(x)$ for second derivatives and notation such as $\dfrac{d^ny}{dx^n}, \dfrac{d^n}{dx^n}f(x)$, and $f^{(n)}(x)$ for nth derivatives.

Raising Derivatives to New Heights

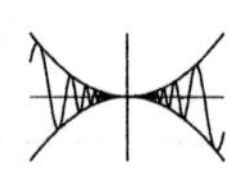

If $x^3 + y^3 = 1$, find y'' by implicit differentiation.

Lynette's Solution

Implicit differentiation with respect to x of the equation $x^3 + y^3 = 1$ yields

$$3x^2 + 3y^2y' = 0$$

Solving for y' we get

$$y' = -\frac{x^2}{y^2}$$

With the insertion point in the equation $x^3 + y^3 = 1$, choose Calculus + Implicit Differentiation (Differentiation Variable x). With the insertion point in the equation $3x^2 + 3y^2 y' = 0$, choose Solve + Exact (Variable(s) to Solve For: y').

This can be differentiated again with respect to x to yield

$$y'' = -2\frac{x}{y^2} + 2\frac{x^2}{y^3}y'$$

Replacing y' by its equivalent form $-\frac{x^2}{y^2}$ gives

$$\begin{aligned} y'' &= -2\frac{x}{y^2} + 2\frac{x^2}{y^3}\left(-\frac{x^2}{y^2}\right) \\ &= -2\frac{x}{y^2} - 2\frac{x^4}{y^5} \end{aligned}$$

With the insertion point in the equation $y' = -\frac{x^2}{y^2}$, choose Calculus + Implicit Differentiation (Differentiation Variable x). To replace y' with $\left(-\frac{x^2}{y^2}\right)$ in the equation $y'' = -2\frac{x}{y^2} + 2\frac{x^2}{y^3}y'$, select $-\frac{x^2}{y^2}$ with the mouse and click the Parentheses button on the Math toolbar; select $\left(-\frac{x^2}{y^2}\right)$ with the mouse and choose Edit + Copy; then, select y' with the mouse and choose Edit + Paste. With the insertion point in the expression $-2\frac{x}{y^2} + 2\frac{x^2}{y^3}\left(-\frac{x^2}{y^2}\right)$, choose Evaluate.

Splines Are Smooth

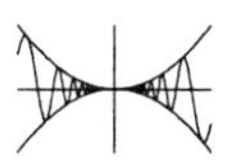

The word *spline* comes from the art of shipbuilding. Pieces of wood were soaked and reshaped smoothly to form the hull of a ship. Today, *spline* refers to a piecewise-defined function where the pieces are polynomials that fit together smoothly. In particular, a *cubic spline* is a piecewise-defined function f where the pieces are polynomials of degree 3 and such that $f''(x)$ is a continuous function.

Find a pair of cubic polynomials $p(x) = ax^3 + bx^2 + cx + d$ and $q(x) = rx^3 + sx^2 + tx + u$ that fit together to form a cubic spline

$$f(x) = \begin{cases} p(x) & \text{if} \quad x < 0 \\ q(x) & \text{if} \quad x \geq 0 \end{cases}$$

that has each of the following properties:

$$\begin{aligned} f(-1) &= 0 \\ f(0) &= 1 \\ f(1) &= 0 \end{aligned}$$

$$\begin{aligned} f'(-1) &= 0 \\ f'(1) &= 0 \end{aligned}$$

(A spline that specifies the derivative $f'(0)$ and $f'(2)$ at the endpoints is called a *clamped spline.*) Plot the graphs of p and q, and show how they fit together smoothly to form a spline.

Levi's Solution

We need to find a pair of cubic polynomials $p(x) = ax^3 + bx^2 + cx + d$ and $q(x) = rx^3 + sx^2 + tx + u$ that fit together to form a cubic spline

$$f(x) = \begin{cases} p(x) & \text{if } x < 0 \\ q(x) & \text{if } x \geq 0 \end{cases}$$

that has each of the following properties:

$$\begin{aligned} f(-1) &= 0 \\ f(0) &= 1 \\ f(1) &= 0 \\ f'(-1) &= 0 \\ f'(1) &= 0 \end{aligned}$$

The required properties may be stated in terms of p and q as follows:

$$\begin{array}{ll} p(-1) = 0 & p'(-1) = 0 \\ p(0) = 1 & q'(1) = 0 \\ q(0) = 1 & p'(0) = q'(0) \\ q(1) = 0 & p''(0) = q''(0) \end{array}$$

Note there are a total of eight conditions, corresponding to eight unknowns. The solution to this system of equation is:

$$\{d = 1, u = 1, a = -2, r = 2, b = -3, c = 0, t = 0, s = -3\}$$

With the insertion point in the equation $p(x) = ax^3 + bx^2 + cx + d$, choose Define + New Definition. With the insertion point in the equation $q(x) = rx^3 + sx^2 + tx + u$, choose Define + New Definition. Type $p(-1) = 0$ into a display; then press ENTER and type the remaining seven equations, one at a time. With the insertion point in the display, choose Solve + Exact.

Graphs of $p(x) = -2x^3 - 3x^2 + 1$ (using the thick pen) and $q(x) = 2x^3 - 3x^2 + 1$ (using a thin pen) are given in the following figure:

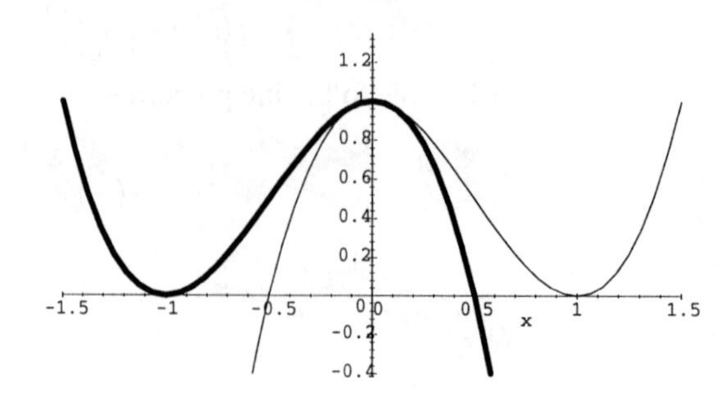

Define each of the coefficients by selecting the equation and choosing Define + New Definition. With the insertion point in the expression $p(x)$, choose Plot 2D + Rectangular. Select the expression $q(x)$ and drag it to the frame. To change the line thickness, choose Edit + Properties and revise the Plot Components page by changing the Thickness for Item Number 1 to Medium.

Here is a graph of the cubic spline.

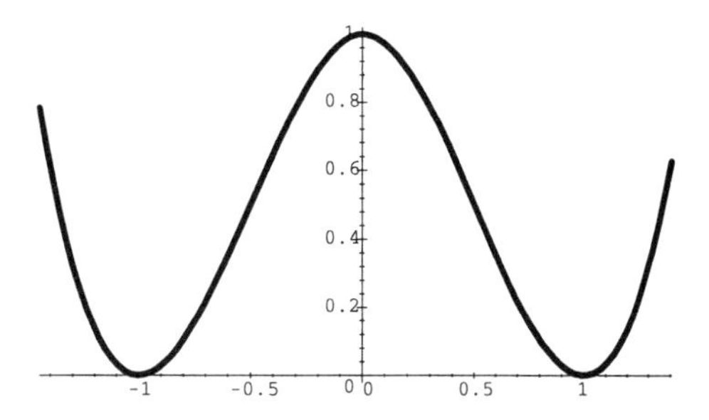

Type $f(x) = \begin{cases} p(x) & \text{if} \quad x < 0 \\ q(x) & \text{if} \quad x \geq 0 \end{cases}$ by choosing a left brace and empty right bracket; then choose Insert + Matrix (2 rows, 3 columns), typing expressions into the left column, if into the middle column, and the inequalities $x < 0$ and $x \geq 0$ into the right column. With the insertion point in the expression $f(x) = \begin{cases} p(x) & \text{if} \quad x < 0 \\ q(x) & \text{if} \quad x \geq 0 \end{cases}$, choose Define + New Definition. With the insertion point in the expression $f(x)$, choose Plot 2D + Rectangular. To change the line thickness, choose Edit + Properties and revise the Plot Components page by changing the Thickness to Medium.

Notice how smooth the graph of f looks. It is very similar to the graph of the fourth-degree polynomial

$$g(x) = x^4 - 2x^2 + 1 = \left(x^2 - 1\right)^2$$

whose graph follows:

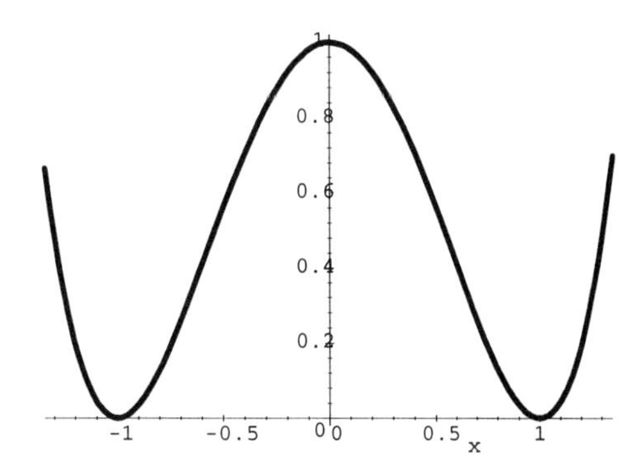

Activities

1. Let $f(x) = \sin x$. Explain how to find $f^{(n)}(x)$ without first computing all the intermediate derivatives $f'(x)$, $f''(x)$, ..., $f^{(n-1)}(x)$. Identify $y = f(x)$, $y = f'(x)$, and $y = f''(x)$ from plots and explain how the graphs of the first and second derivative describe features of the graph of f.

2. Find a cubic spline f defined on the interval $0 \leq x \leq 2$ that satisfies each of the following conditions:
$$f(0) = 0$$
$$f(1) = 1$$
$$f(2) = 1$$
$$f''(0) = 0$$
$$f''(2) = 0$$
(A cubic spline with the endpoint conditions $f''(0) = 0$ and $f''(2) = 0$ is called a *free spline*.)

2.8 Related Rates

Given the rate of change of one function, can you determine the rate of change of a related function?

The Fastest Ladder in the West

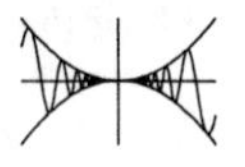

A ladder 13 feet long is leaning against a vertical wall. Sarah has finished painting for the day and decides to pull the foot of the ladder away from the wall to lay it on the ground. The ladder is initially 5 feet from the wall. If she moves the foot of the ladder away from the wall at a rate of 1 foot/second, how fast is the top of the ladder sliding down the wall at an instant when the foot of the ladder is x feet from the wall? Draw a set of ladders with the foot of the ladder at distances 5, 6, 7, 8, 9, 10, 11, and 12 feet from the wall. Make a table of vertical distances from the tip of the ladder to the ground. Describe how these distances are changing. Will the top of the ladder remain against the wall, or will it pull away from the wall before it hits the ground?

Bill's Solution

The following figure shows several positions of the ladder as the foot of the ladder is being pulled away from the wall. Notice that equal spacing along the horizontal axis produces increased spacing along the vertical axis.

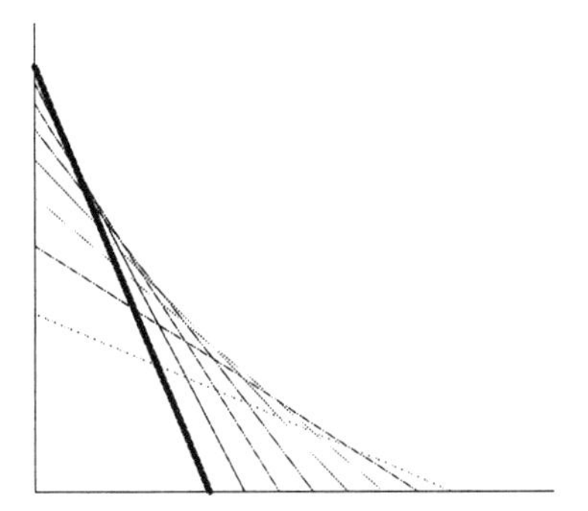

Foot of ladder sliding to the right

To make the wall and the ground, with the insertion point in the list $(0, 13, 0, 0, 31, 0)$, choose Plot 2D + Rectangular. Choose Insert + Properties, Axes page. Check Equal Scaling Along Each Axis and set Axes to None. Choose OK. To add the ladders, select and drag each of the following to the frame. $\left(5, 0, 0, \sqrt{13^2 - 5^2}\right)$, $\left(6, 0, 0, \sqrt{13^2 - 6^2}\right)$, $\left(7, 0, 0, \sqrt{13^2 - 7^2}\right)$, $\left(8, 0, 0, \sqrt{13^2 - 8^2}\right)$, $\left(9, 0, 0, \sqrt{13^2 - 9^2}\right)$, $\left(10, 0, 0, \sqrt{13^2 - 10^2}\right)$, $\left(11, 0, 0, \sqrt{13^2 - 11^2}\right)$, $\left(12, 0, 0, \sqrt{13^2 - 12^2}\right)$. To add colors to the graph, choose Edit + Properties, Plot Components page, and for each of Item Number 1 to Item Number 9, select the color of your choice.

The average velocity can be calculated during each 1-second interval. Notice that the top of the ladder is moving faster and faster as the foot of the ladder is being pulled away from the wall. Let $x_1(t) = \frac{x(t+1)-x(t)}{1}$ and $y_1(t) = \frac{y(t+1)-y(t)}{1}$.

t	0	1	2	3	4	5	6	7	8
$x(t)$	5	6	7	8	9	10	11	12	13
$y(t)$	12	11.53	10.95	10.2	9.38	8.31	6.93	5.0	0
$x_1(t)$	1	1	1	1	1	1	1	1	1
$y_1(t)$	$-.47$	$-.58$	$-.71$	$-.87$	-1.07	-1.38	-1.93	-5.0	

Choose Insert + Table and select 10 columns and 5 rows. Entries in the first row represent time for $t = 0, 1, \ldots, 8$ seconds. The second row is determined by the initial position 5 for x and the specification that x increases at the rate of 1 foot/second. The third row is determined by the relationship $x^2 + y^2 = 13^2$. To fill in the numerical values, define $y(t) = \sqrt{13^2 - t^2}$ and type the expressions

$$y(0) \quad y(1) \quad y(2) \quad y(3) \quad y(4) \quad y(5) \quad y(6) \quad y(7) \quad y(8)$$

in the input boxes. Select each expression with the mouse and, while holding down the control key, choose Evaluate + Numerically (in-place replacements). The fourth and fifth rows are self-explanatory. For row 5, define $h(t) = y(t+1) - y(t)$ and follow a procedure similar to that suggested for row 3.

Differentiating the equation

$$x^2 + y^2 = 13^2$$

with respect to time t yields

$$2x\frac{dx}{dt} + 2y\frac{dy}{dt} = 0$$

which implies

$$\frac{dy}{dt} = -\frac{x}{y}\frac{1}{dx/dt} = -\frac{x}{\sqrt{13^2 - x^2}}\frac{1}{dx/dt}$$

N With the insertion point in the equation $x^2 + y^2 = 13^2$, choose Calculus + Implicit Differentiation (Differentiation Variable: t). With the insertion point in the equation $2xx' + 2yy' = 0$, choose Solve + Exact (Variable to Solve For: y'). For clarity, you may wish to replace occurrences of x' by dx/dt and of y' by dy/dt.

Notice that at 8 seconds we have $x = 13$ and hence dy/dt is undefined. In fact, $\lim_{t \to 8^-} dy/dt = -\infty$, which means that if the tip of the ladder remains against the wall, it would achieve an infinite velocity (which certainly would make it the fastest ladder in the west). We must conclude that the tip of the ladder pulls away from the house just before the ladder hits the ground.

N To enter the expression $\lim_{t \to 8^-} dy/dt$, type lim in mathematics. Choose Insert + Subscript and type $t \to 8$ in the input box. Choose Insert + Superscript and type $-$ in the input box. Press SPACEBAR twice to return to the line. Type dy/dt in mathematics. To compute the limit, with the insertion point in the expression $\lim_{t \to 8^-} dy/dt$, choose Evaluate.

Activities

1. A sedan travels south at a rate of 60 miles per hour. At 11:00 p.m. the sedan is located 60 miles north of Crossroads. A sports car travels west at a rate of 80 miles per hour. At 11:00 p.m. the sports car is located 80 miles east of Crossroads. Plot line segments that represent the distance between the two vehicles at 10-minute intervals, beginning at 11:00 p.m. Calculate the average rate of change in the distance between the two vehicles during these 10-minute intervals, and display this information in a table. Use the figure and your table of values to estimate the relative velocity with which the two automobiles were approaching each other at an instant before impact.

2. A moderate-sized two-story frame house has the shape of a cube of edge 18 feet. During the day the wood frame expands as the house is warmed by the sun. Find the rate of change in the volume (in cubic inches per house) inside the house at an instant when the edge is 18 feet and is increasing at a rate of 0.01 inch per hour. Describe a common object whose volume is roughly equal to the hourly rate of change in the house volume. Discuss whether or not you think this model is realistic.

2.9 Differentials; Linear and Quadratic Approximations

The approximation

$$f(x) \approx f(a) + f'(a)(x - a)$$

is called the *linear approximation* or *tangent line approximation* of f at a, and the function

$$L(x) = f(a) + f'(a)(x - a)$$

is called the *linearization* of f at a. The line $L(x)$ is the best approximation of the function f by a line at the point $x = a$, $y = f(a)$.

If $y = f(x)$, the differential dy depends on the two variables x and dx. Under most circumstances, *Scientific Notebook* treats the expressions dy and dx as the product of d and y or x. To use the differentials as variables, define dx and dy as *math names* as follows:

- Choose Insert + Math Name; in the Name box, type dx; choose OK.
- Choose Insert + Math Name; in the Name box, type dy; choose OK.

(The dx and dy should appear in gray on the screen.) Now Define + New Definition applied to $\mathrm{dy} = f'(x)\,\mathrm{dx}$ defines dy as a differential, and if $f(x) = \sqrt{x}$, then Evaluate produces

$$[f'(x)\,\mathrm{dx}]_{x=1.0,\mathrm{dx}=0.1} = .05$$

What a Difference a Differential Makes

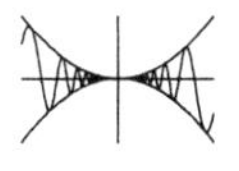

Make "math names" out of dx and dy as described above. Define $f(x) = \sqrt{x}$ and define $\mathrm{dy} = f'(x)\,\mathrm{dx}$. Calculate dy at $x = .2$, $\mathrm{dx} = .1$ by applying Evaluate to

$$[\mathrm{dy}]_{x=.2,\mathrm{dx}=.1}$$

Calculate the difference in the function values $f(.3)$ and $f(.2)$. Visualize the differentials dx and dy by plotting a graph of $y = \sqrt{x}$ and dragging the expressions

$$\left(.2, \sqrt{.2}, .3, \sqrt{.2}, .3, \sqrt{.2} + .1118\right) \quad \text{and} \quad f(.5) + f'(.5)(x - .5)$$

to the frame. Explain the connection between the plot and the equation

$$\mathrm{dy} = f'(x)\,\mathrm{dx}$$

Evelyn's Solution

Define $f(x) = \sqrt{x}$. Evaluating the differential $\mathrm{dy} = f'(x)\,\mathrm{dx}$ at $x = .2$, $\mathrm{dx} = .1$, we get

$$[\mathrm{dy}]_{x=.2,\mathrm{dx}=.1} = .1118$$

With the insertion point in the equation $f(x) = \sqrt{x}$, choose Define + New Definition. Choose Insert + Math Name (type dy, and choose OK). Type $= f'(x)$. Choose Insert + Math Name (type dx, and choose OK). With the insertion point in the equation $\mathrm{dy} = f'(x)\,\mathrm{dx}$, choose Define + New Definition. Choose Insert + Math Name (type dy, and choose OK). Select dy with the mouse, choose Insert + Brackets, and select Square Brackets. Choose Insert + Subscript and type $x = .2$, choose Insert + Math Name (type dx, and choose OK). Type $= 1$. With the insertion point in the expression $[\mathrm{dy}]_{x=.5,\mathrm{dx}=.1}$, choose Evaluate.

Notice that this is roughly the same as the difference in function values $f(.3) - f(.2) = .10051$. In the following figures, the graph of $y = \sqrt{x}$ is drawn with a thick pen, and the tangent line and both the horizontal and vertical line segments are drawn with a thin pen. The differential dx corresponds to the length of the horizontal leg of the right triangle, and the differential dy corresponds to the length of the vertical leg of the right triangle. We know that $f'(.2)$ is the slope of the tangent line at $x = .2$, and indeed the slope of the tangent line is the length dy of the opposite side divided by the length dx of the adjacent side of the right triangle. Notice that the vertical leg of the "triangular" region bounded above by the graph of $y = \sqrt{x}$ has length $f(.3) - f(.2) = 0.0051$, which is slightly shorter than the length of dy, which is $f'(.2)\,(.1) = 0.1118$.

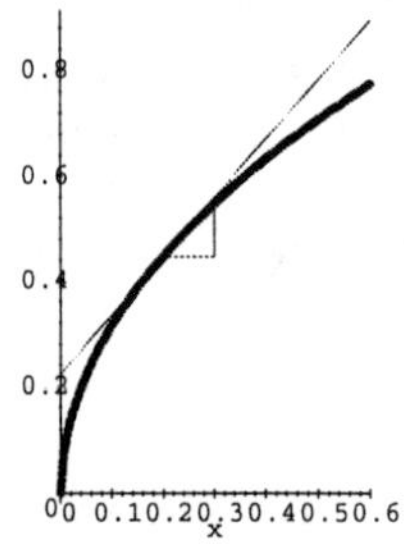

$f(x) = \sqrt{x}$, $\mathrm{dy} = f'(x)\,\mathrm{dx}$

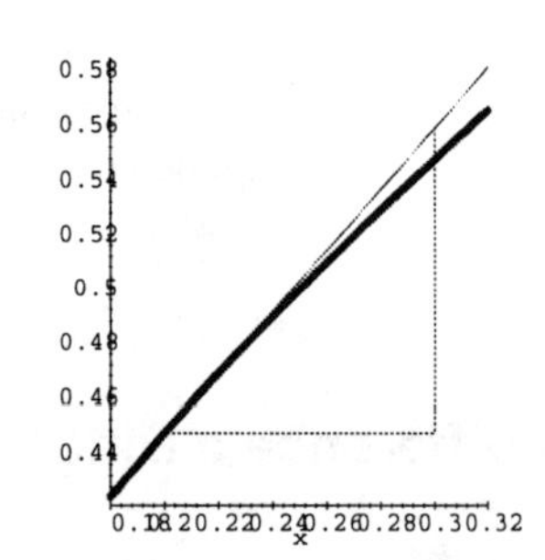

$f(x) = \sqrt{x}$, $\mathrm{dy} = f'(x)\,\mathrm{dx}$

With the insertion point in the expression $f(.3) - f(.2)$, choose Evaluate. With the insertion point in the expression $\sqrt{x}$, choose Plot 2D + Rectangular. With the insertion point in the expression $(.2, f(.2), .3, f(.2), .3, f(.2) + .1118)$, choose Evaluate. Select and drag to the frame $(.2, .44721, .3, .44721, .3, .55901)$. With the insertion point in the expression $f(.2) + f'(.2)(x - .2)$, choose Evaluate. Select and drag to the frame $.22361 + 1.118x$. For the two views shown, change the Domain Interval to $0 < x < .6$ and $.18 < x < .32$, respectively, and check Equal Scaling Along Each Axis.

Activities

1. Define $f(x) = x^3$ and define $\mathrm{dy} = f'(x)\,\mathrm{dx}$. Calculate dy at $x = 1$, $\mathrm{dx} = .1$ by applying Evaluate to
$$[\mathrm{dy}]_{x=1,\mathrm{dx}=.1}$$
Visualize the differentials dx and dy by plotting a graph of $y = x^3$ and dragging the expressions $[1, 1, 1.1, 1, 1.1, 1.3]$ and $f(1) + f'(1)(x - 1)$ to the frame. Name the horizontal and vertical legs of the right triangle and explain the connection between

the plot and the equation dy $= f'(x)$ dx.

2. Define $f(x) = \sqrt{x}$. Define $P(x) = A(x-1)^2 + B(x-1) + C$ and solve the system

$$P(1) = f(1)$$
$$P'(1) = f'(1)$$
$$P''(1) = f''(1)$$

for A, B, and C. Visualize the quadratic approximation to f at 1 by plotting the graph of f together with the graph of P. Zoom in slightly so that the viewing rectangle is restricted to $-1 \leq x \leq 5$. Over what approximate interval does the quadratic approximation provide a reasonably good fit?

3. Define $f(x)$ to be a generic function and define

$$P(x) = A(x-a)^2 + B(x-a) + C$$

If $P(x)$ is the quadratic approximation to f at a, find a formula for P in terms of $f(a)$, $f'(a)$, and $f''(a)$ by solving the system

$$P(a) = f(a)$$
$$P'(a) = f'(a)$$
$$P''(a) = f''(a)$$

for A, B, and C.

4. A section of highway that crosses a gully has the shape of a parabola $f(x) = ax^2 + bx + c$, where x measures horizontal distance from a reference point and y measures elevation relative to the reference point. The reference point is chosen as the low point as the highway crosses the gully. The radius of curvature at the reference point is 2000 feet. This means that a circle of radius 2000 feet resting on the reference point would have the same second derivative as the parabola at the reference point. Determine a, b, and c, and plot the parabolic highway together with the graph of a circle of radius 2000 feet that rests on the reference point.

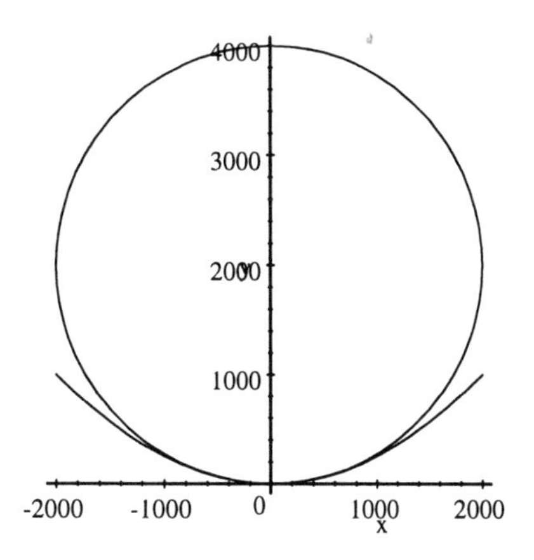

Hint Choose Equal Scaling Along Each Axis to make the circle look like a circle.

2.10 Newton's Method

Solve + Exact and Solve + Numeric provide tools for locating x-intercepts of a curve. Occasionally not all x-intercepts are located by these tools. By plotting a graph of a curve the x-intercepts can be located approximately. This information can be used to provide more precise estimates for the x-intercepts.

Newton Had a Method

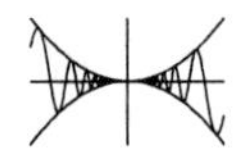

Newton's method uses tangent lines in an attempt to locate x-intercepts. Let $f(x) = x^3 + 2x^2 - 5x - 6$. Use $x_1 = 1$ as an initial approximation, and calculate

$$\begin{aligned} x_2 &= x_1 - \frac{f(x_1)}{f'(x_1)} \\ x_3 &= x_2 - \frac{f(x_2)}{f'(x_2)} \\ x_4 &= x_3 - \frac{f(x_3)}{f'(x_3)} \end{aligned}$$

Plot a graph of $y = f(x)$ on the interval $0 \leq x \leq 5$, attach tangent lines at each of the points $(x_1, f(x_1))$, $(x_2, f(x_2))$, and $(x_3, f(x_3))$, and attach the vertical line segments connecting $(x_1, 0)$ to $(x_1, f(x_1))$, $(x_2, 0)$ to $(x_2, f(x_2))$, and $(x_3, 0)$ to $(x_3, f(x_3))$.

Norman's Solution

Let $f(x) = x^3 + 2x^2 - 5x - 6$. Use $x_1 = 1$ as an initial approximation, and calculate

$$\begin{aligned} x_2 &= x_1 - \frac{f(x_1)}{f'(x_1)} \\ x_3 &= x_2 - \frac{f(x_2)}{f'(x_2)} \\ x_4 &= x_3 - \frac{f(x_3)}{f'(x_3)} \end{aligned}$$

With the insertion point in the equation $f(x) = x^3 + 2x^2 - 5x - 6$, choose Define + New Definition. With the insertion point in the equation $x_1 = 1$, choose Define + New Definition. With the insertion point in the expression $x_1 - f(x_1)/f'(x_1)$, choose Evaluate Numerically. With the insertion point in the equation $x_2 = 5.0$, choose Define + New Definition. With the insertion point in the expression $x_2 - f(x_2)/f'(x_2)$, choose Evaluate Numerically. With the insertion point in the equation $x_3 = 3.4$, choose Define + New Definition. With the insertion point in the expression $x_3 - f(x_3)/f'(x_3)$, choose Evaluate Numerically. With the insertion point in the equation $x_4 = 2.4891$, choose Define + New Definition.

Here is a graph of $y = f(x)$ together with line segments that illustrate the first three iterations of Newton's method.

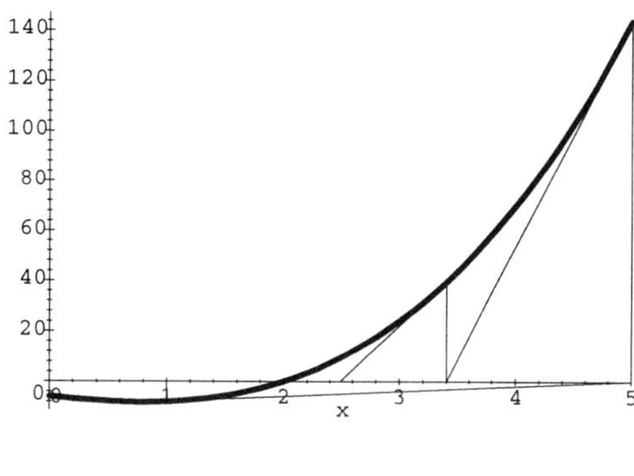

$$f(x) = x^3 + 2x^2 - 5x - 6$$

With the insertion point in the expression $f(x)$, choose Plot 2D + Rectangular. Choose Edit + Properties, Plot Components page, and change Domain Interval to $0 < x < 5$. Choose OK. Select the list of points

$$(x_1, 0, x_1, f(x_1), x_2, 0, x_2, f(x_2), x_3, 0, x_3, f(x_3), x_4, 0)$$

and drag to the frame.

Getting Off to a Good Start

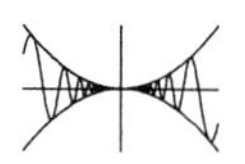

This activity is designed to demonstrate the significance of the initial approximation x_1 in Newton's method. Choosing x_1 close to an x-intercept of $y = f(x)$ will usually cause Newton's method to give excellent approximations quickly. This activity shows the typical behavior of Newton's method for an interval of starting x-coordinates. Define f by the equation $f(x) = x^3 + 2x^2 - 5x - 6$ and define g by the equation

$$g(x) = x - \frac{f(x)}{f'(x)}$$

If x_1 represents an initial approximation for Newton's method, explain why the second approximation is given by $x_2 = g(x_1)$. Explain why the term *Newton Iteration Function* is an appropriate name for the function g.

Plot a graph of $y = f(x)$ and superimpose a graph of $y = g(x)$ by selecting $g(x)$ and dragging it to the plot frame. Use the Plot Properties dialog to change the View Intervals to $-5 \leq x \leq 5$ and $-8 \leq y \leq 5$. Explain why the graph of $y = g(x)$ appears to have vertical asymptotes near $x = -2$ and $x = 1$.

Use **Solve + Numeric** to solve the equation $f'(x) = 0$. What do these solutions represent?

Plot a new graph of $y = f(x)$ and superimpose a graph of $y = g(g(x))$. If x_1 represents an initial approximation for Newton's method, explain why the third approximation is given by $x_3 = g(g(x_1))$. Parts of the graph of $y = g(g(x))$ appear to be nearly horizontal. Describe what you think causes these parts to be nearly horizontal. At least two new vertical asymptotes have appeared. To help understand them, start with the solutions $x = a_1$ and $x = a_2$ that you found earlier as solutions to the equation $f'(x) = 0$. Then solve $g(x) = a_1$ and $g(x) = a_2$. What do these solutions represent?

Stephanie's Solution

Define f by the equation $f(x) = x^3 + 2x^2 - 5x - 6$ and define g by the equation

$$g(x) = x - \frac{f(x)}{f'(x)}$$

With the insertion point in the equation $f(x) = x^3 + 2x^2 - 5x - 6$, choose Define + New Definition. With the insertion point in the equation $g(x) = x - \frac{f(x)}{f'(x)}$, choose Define + New Definition.

Newton's method states that if x_1 represents an initial approximation for Newton's method, then the second approximation is given by $x_2 = x_1 - \frac{f(x_1)}{f'(x_1)} = g(x_1)$. The term *Newton Iteration Function* is an appropriate name for the function g because it gives the next iteration of Newton's method starting at any number x. The following plot shows graphs of f and g. The graph of g and its asymptotes appear as a dotted line. Notice that the vertical asymptotes of g appear where $f'(x) = 0$.

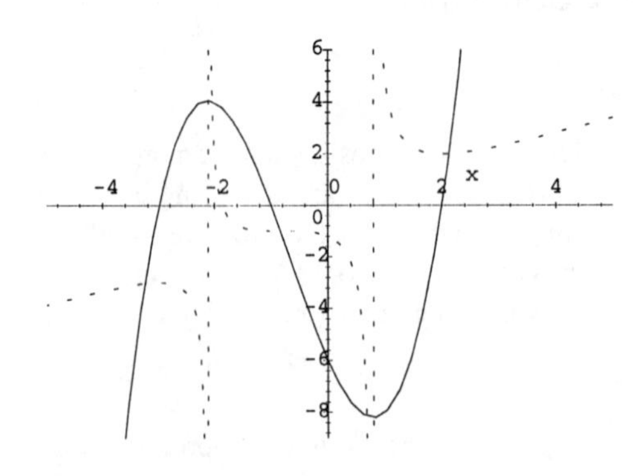

$f(x)$ and $g(x)$

With the insertion point in the expression $f(x)$, choose Plot 2D + Rectangular. Select and drag $g(x)$ to the plot. Choose Edit + Properties, Plot Components page. For Item Number 2, change Line Style to Dots. Choose the View page. Change View Intervals to $-5 < x < 5$ and $-9 < y < 6$. Choose OK.

The roots of $f'(x)$ are given by $x = -\frac{2}{3} - \frac{1}{3}\sqrt{19} = -2.1196$ and $x = -\frac{2}{3} + \frac{1}{3}\sqrt{19} = .7863$, which are relatively close to -2 and 1.

 With the insertion point in the equation $f'(x) = 0$, choose Solve + Numeric.

Here are the graphs of $f(x)$ and $g(g(x))$. If x_1 represents an initial approximation for Newton's method, then

$$x_3 = x_2 - \frac{f(x_2)}{x_2} = g(x_2) = g\left(x_1 - \frac{f(x_1)}{f'(x_1)}\right) = g(g(x_1))$$

Parts of the graph of $y = g(g(x))$ appear to be nearly horizontal. These horizontal pieces represent intervals where Newton's method gives very good approximations to roots of f after only two iterations.

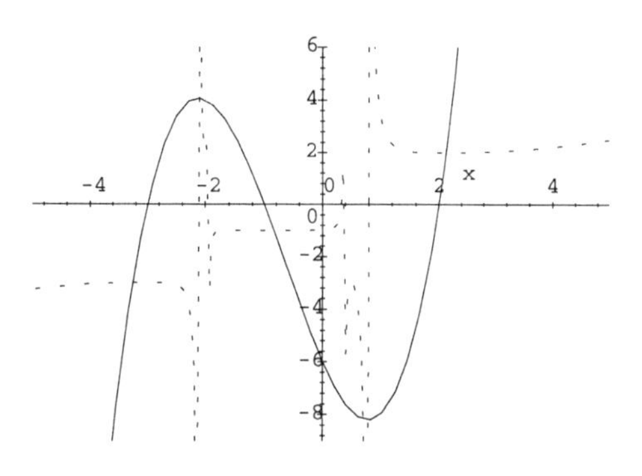

$f(x)$ and $g(g(x))$

Follow the same directions as for the previous graph, except select and drag $g(g(x))$ to the plot.

Solving $g(x) = -2.1196$, we get $x = .38378$. This means that if we start with $x_1 = .38378$, then $x_2 = -2.1196$ and $f'(-2.1196) = 0$, and hence $g(g(x_1))$ is undefined, and $x = .38378$ is a vertical asymptote of $g(g(x))$. Similarly, the equation $g(x) = .7863$ has the solution $x = -1.9438$, which indicates that $x = -1.9438$ is a vertical asymptote of $g(g(x))$.

With the insertion point in the equation $g(x) = -2.1196$, choose Solve + Exact. With the insertion point in the equation $g(x) = .7863$, choose Solve + Exact.

Multiple iterations such as $g(g(g(g(x))))$ would generate additional vertical asymptotes, indicating all possible starting points that would lead to a number x where $f'(x) = 0$ in four or fewer iterations of Newton's method.

Activities

1. *Divide and Average* is an algorithm for finding square roots that is based upon the following observations. If $x = \sqrt{a}$, then $x^2 = a$ and hence $x = a/x$. If x is *close* to $\sqrt{a}$, then so is a/x. Given two approximations, a better one is often available by averaging the first two. In this case the average is
$$\frac{x + \frac{a}{x}}{2}$$
Divide and Average is based on iterating the function
$$s(x) = \frac{x + \frac{a}{x}}{2}$$
Show that Divide and Average is really Newton's method (applied to the function $f(x) = x^2 - a$) in disguise. Show that Divide and Average is also equivalent to the following recipe for approximating $\sqrt{a}$:
   a. Start with a positive number x.
   b. Calculate the square x^2 of x.
   c. If $x^2 \approx a$, then stop.
   d. Add x^2 to a.
   e. Divide by x.
   f. Divide by 2.
   g. Call your answer x and go to step b.

   Use ten iterations of Divide and Average to estimate $\sqrt{2}$, starting with $x_1 = 1$.

2. Divide and Average can be modified to calculate cube roots. If $x \approx \sqrt[3]{a}$, then $x^3 \approx a$ and hence $x \approx a/x^2$. Calculate a weighted average of x and a/x^2—namely,
$$\frac{2x + \frac{a}{x^2}}{3} = \frac{2x^3 + a}{3x^2}$$
Show that iteration on the function
$$c(x) = \frac{2x^3 + a}{3x^2}$$
is equivalent to Newton's method. Use this variation of Divide and Average to estimate $\sqrt[3]{2}$.

3. Find a Divide and Average algorithm for calculating nth roots. Show that your algorithm is equivalent to Newton's method.

4. Newton's method is based on finding the x-intercept of the linearization of a function f at a. Develop an analogous method that is based on finding an x-intercept of the quadratic approximation to $f(x)$ near a. Apply your method to find an x-intercept of $y = \sin(\sin(x + \cos x)) + \frac{1}{2}$ near $x = 4$.

3 INVERSE FUNCTIONS

In this chapter, you will explore several families of functions and the inverse relations associated with these functions.

3.1 Exponential Functions and Their Derivatives

Exponential functions provide mathematical models for diverse applications.

Getting Into Exponential Shape

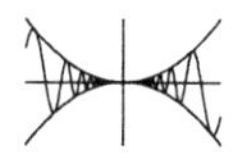

Plot graphs of $y = b^x$ for $b = \frac{1}{4}, \frac{1}{2}, 1, \frac{3}{2}, 2, 4$, and 10, all in the same viewing rectangle $-2 \leq x \leq 2$. Discuss the similarities and differences in the curves. What is special about $b = 1$? Which of these graphs are one-to-one?

Elizabeth's Solution

The plots of $y = b^x$ for the values $b = \frac{1}{4}, \frac{1}{2}, 1, \frac{3}{2}, 2, 4$, and 10 are given in the following figure:

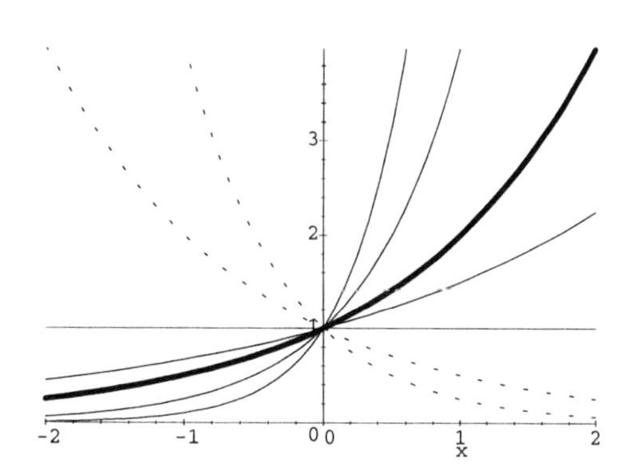

The curve $y = 2^x$ is drawn with a fat pen, the other curves $y = b^x$ with $b > 1$ are drawn with a regular pen, and the curves $y = b^x$ with $0 < b < 1$ are dotted curves. The curves all go through the point $(0, 1)$.

With the insertion point in the expression $\left(\frac{1}{4}\right)^x$, choose Plot 2D + Rectangular. Select and drag to the frame each of $\left(\frac{1}{2}\right)^x, 1^x, \left(\frac{3}{2}\right)^x, 2^x, 4^x, 10^x$. Choose Edit + Properties, and click Plot Components tab. Change Domain Interval to $-2 \leq x \leq 2$; click Item Number, and choose Line Style, Line Color, and Thickness for each item. Click the Frame tab; change Placement to Displayed. Choose OK.

- If $b < 1$, then the slope is negative, the curve is asymptotic to the positive x-axis, and the curve models exponential decay. Note that the graph of $y = \left(\frac{1}{4}\right)^x$ is steeper than the graph of $y = \left(\frac{1}{2}\right)^x$ at $(0, 1)$.
- If $b > 1$, then the slope is positive, the curve is asymptotic to the negative x-axis, and the curve models exponential growth. In this case, as b increases so does the slope of the tangent line at the point $(0, 1)$.
- If $b = 1$, then the graph of $y = b^x$ is a horizontal line with y-intercept 1.
- All of these functions except $y = 1^x$ appear to be one-to-one, as any horizontal line would intercept the graph at most once. The function $y = 1^x$ is not one-to-one because it fails this test for the horizontal line $y = 1$.

Activities (Exponential Functions)

1. Find three real numbers that satisfy the equation $x^2 = 2^x$. Describe the connection between these numbers and the graphs of $y = x^2$ and $y = 2^x$.

2. Define f by the equation
$$f(x) = \frac{e^{x+h} - e^x}{h}$$
and set $h = 0.1$. Describe the connection between f and the derivative of the exponential function e^x. Plot a graph of f. On the basis of your graph, explain why it is reasonable to expect that
$$\frac{d}{dx}e^x = e^x$$

3. Define f by the equation
$$f(x) = \frac{10^{x+h} - 10^x}{h}$$
and set $h = 0.1$. Describe the connection between f and the derivative of the exponential function 10^x. Plot a graph of f together with a graph of $y = 10^x$. On the basis of your graph, compare the functions
$$\frac{d}{dx}10^x \qquad \text{and} \qquad 10^x$$

4. Determine appropriate values for b in the following graphs of $y = b^x$. Describe the

clues you used to find these values. Verify your solutions visually by plotting the graph of $y = b^x$ with your choices of b.

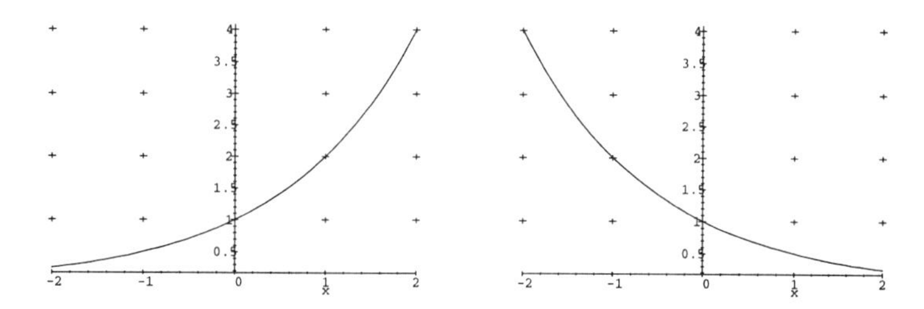

3.2 Inverse Functions

The inverse graph of a function f is obtained by reflecting the graph of $y = f(x)$ about the line $y = x$. This reflected curve is the *inverse relation* of f. If the function f is one-to-one, the inverse relation of f is again a function. In this case, it is denoted by f^{-1} and referred to as "f *inverse.*"

A curve can be described by expressing the x- and y-coordinates as functions of a third variable (or parameter) such as t. In the special case that the curve is the graph of a function $y = f(x)$, you can take $t = x$ and write the coordinates as $(x, f(x))$; for its inverse relation, you can write the coordinates as $(f(x), x)$.

To graph a curve described by parametric coordinates, enclose the coordinate functions inside parentheses (), separated by a (red) comma, and apply Plot 2D + Parametric.

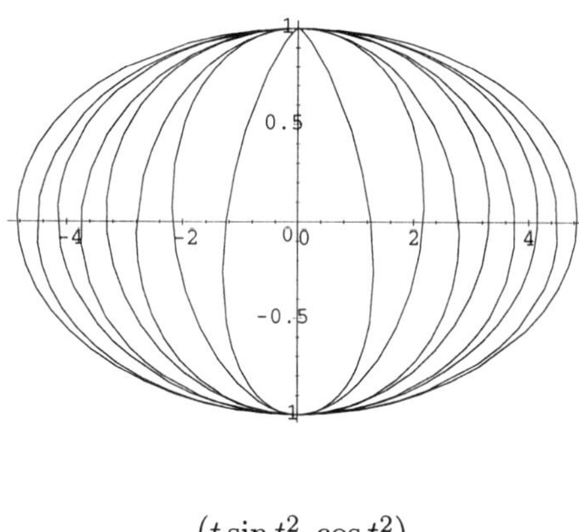

$$(t \sin t^2, \cos t^2)$$

The parametric form easily plots the inverse of a function. For example, here are plots of $y = x^3$ and its inverse function.

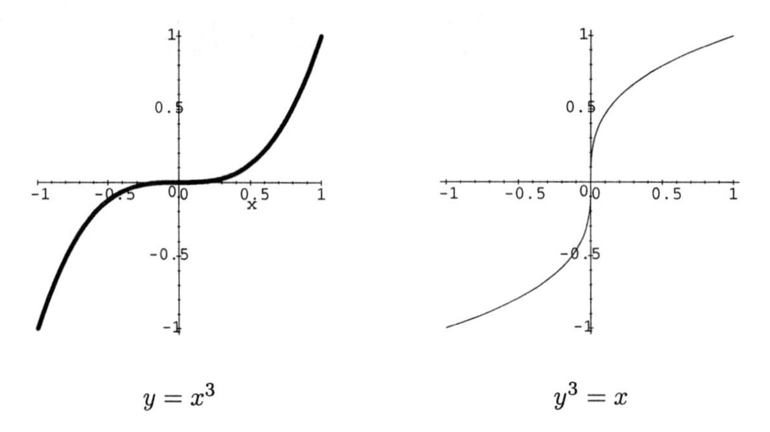

$y = x^3$ $y^3 = x$

Multiple curves can be plotted together by selecting other ordered pairs and dragging them to the frame. In particular, a function, its inverse, and the line $y = x$ can be plotted together to yield a plot similar to the following. (Use the **Plot Components** page of the **Plot Properties** dialog to set the **Line Thickness, Style,** and **Color,** and use the **Labeling** page to add the caption.)

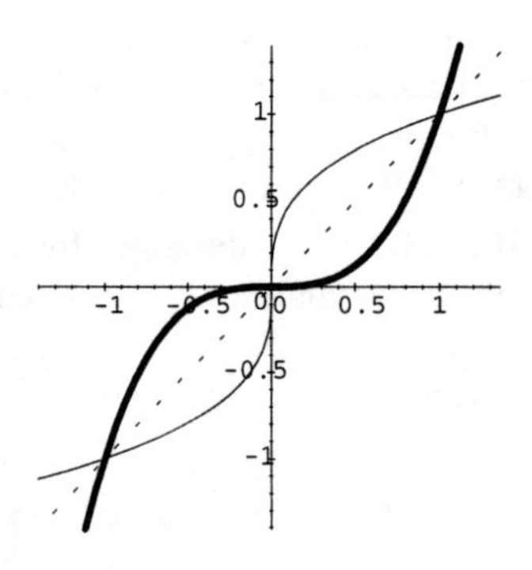

$y = x^3, y^3 = x, y = x$

Speaking Inverse Does Not A Poet Make

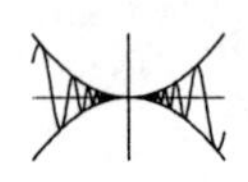

Investigate the inverse of the function $f(x) = e^x$ by using **Plot 2D + Parametric** to study its properties. Plot f and f^{-1} together with the line $y = x$. Verify visually that f^{-1} is a certain well-known function.

Dan's Solution

The following figure shows the graph of $y = e^x$ (drawn with a fat pen) together with the graph of its inverse. Notice the symmetry about the dotted line $y = x$.

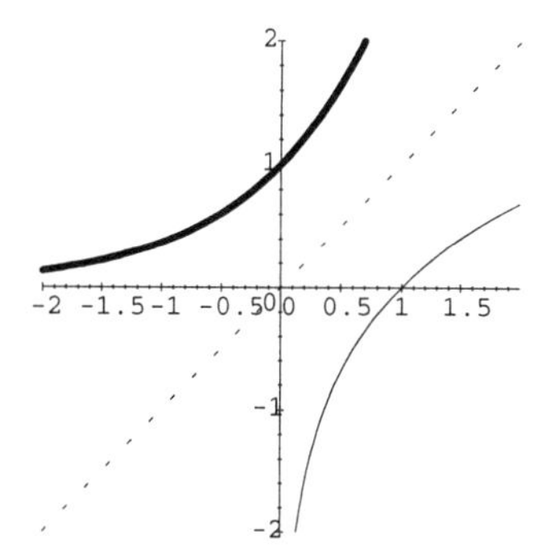

With the insertion point in the expression (x, e^x), choose Plot 2D + Parametric. Select and drag to the frame (e^x, x). Select and drag to the frame (x, x). Choose Edit + Properties, Plot Components page; set Line Style, Line Color, Thickness. Choose the Axes page: check Equal Scaling Along Each Axis. Choose OK.

Plotting the inverse of e^x together with the graph of $y = \ln x$, it appears that $\ln x$ is the inverse of the exponential function. The figure appears to be the graph of a single curve.

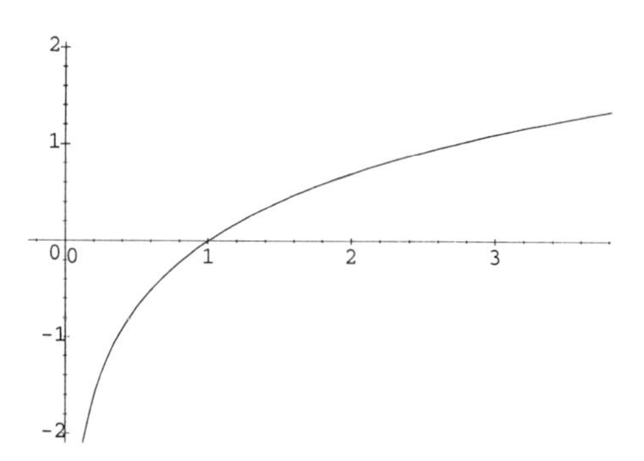

With the insertion point in the expression $(x, \ln x)$, choose Plot 2D + Parametric. Select and drag to the frame (e^x, x).

Activities

1. Study the inverse of $f(x) = \frac{1}{x}$ by using Plot 2D + Parametric. Explain why only a single curve appears. Add a graph of $y = x$ to your plot and discuss symmetry with respect to this line. Find a formula for the inverse function by solving the equation $x = \frac{1}{y}$ for y. Find another example of a function that is equal to its own inverse.

2. Study the inverse of $f(x) = x^3 + x$ by using Plot 2D + Parametric. Explain why this function is one-to-one. Apply Solve + Exact to solve the equation $x = y^3 + y$ for y in terms of x. You should see three possible formulas. Explain why one formula is more appropriate than the other two. Verify your choice by applying Plot 2D + Parametric to the expressions $(x^3 + x, x)$ and $(x, g(x))$, where $g(x)$ is the formula returned by Solve + Exact.

3.3 Logarithmic Functions

The logarithmic function base a, denoted by $y = \log_a x$, is the inverse of the exponential function $y = a^x$.

Getting Into Logarithmic Shape

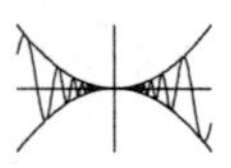

Plot graphs of $y = \log_b x$ for $b = \frac{1}{4}, \frac{1}{2}, \frac{3}{2}, 2, 4$, and 10, all in the same viewing rectangle $-2 \leq x \leq 2$. Discuss the similarities and differences in the curves. Explain why $b = 1$ is not an appropriate base for a logarithm.

Lorenzo's Solution

The graphs of $y = \log_b x$ for $b = \frac{1}{4}, \frac{1}{2}, \frac{3}{2}, 2$, and 10 are given in the following figure:

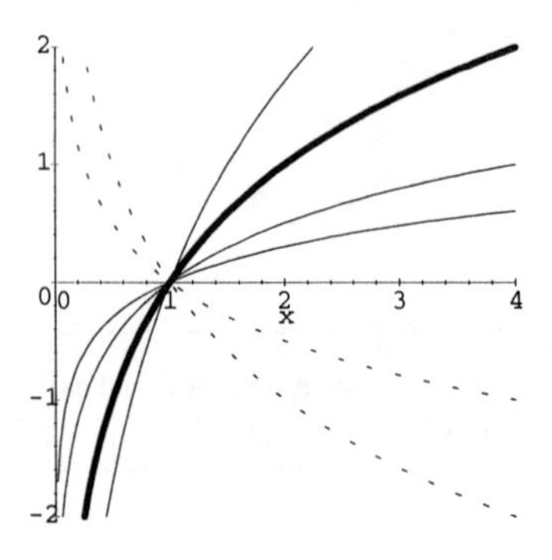

N With the insertion point in the expression $\log_{1/4} x$, choose Plot 2D + Rectangular. Select and drag each of the following to the frame: $\log_{1/2} x$, $\log_{3/2} x$, $\log_2 x$, $\log_{10} x$. Choose Edit + Properties, Plot Components page. Reset Domain Interval, and specify Line Style and Thickness for each Item Number. Choose OK.

In the figure, the graph of $y = \log_b x$ appears as a dashed curve for $b < 1$, and the curve $y = \log_2 x$ is drawn with a fat pen. If $b < 1$, then the graph is decreasing; the positive y-axis is a vertical asymptote; and the closer b is to 1, the steeper the graph at $x = 1$. If $b > 1$, then the graph is increasing; the negative y-axis is a vertical asymptote; and the closer b is to 1, the steeper the graph at $x - 1$. The choice $b = 1$ would yield a vertical line $x = 1$, which is not the graph of a function.

Activities

1. Study the inverse of $f(x) = 2^x$ by applying Plot 2D + Parametric to the two pairs of expressions $(x, 2^x)$ and $(2^x, x)$. Find a formula for $f^{-1}(x)$ by solving the equation $x = 2^y$ for y in terms of x. Show that the inverse is a logarithm base b for some b.

2. Compare parametric graphs of (e^x, x) and $(x, \ln x)$.

3. Determine values of b corresponding to the following graphs of $y = \log_b x$. Describe the strategy you used to determine each value of b. Verify your solutions by graphing $y = \log_b x$ in appropriate viewing rectangles.

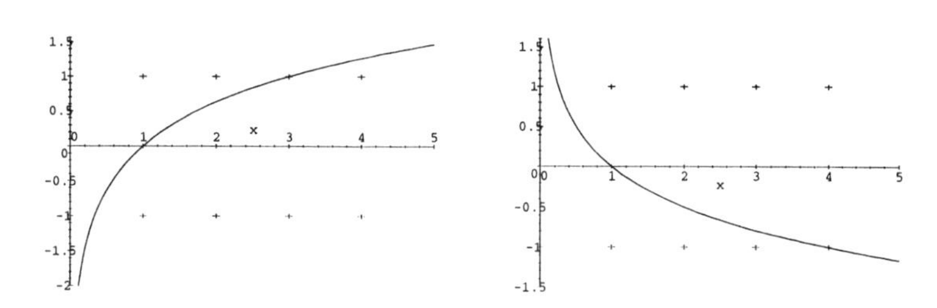

3.4 Derivatives of Logarithmic Functions

How do you describe the rate of change of logarithmic functions? In this activity, you use the definition of a derivative to find approximations to the derivative.

Filling in the Gap

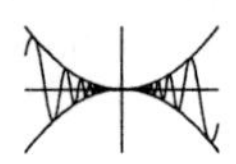

Define g by the equation

$$g(x) = \frac{\ln(x+h) - \ln x}{h}$$

and set $h = 0.1$. Describe the connection between g and the derivative of the natural logarithm $\ln x$. Plot a graph of g. On the basis of your graph, explain why it is reasonable to expect that

$$\frac{d}{dx} \ln x = \frac{1}{x}$$

Evaluate the limit $\lim_{h \to 0} \frac{\ln(x+h)-\ln x}{h}$ and discuss the result. What is the relationship between the derivative of the natural logarithm and the power rule? (*Hint*: Look at the name of this activity.)

Elva's Solution

The derivative of the natural logarithm $\ln x$ is defined to be the limit as h approaches zero of the expression $\frac{\ln(x+h)-\ln x}{h}$. To investigate this limit, we define a function g by the equation

$$g(x) = \frac{\ln(x+h) - \ln x}{h}$$

and set $h = 0.1$.

 Choose Define + Clear Definitions. With the insertion point in the equation

$$g(x) = \frac{\ln(x+h) - \ln x}{h}$$

choose Define + New Definition. With the insertion point in the equation $h = 0.1$, choose Define + New Definition.

On the interval $0 < x < 10$, the graph of g looks like

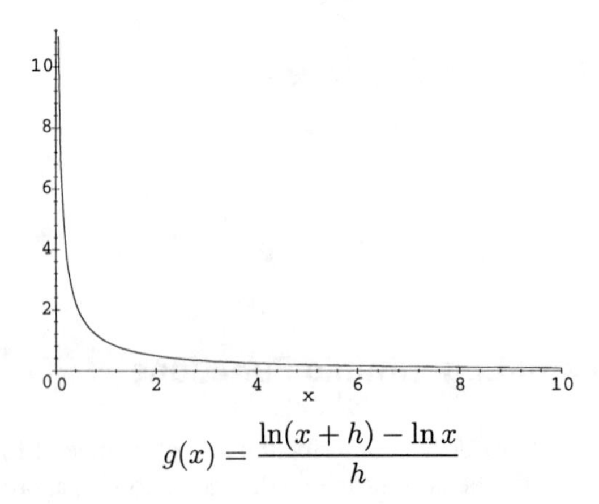

$$g(x) = \frac{\ln(x+h) - \ln x}{h}$$

With the insertion point in the expression $g(x) = \frac{\ln(x+h)-\ln x}{h}$, choose Plot 2D + Rectangular. Choose Edit + Properties. On the Plot Components page, change the Domain Interval to $0 < x < 10$. On the Frame page, under Placement Options, check Displayed. Choose OK.

This graph resembles the graph of $\frac{1}{x}$, and we check this resemblance by dragging $\frac{1}{x}$ to the frame, and plotting the graphs of g and $\frac{1}{x}$ together.

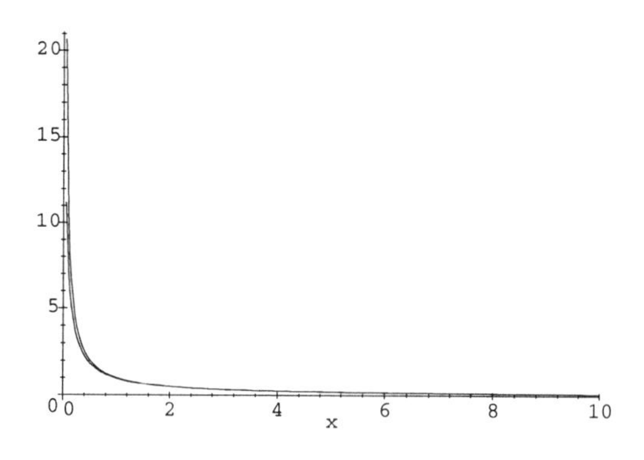

$$\frac{\ln(x+h)-\ln x}{h} \text{ and } \frac{1}{x}$$

 Select and drag to the plot the expression $\frac{1}{x}$.

The striking similarity of these two plots makes it reasonable to conjecture that

$$\frac{d}{dx}\ln x = \frac{1}{x}$$

Evaluating the limit in the definition of this derivative yields

$$\lim_{h\to 0}\frac{\ln(x+h)-\ln x}{h} = \frac{1}{x}$$

which gives further evidence that our conjecture is correct.

 With the insertion point in the expression $\lim_{h\to 0}\frac{\ln(x+h)-\ln x}{h}$, choose Evaluate.

The power rule

$$\frac{d}{dx}x^n = nx^{n-1}$$

yields a multiple of all integer powers of x (except x^{-1}) as the derivative of an integer power of x:

y	$\cdots$	x^{-3}	x^{-2}	x^{-1}	$\ln x$	x^0	x^1	x^2	x^3	x^4	$\cdots$
$\frac{dy}{dx}$	$\cdots$	$-3x^{-4}$	$-2x^{-3}$	$-x^{-2}$	x^{-1}	0	x^0	$2x^1$	$3x^2$	$4x^3$	$\cdots$

Surprisingly, the natural logarithm fills in the gap!

Activities (Logarithmic Differentiation)

Historically, logarithms were developed in order to simplify computations involving large numbers based on the fact that the algorithm for addition is computationally simpler than the algorithm for multiplication. The identity

$$\ln xy = \ln x + \ln y$$

serves to replace computation of the product xy with computation of the sum $\ln x + \ln y$ followed by an application of the inverse function e^x. This same simplification can help you compute derivatives of complicated functions, as illustrated in this activity. By *logarithmic differentiation* of a function $y = f(x)$, we mean first taking the natural logarithm of both sides of the equation to get $\ln y = \ln f(x)$, then taking the derivative of both sides to get $\frac{d}{dx}(\ln y) = \frac{d}{dx}(\ln f(x))$, and then solving for y' to get the derivative of the original function.

1. Differentiate
$$y = \frac{(2x-1)^3\sqrt{x^2+2}}{(5x+3)^{5/3}}$$
by using logarithmic differentiation. Fill in all the missing steps. Verify your answer by comparing it with direct differentiation.

2. Discuss reasons for using the technique of logarithmic differentiation when taking derivatives of complicated functions by hand (without the aid of technology), and discuss the role this technique might play in doing mathematics with the aid of a symbolic computation system such as you have with *Scientific Notebook*.

Hint Note that the command Expand gives the result

$$\ln\left(\frac{(2x-1)^3\sqrt{x^2+2}}{(5x+3)^{5/3}}\right) = 3\ln(2x-1) + \frac{1}{2}\ln\left(x^2+2\right) - \frac{5}{3}\ln(5x+3)$$

3.5 Exponential Growth and Decay

Exponential functions are often used to model population growth.

Revisiting Rabbit Island

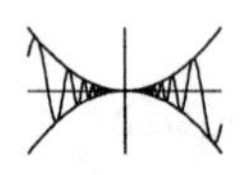

Rabbits living on a small island have a population that grows at a rate proportional to the number of rabbits, but is limited by the size of the island. The population growth is modeled by the differential equation

$$\frac{dR}{dt} = cR\left(1 - \frac{R}{M}\right)$$

where M is the maximum population the small island can sustain. A solution to this ordinary differential equation (ODE) is a function $R(t)$ that satisfies the equation. You will learn techniques for solving such differential equations in a later course. For now, you can use Solve ODE + Exact to find a function $R(t)$ that satisfies this differential equation. The monthly population $R(t)$ was estimated at $R(0) = 500$, $R(10) = 800$, and $R(20) = 1000$. Predict the maximum population that the island will sustain. Plot the rabbit population and attach the horizontal asymptote $y = M$.

Don's Solution

The solution to the differential equation

$$\frac{dR}{dt} = cR\left(1 - \frac{R}{M}\right)$$

is given by

$$\frac{1}{R(t)} = \frac{1 + e^{-ct}C_1 M}{M}$$

 With the insertion point in the equation $\frac{dR}{dt} = cR\left(1 - \frac{R}{M}\right)$, choose Solve ODE + Exact.

Inverting both sides and clearing the negative exponent, we define R by the equation

$$R(t) = \frac{e^{ct}M}{e^{ct} + C_1 M}$$

This solution is compatible with the fact that exponential functions are often used to model population growth.

Select the expression $R(t)$ with the mouse and choose Define + New Definition. With the insertion point in the equation $\frac{1}{R(t)} = \frac{1+e^{-ct}C_1 M}{M}$, choose Solve + Exact (Variable to Solve For: $R(t)$). With the insertion point in the equation $R(t) = \frac{e^{ct}M}{e^{ct}+C_1 M}$, choose Define + New Definition.

To find the parameters c, M, and C_1 we solve the system

$$\begin{aligned} R(0) &= 500 \\ R(10) &= 800 \\ R(20) &= 1000 \end{aligned}$$

which gives us

$$\begin{aligned} c &= \frac{1}{10}\ln 3 \\ C_1 &= \frac{9}{8000} \\ M &= \frac{8000}{7} \end{aligned}$$

Choose Insert + Display. Type $R(0) = 500$ in the input box. Press ENTER and type $R(10) = 800$ in the input box. Press ENTER and type $R(20) = 1000$ in the input box. With the insertion point in the system of equations, choose Solve + Exact.

The graph of R is then given by

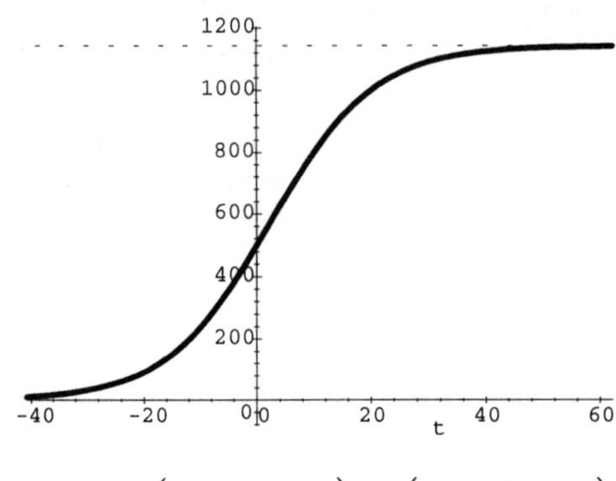

$$R(t) = \left(8000 \times 3^{\frac{1}{10}t}\right) \div \left(7 \times 3^{\frac{1}{10}t} + 9\right)$$

Select the equation $c = \frac{1}{10} \ln 3$ with the mouse (that is, "highlight" it) and choose Define + New Definition. Select the equation $C_1 = \frac{9}{8000}$ with the mouse and choose Define + New Definition. Select the equation $M = \frac{8000}{7}$ with the mouse and choose Define + New Definition. With the insertion point in the expression $R(t)$, choose Plot 2D + Rectangular. Select the graph and choose Edit + Properties. Choose Plot Components page and change Plot Thickness to Medium. Change Domain Interval to $-40 < x < 60$. Choose OK.

The horizontal dashed line shows the limiting population of $\frac{8000}{7} \approx 1143$.

Select $\frac{8000}{7}$ and drag it to the plot. Choose Edit + Properties, Plot Components page. Choose Item Number 2. Change Line Style to Dots. Change Line Color to Red. Choose OK.

Activity (Murder Investigation)

A police officer arrived on the scene at 6:30 a.m. and noted that the body temperature of the murder victim was 77° Fahrenheit. At 7:15 the body temperature had dropped to 75°. The room temperature was a steady 68°. A suspect was seen leaving the building at 3:00 a.m., but had a solid alibi for the hours between 7:00 p.m. and 2:00 a.m. Should the suspect be held for further questioning, or does evidence indicate that the suspect could not have committed this crime?

Hint Falling temperature is often modeled by an exponential function.

3.6 Inverse Trigonometric Functions

How do you find an angle having a specified sine or cosine? In this activity you will plot the inverse sine and cosine functions.

Inverse Sine and Cosine Functions

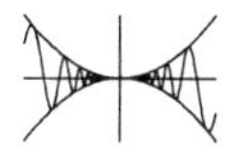

Use Plot 2D + Parametric to plot $(x, \sin x)$ and $(\sin x, x)$. Do both of these graphs satisfy the vertical-line test for functions? Restrict the domain of $f(x) = \sin x$ so that f is one-to-one. Plot $(x, \sin x)$ and $(\sin x, x)$ using the restricted domain. Compare and contrast this graph of $(\sin x, x)$ with a plot of $y = \sin^{-1} x$.

Scott's Solution

The graphs of $y = \sin x$ and its inverse relation are depicted in the following two figures:

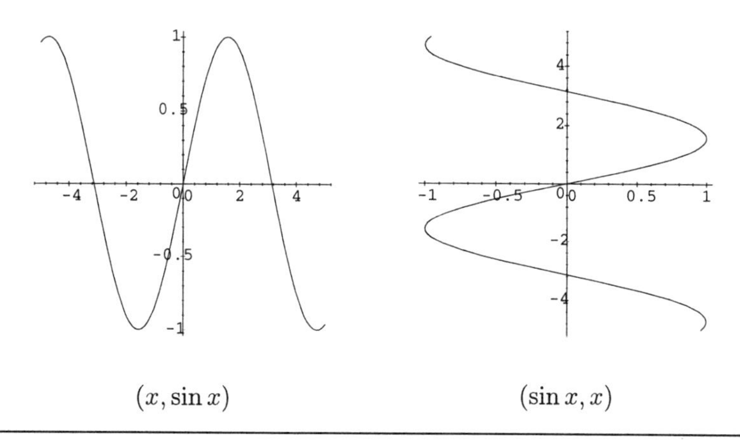

$(x, \sin x)$ $(\sin x, x)$

With the insertion point in the expression $(x, \sin x)$, choose Plot 2D + Parametric. With the insertion point in the expression $(\sin x, x)$, choose Plot 2D + Parametric.

Because the plot on the right fails the vertical-line test for functions, we restrict the domain of $\sin x$ to $-\frac{\pi}{2} \leq x \leq \frac{\pi}{2}$. This gives the following plots of $y = \sin x$ on the restricted domain and its inverse function:

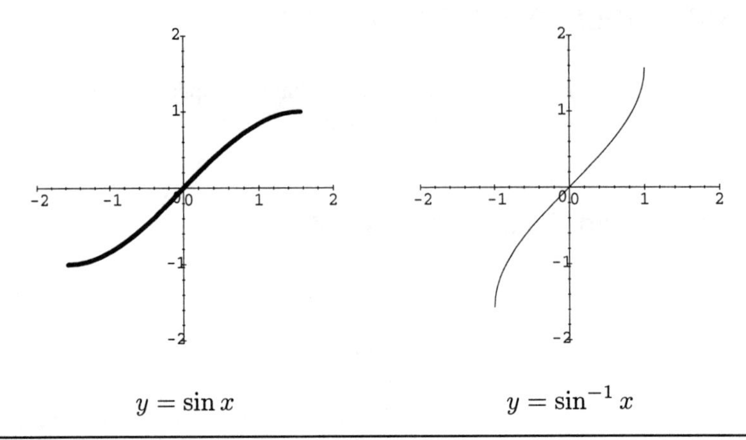

$y = \sin x$ $\qquad$ $y = \sin^{-1} x$

Select each of the two plots, and choose Edit + Properties, Plot Components page. Change Domain Intervals to $-1.57 < x < 1.57$. Choose OK.

We plot the inverse together with the curve $y = \sin^{-1} x$. The figure indicates the graph of a single curve. Plotting the inverse of the sine function together with the curve $y = \sin^{-1} x$ appears to yield the graph of a single curve.

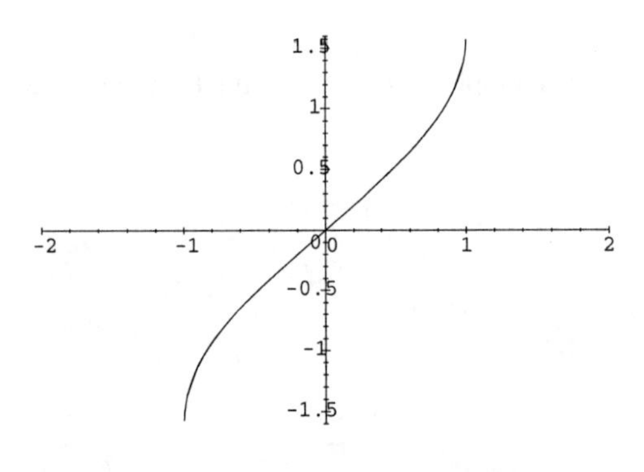

$x = \sin y,\, y = \sin^{-1} x$

With the insertion point in the list $(x, \sin^{-1} x)$, $(\sin x, x)$, choose Plot 2D + Parametric. Choose Edit + Properties, Plot Components page, and change Domain Intervals to $-1.57 < x < 1.57$. Choose OK.

This is a reasonable result, because $\sin^{-1} x$ is called the inverse sine function!

Activities

1. Plot $y = \cos^{-1}(\cos x)$ and determine the numbers x where $\cos^{-1}(\cos x) = x$. Explain how the remaining parts of the graph relate to the definition of inverse function.

2. Plot $\cos\left(\cos^{-1} x\right)$ and explain what is wrong with this graph.

3.7 Hyperbolic Functions

The hyperbolic functions and their inverses are used to solve a variety of problems in the physical sciences and engineering.

Not Just Hyperbole

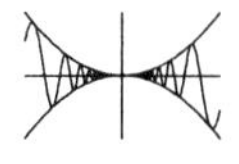

Experiment with the hyperbolic sine

$$\sinh x = \frac{e^x - e^{-x}}{2}$$

and its inverse. Plot a graph of $y = \sinh x$ and its inverse $y = \sinh^{-1} x$ by applying Plot 2D + Parametric to $(x, \sinh x)$ and $(\sinh x, x)$. Find a formula for $\sinh^{-1} x$ in terms of the natural logarithm function by solving the equation

$$x = \frac{e^y - e^{-y}}{2}$$

for y in terms of x. Verify your formula by plotting it together with $\sinh^{-1} x$. Carry out similar experiments with the hyperbolic cosine

$$\cosh x = \frac{e^x + e^{-x}}{2}$$

Maria's Solution

The graph of $y = \sinh x$ (drawn with a fat pen) and its inverse are given by

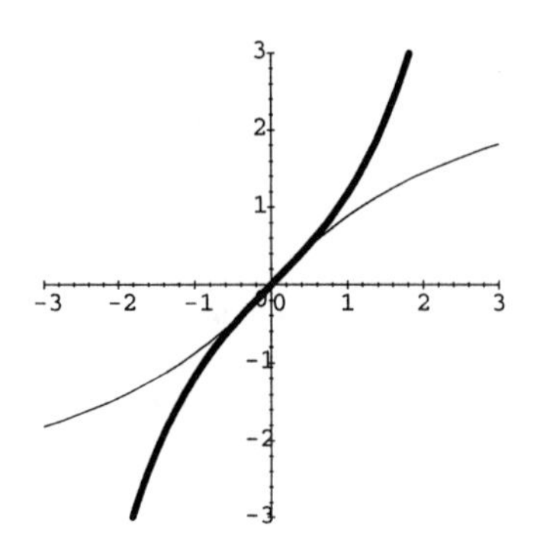

$y = \sinh x, y = \sinh^{-1} x$

With the insertion point in the pair $(x, \sinh x)$, choose Plot 2D + Parametric. Select and drag to the frame $(\sinh x, x)$. Choose Edit + Properties. Choose the Plot Components page, and change Line Thickness to Medium for Item 1. Change Domain Interval to $-3 < t < 3$. Choose the Axes page, and check Equal Scaling on Each Axis. Choose the View page and, under View Intervals, click Default (to remove the check) and change the intervals to $-3 < x < 3$ and $-3 < y < 3$. Choose OK.

The inverse can also be found by solving the equation

$$x = \frac{e^y - e^{-y}}{2}$$

for y in terms of x. Faced with the two possible solutions $y = \ln\left(x - \sqrt{x^2+1}\right)$ and $y = \ln\left(x + \sqrt{x^2+1}\right)$, we choose the second solution because $x - \sqrt{x^2+1} < 0$ for all x, making $y = \ln\left(x - \sqrt{x^2+1}\right)$ undefined. One way to see this is to graph the two expressions $x - \sqrt{x^2+1}$ and $x + \sqrt{x^2+1}$.

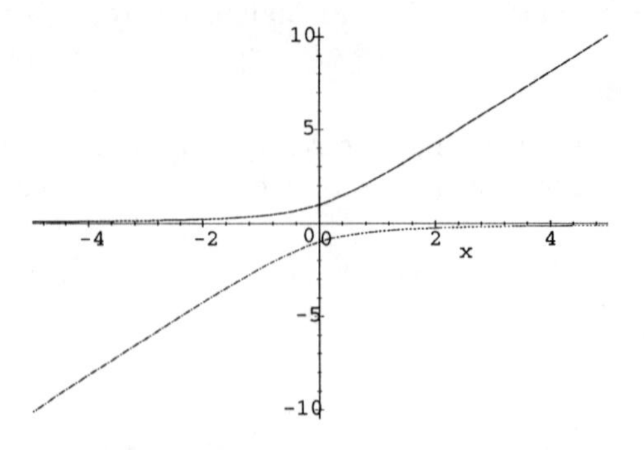

$x - \sqrt{x^2+1}$ and $x + \sqrt{x^2+1}$

With the insertion point in the equation $x = \frac{e^y - e^{-y}}{2}$, choose Solve + Exact (Variable to Solve For: y). Select the solution $\ln\left(x-\sqrt{x^2+1}\right)$ with the mouse and choose Plot 2D + Rectangular. Select and drag the other solution $\ln\left(x+\sqrt{x^2+1}\right)$ to the plot.

As a visual check, observe the similarity between the graphs of $y=\ln\left(x+\sqrt{x^2+1}\right)$ and $y=\sinh^{-1}x$.

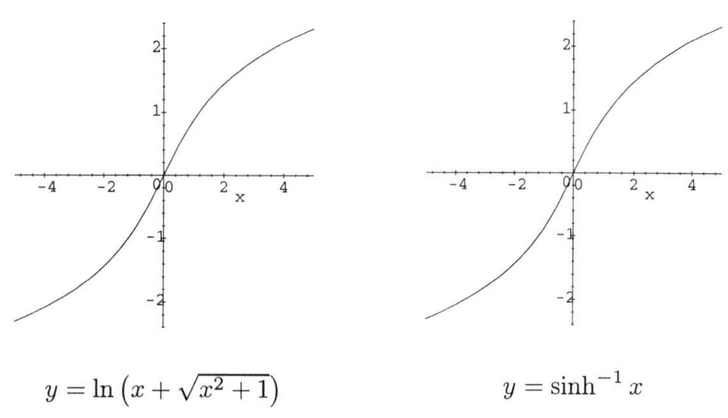

$y=\ln\left(x+\sqrt{x^2+1}\right)$ $\qquad$ $y=\sinh^{-1}x$

The two plots appear to be identical. Dragging one formula to the other frame produces no new visible points.

With the insertion point in the expression $\ln\left(x+\sqrt{x^2+1}\right)$, choose Plot 2D + Rectangular. With the insertion point in the expression $\sinh^{-1}x$, choose Plot 2D + Rectangular. Select one expression with the mouse and drag it to the other plot.

The graph of $y=\cosh x$ (drawn with a fat pen) and its inverse relation are shown in the following parametric plot of $(x,\cosh x)$ and $(\cosh x, x)$:

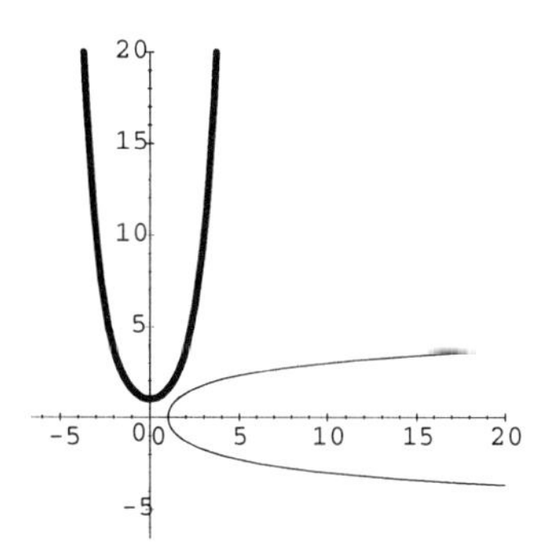

With the insertion point in the pair $(x,\cosh x)$, choose Plot 2D + Parametric. Select and drag to the frame $(\cosh x, x)$. Choose Edit + Properties, Plot Components page, and change Line Thickness to Medium for Item Number 1. Choose Axes page, and check Equal Scaling on Each

Axis. Choose View page and, under View Intervals, change to $-6 < x < 20$ and $-6 < y < 20$. Choose OK.

Because the hyperbolic cosine function is not one-to-one, its inverse relation is not a function. The inverse hyperbolic cosine function $y = \cosh^{-1} x$ is defined to be the upper half of the inverse relation. The inverse can also be found by solving the equation

$$x = \frac{e^y + e^{-y}}{2}$$

for y in terms of x. Faced with the two possible solutions $y = \ln\left(x - \sqrt{x^2 - 1}\right)$ and $y = \ln\left(x + \sqrt{x^2 - 1}\right)$, we choose the second solution because $x + \sqrt{x^2 - 1} > x - \sqrt{x^2 - 1}$, which implies $\ln\left(x + \sqrt{x^2 - 1}\right) > \ln\left(x - \sqrt{x^2 - 1}\right)$, and we want the upper half of the inverse relation.

With the insertion point in the equation $x = \frac{e^y + e^{-y}}{2}$, choose Solve + Exact (Variable to Solve For: y).

As a visual check, observe the similarity between the graphs of $y = \ln\left(x + \sqrt{x^2 - 1}\right)$ and $y = \cosh^{-1} x$:

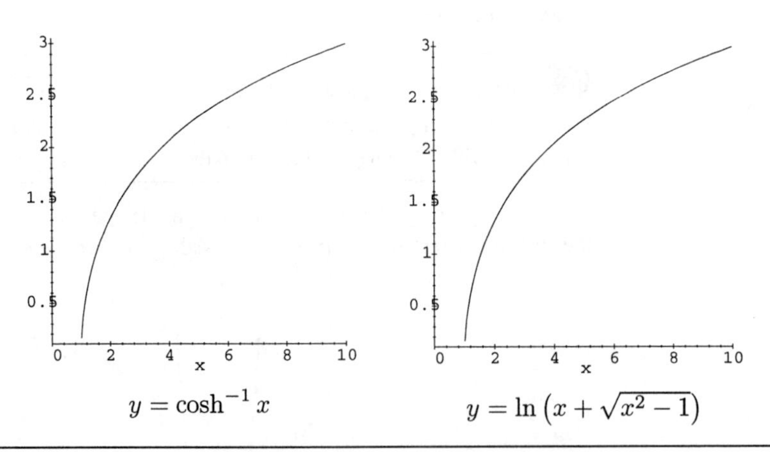

$y = \cosh^{-1} x$ $\qquad$ $y = \ln\left(x + \sqrt{x^2 - 1}\right)$

With the insertion point in the expression $\ln\left(x + \sqrt{x^2 - 1}\right)$, choose Plot 2D + Rectangular. With the insertion point in the expression $\cosh^{-1} x$, choose Plot 2D + Rectangular.

Activities

1. All of the internal forces are in equilibrium when a cable hangs freely. Consideration

of these forces leads to the differential equation

$$a\frac{d^2y}{dx^2} = \sqrt{1 + \left(\frac{dy}{dx}\right)^2}$$

which must be satisfied by the equation of the curve formed by a hanging cable whose ends are supported from the same height. This curve is called a *catenary*. (We heard that this curve is called a "catenary" because the curve resembles the smile of the Cheshire cat, suspected of being the cat (that ate the ca)nary. Others claim the name comes from the Latin word *catena* for chain.)

> **Hint** Use SolveODE + Exact to find functions $y(x)$ that satisfy the differential equation; then use Combine + Trig Functions to simplify the expression.

You will find that the formula for a catenary involves the hyperbolic cosine function. Both catenaries and inverted catenaries occur in a number of important applications. For example, constructing an arch in the shape of an inverted hyperbolic cosine creates a structure for which there are no transverse forces that might cause the arch to collapse, giving such a structure an inherent stability.

When a cable is hung between two 30-foot-high poles that are 36 feet apart, and the lowest point of the cable is 24 feet off the ground, how high will the cable be above a point on the ground between the poles that is 12 feet from the nearest pole?

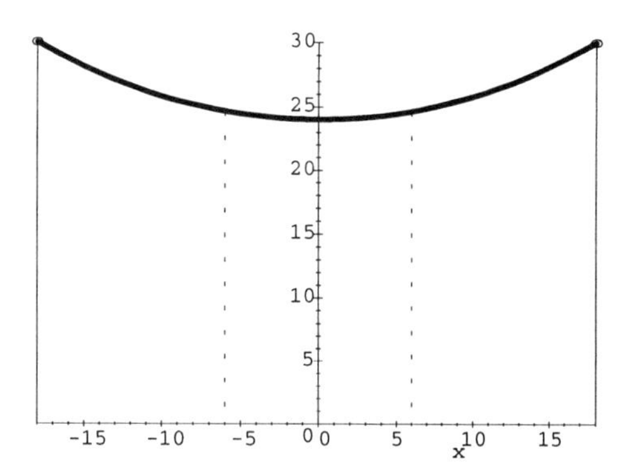

> **Hint** Placing the center of the cable above the origin lets you set one of the parameters to zero, leaving you with a system of two equations in two unknowns. Use Solve + Numeric to get a solution to this system of equations.

2. Study the function $\tanh x = \frac{\sinh x}{\cosh x}$ and its derivative by plotting appropriate graphs and explaining how the graphs are related. Study the function $\tanh x$ and its inverse by plotting appropriate graphs and explaining how the graphs are related. Verify the

formula $\tanh^{-1} x = \frac{1}{2} \ln\left(\frac{1+x}{1-x}\right)$ visually by plotting appropriate graphs.

3. Verify the following formulas visually by plotting appropriate graphs.

   a. $\dfrac{d}{dx} \sinh^{-1} x = \dfrac{1}{\sqrt{1+x^2}}$
   b. $\dfrac{d}{dx} \cosh^{-1} x = \dfrac{1}{\sqrt{x^2-1}}$
   c. $\dfrac{d}{dx} \tanh^{-1} x = \dfrac{1}{1-x^2}$

4. Verify the following formulas visually by plotting appropriate graphs.

   a. $\cosh^2 x = 1 + \sinh^2 x$

   Hint Look at the graph of $\cosh^2 x - \sinh^2 x$.

   b. $1 - \tanh^2 x = \operatorname{sech}^2 x$, where $\operatorname{sech} x = \dfrac{1}{\cosh x}$

   Hint Look at the graph of $\operatorname{sech}^2 x + \tanh^2 x$.

   c. $\sinh(x+y) = \sinh x \cosh y + \cosh x \sinh y$
   d. $\sinh(x+y) = \cosh x \cosh y + \sinh x \sinh y$

5. Make a parametric plot for $x = \cosh t$, $y = \sinh t$. Add the graph of $x = -\cosh t$, $y = \sinh t$ to the plot. To get a smooth view, set the domain interval to $-2 < t < 2$ and the view intervals to $-5 < x < 5$ and $-5 < y < 5$. Describe the result and compare the graph with the implicit plot of $x^2 - y^2 = 1$. How do these graphs help to explain the origin for the name of the hyperbolic functions?

3.8 Indeterminate Forms and l'Hospital's Rule

Suppose $\lim_{x\to c} f(x)$ and $\lim_{x\to c} g(x)$ both exist. If the latter is not zero, then the limit of the quotient exists and is equal to the quotient of the limits.

$$\lim_{x\to c} \frac{f(x)}{g(x)} = \frac{\lim_{x\to c} f(x)}{\lim_{x\to c} g(x)}$$

However, there are many cases in which the limit of the quotient exists but the quotient of the limits does not make sense. For example, $\lim_{x\to c} \frac{x-c}{x-c} = 1$ and $\lim_{x\to 0} \frac{\sin x}{x} = 1$. A common procedure for computing limits of quotients, called *l'Hospital's rule*, is based on the fact that, in certain cases where $\lim_{x\to c} \frac{f(x)}{g(x)}$ is indeterminate of type $\frac{0}{0}$ or $\frac{\infty}{\infty}$ (see page 38), the limit of a quotient of functions is equal to the quotient of the limits of their derivatives.

Lining Up l'Hospital's Rule

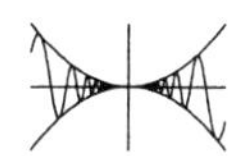

Consider the limit

$$\lim_{x \to 1} \frac{x^3 - 1}{x^2 - 1}$$

Plot the functions $f(x) = x^3 - 1$ and $g(x) = x^2 - 1$ in the same viewing window. Zoom in on the x-intercept until the curves start to look almost linear. Find the linearizations $L(x) = f(1) + f'(1)(x - 1)$ and $K(x) = g(1) + g'(1)(x - 1)$. (See page 77.) In a second viewing window, plot the linearizations L and K. Explain why it is reasonable that

$$\lim_{x \to 1} \frac{f(x)}{g(x)} = \lim_{x \to 1} \frac{f'(x)}{g'(x)}$$

Michelle's Solution

Define f and g by the equations

$$\begin{aligned} f(x) &= x^3 - 1 \\ g(x) &= x^2 - 1 \end{aligned}$$

Choose Define + Clear Definitions. Select the equation $f(x) = x^3 - 1$ with the mouse and choose Define + New Definition. Select the equation $g(x) = x^2 - 1$ with the mouse and choose Define + New Definition.

The graphs near $x = 1$ are given by

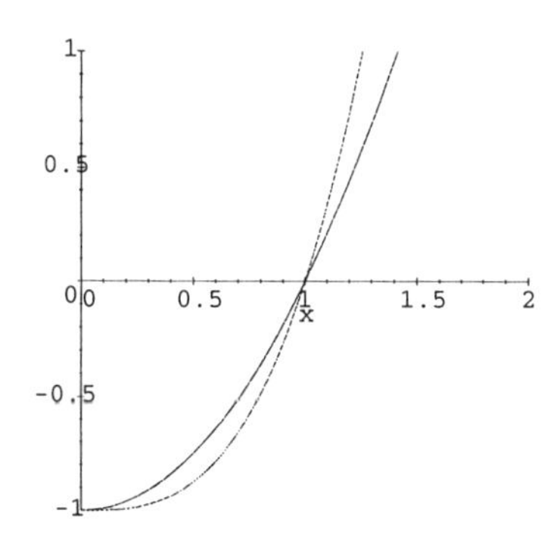

$x^3 - 1, x^2 - 1$

With the insertion point in the expression $x^3 - 1$, choose Plot 2D + Rectangular. Select $x^2 - 1$ with the mouse and drag to the plot. Choose Edit + Properties, Plot Components page. Change Domain Intervals to $0 < x < 2$. Choose the Axes page and check Equal Scaling on

Both Axes. Choose the View page; under View Intervals, check Default (to turn it off) and change intervals to $0 < x < 2$ and $-1 < y < 1$. Choose OK.

Notice that both curves cross the x-axis at 1.
We zoom in for a closer look.

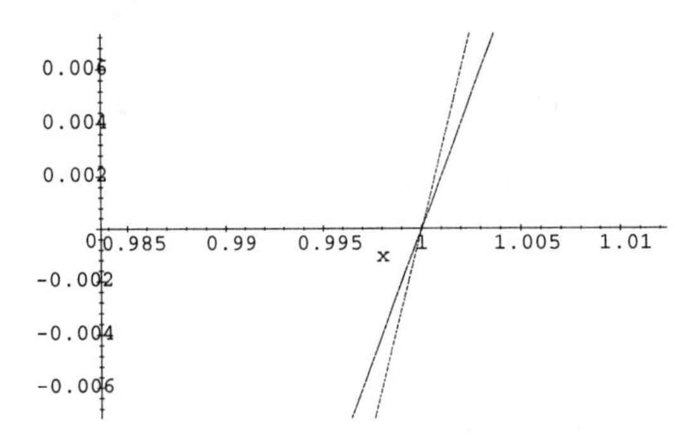

Double-click the plot. Choose the zoom-in tool (the large mountain). Place the mouse cursor at a point in the plot above and to the left of $(1, 0)$. Press the (left) mouse button, drag across and down to create a box, and release the mouse button. The plot inside the box will fill the frame. Repeat until you have the desired view.

With this viewing rectangle, the curves appear to be straight lines.

Consider the following linearizations (see page 77) at $x = 1$:

$$\begin{aligned} L(x) &= f'(1)(x-1) \\ K(x) &= g'(1)(x-1) \end{aligned}$$

Graphs of the linearizations at $x = 1$ look identical to the graphs of f and g in the given viewing rectangle. For fixed x very near to 1, the ratio of function values is very close to the ratio of values on the linearizations.

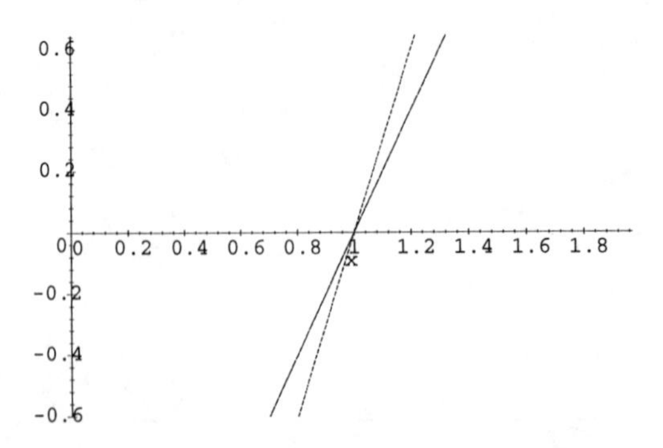

With the insertion point in the list $f'(1)(x-1), g'(1)(x-1)$, choose Plot 2D + Rectangular. Choose Edit + Properties, Axes page, and check Equal Scaling on Both Axes. Choose the View page and, under View Intervals, click Default (to turn it off). Choose OK. Double-click the plot and choose the zoom-in tool. Press the mouse button at a place above and to the left of the point $(1, 0)$, drag to the right and down to make a box symmetric about the point $(1, 0)$, and release the mouse button. Repeat until you have the desired view.

Activities

1. Verify that the right-hand limit $\lim_{x\to 0^+} x^x$ is an indeterminate form of type 0^0. (See page 38.) Justify each of the following steps:

$$\begin{aligned}
\ln\left(\lim_{x\to 0+} x^x\right) &= \lim_{x\to 0+} (\ln x^x) \\
&= \lim_{x\to 0+} x \ln x \\
&= \lim_{x\to 0+} \frac{\ln x}{\frac{1}{x}} \\
&= \lim_{x\to 0+} \frac{\frac{d}{dx}(\ln x)}{\frac{d}{dx}\left(\frac{1}{x}\right)} \\
&= \lim_{x\to 0+} \frac{\frac{1}{x}}{-\frac{1}{x^2}} \\
&= \lim_{x\to 0+} -x \\
&= 0
\end{aligned}$$

and conclude that $\lim_{x\to 0^+} x^x = e^0 = 1$. In justifying the fourth step, notice that $\lim_{x\to 0^+} \dfrac{\ln x}{\frac{1}{x}}$ is indeterminate of type $\dfrac{-\infty}{\infty}$ and explain why l'Hospital's rule still applies. Verify the limit visually by looking closely at the graph of $y = x^x$ for x near 0.

2. Verify that $\lim_{x\to\infty} \left(1 + \frac{1}{x}\right)^x$ is an indeterminate form of type 1^∞. Write $\left(1 + \frac{1}{x}\right)^x$ as an exponential by using the formula $[f(x)]^{g(x)} = e^{g(x)\ln f(x)}$. Rewrite the exponent as an indeterminate form of type $\frac{0}{0}$ and apply l'Hospital's rule to find an exact value for the limit.

4 THE MEAN VALUE THEOREM AND CURVE SKETCHING

Mean refers to "average"—not cruel, stingy, or inferior.

4.1 Maximum and Minimum Values

Calculus helps you find the biggest, the smallest, the cheapest, the most economical, the best. First you model a situation with a mathematical function, then find maximum and minimum values of the function. In this activity, you start with the maximum and minimum values, and find the function!

Dealing with Extreme Conditions

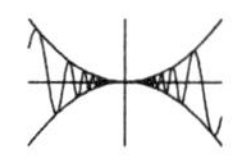

Find a polynomial of degree 3 that has a local maximum at $(-1, 3)$ and a local minimum at $(5, -11)$. Plot the graph and mark the points $(-1, 3)$ and $(5, -11)$ with small circles to verify your formula. Find a second polynomial of degree 3 that has no extreme values.

Hector's Solution

Consider the function

$$f(x) = ax^3 + bx^2 + cx + d$$

with undetermined coefficients. In order to have extreme values at the points $(-1, 3)$ and $(5, -11)$, the following four conditions must be satisfied:

$$\begin{aligned} f(-1) &= 3 \\ f(5) &= -11 \\ f'(-1) &= 0 \\ f'(5) &= 0 \end{aligned}$$

The solution to this system of equations is given by $b = -\frac{7}{9}$, $d = \frac{53}{27}$, $a = \frac{7}{54}$, $c = -\frac{35}{18}$.

N Choose Define + Clear Definitions. With the insertion point in the equation $f(x) = ax^3 + bx^2 + cx + d$, choose Define + New Definition. Choose Insert + Display and enter the conditions, pressing ENTER after each equation. With the insertion point in the system of equations, choose Solve + Exact.

Plotting the curve

$$y = \frac{7}{54}x^3 - \frac{7}{9}x^2 - \frac{35}{18}x + \frac{53}{27}$$

gives visual confirmation that the extreme values are located at the desired points, which are marked with small circles in the following plot:

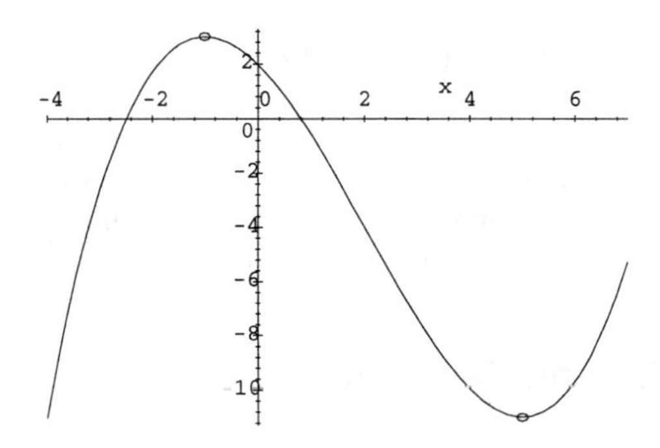

With the insertion point in the expression $\frac{7}{54}x^3 - \frac{7}{9}x^2 - \frac{35}{18}x + \frac{53}{27}$, choose Plot 2D + Rectangular. Select and drag to the frame $(-1, 3, 5, -11)$. Choose Edit + Properties, click the Plot Components tab, and choose Item Number 2. Set Plot Style to Point, and set Point Symbol to Circle. Choose OK.

We construct a cubic with no extreme values by generating a cubic whose derivative has no zeros. Define $g(x) = px^3 + qx^2 + rx$ and let $h(x) = x^2 + 1$. To ensure that $g'(x) = h(x)$, we solve the system

$$\begin{aligned} g'(-1) &= h(-1) \\ g'(0) &= h(0) \\ g'(1) &= h(1) \end{aligned}$$

With the insertion point in the equation $g(x) = px^3 + qx^2 + rx$, choose Define + New Definition. With the insertion point in the equation $h(x) = x^2 + 1$, choose Define + New Definition. Choose Insert + Display and type $g'(-1) = h(-1)$ in the input box. Press ENTER and type $g'(0) = h(0)$ in the input box. Press ENTER and type $g'(1) = h(1)$ in the input box. With the insertion point in the system of equations, choose Solve + Exact.

This yields the polynomial $g(x) = \frac{1}{3}x^3 + x$, whose graph follows.

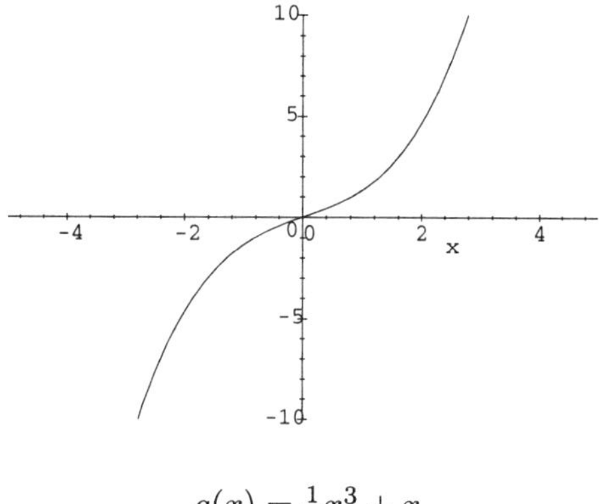

$g(x) = \frac{1}{3}x^3 + x$

 With the insertion point in the expression $\frac{1}{3}x^3 + x$, choose Plot 2D + Rectangular.

Activities

1. Construct three different polynomials f, g, and h, each of degree 5, such that $f(x)$ has four extreme values, $g(x)$ has two extreme values, and $h(x)$ has no extreme values by solving systems of equations that ensure that $f'(x) = (x+2)(x+1)(x-1)(x+2)$, $g'(x) = (x+2)(x-2)(x^2+1)$, and $h'(x) = (x^2+1)^2$. Explain how you know that this method will always work. Verify your solutions visually by plotting graphs of f, g, and h.

2. Construct three different polynomials f, g, and h, each of degree 5, such that $f(x)$ has four extreme values, $g(x)$ has two extreme values, and $h(x)$ has no extreme values by writing f as a product of five linear factors, g as a product of three linear factors times an irreducible polynomial of degree 2, and h as a linear factor times a product of two irreducible polynomials each of degree 2. Explain how you know that this method will always work. Verify your solutions visually by plotting graphs of f, g, and h.

3. Suppose $f(x)$ is a polynomial of degree 8. Discuss the possible number of extreme values and justify your conclusions. Construct one such polynomial that has the maximum possible number of extreme values. Verify your solution visually by plotting its graph.

4.2 The Mean Value Theorem

The *Mean Value Theorem* states that if f is a function continuous on the closed interval

$[a, b]$ and differentiable on the open interval (a, b), then there is a number c between a and b such that the slope of the tangent line for $x = c$ is equal to this "average" slope—that is,

$$f'(c) = \frac{f(b) - f(a)}{b - a}$$

This says that the tangent line to the curve at the point $(c, f(c))$ has slope equal to the slope of the secant line connecting $(a, f(a))$ and $(b, f(b))$. In the following graph, the secant line is shown as a dashed line and the corresponding tangent line is shown as a dotted line.

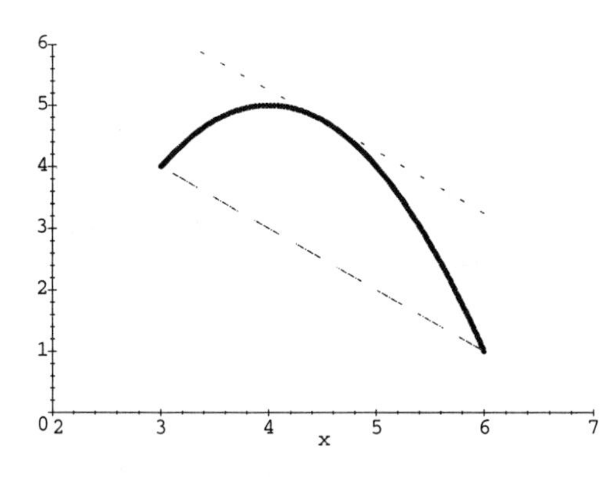

$f(x) = -x^2 + 8x - 11$ and $(a, b) = (3, 6)$

The number c appears to be close to 4.5 in this example. The number c is the solution to the equation $f'(c) = \frac{f(6)-f(3)}{6-3}$, where $f(x) = -x^2 + 8x - 11$, and, as you can easily verify, it is exactly 4.5.

The Mean Value Theorem Game

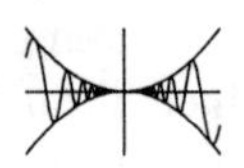

The Mean Value Theorem game is played by observing that the hypotheses of the Mean Value Theorem are satisfied, and then finding a number c that satisfies the conclusion of the theorem. This is a solitaire game, and the winner gets a passing score. Define f by the equation $f(x) = 1 - x^2$, and define $a = 0$, $b = 3$. Let m be the slope of the secant line connecting the points $(a, f(a))$ and $(b, f(b))$. Solve the equation $f'(c) = m$ for c. Plot the graph of f together with the tangent line at $(c, f(c))$ and the secant line connecting the points $(a, f(a))$ and $(b, f(b))$. Explain how your plot illustrates the Mean Value Theorem.

Sarah's Solution

Define $f(x) = 1 - x^2$ and set $a = 0$ and $b = 3$. The function $f(x)$ is continuous and differentiable on $(-\infty, \infty)$, and in particular is continuous and differentiable on the interval $[0, 3]$. The slope of the secant line joining $(a, f(a))$ and $(b, f(b))$ is given by

$$\frac{f(b) - f(a)}{b - a} = -3$$

We solve the equation $f'(c) = -3$, obtaining $c = \frac{3}{2}$.

With the insertion point in the equation $f(x) = 1 - x^2$, choose Define + New Definition. Select the equation $a = 0$ with the mouse and choose Define + New Definition. Select the equation $b = 3$ with the mouse and choose Define + New Definition. With the insertion point in the expression $\frac{f(b)-f(a)}{b-a}$, choose Evaluate. With the insertion point in the equation $f'(c) = -3$, choose Solve + Exact.

The following plot gives visual confirmation that the tangent line at $(c, f(c)) = \left(\frac{3}{2}, f\left(\frac{3}{2}\right)\right) = \left(\frac{3}{2}, -\frac{5}{4}\right)$ is parallel to the secant line joining $(a, f(a)) = (0, f(0)) = (0, 1)$ and $(b, f(b)) = (3, f(3)) = (3, -8)$:

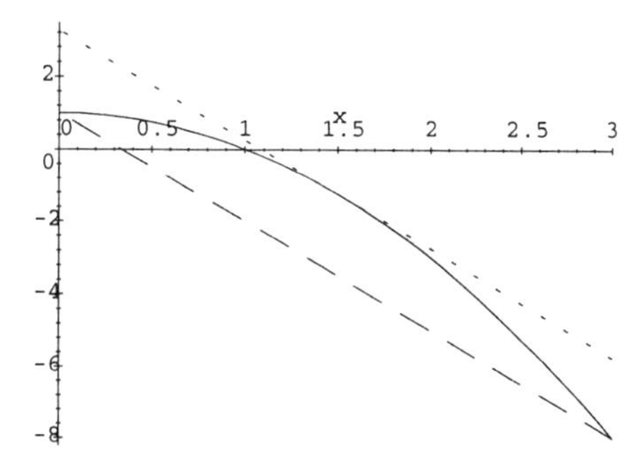

With the insertion point in the expression $f(x)$, choose Plot 2D + Rectangular. Select and drag to the plot each of $f(c) + f'(c)(x - c)$ and $f(a) - 3(x - a)$. Choose Edit + Properties, and click the Plot Components tab. Change Domain Intervals to $0 < x < 3$. Select Item Number 2 and Item Number 3, change Line Style to Dots or Dash, and change Line Color as desired. Choose OK.

The Mean Value Theorem Speed Trap

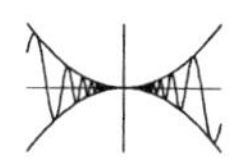

"But officer," replied the motorist, "my speedometer was under 70 mph every time the radar-warning light came on. How can you accuse me of speeding?" "Your car and license number were recorded on video less than an hour ago in Montrose," explained the officer, "and that's over 80 miles away. You certainly deserve a speeding ticket in this 75 mph zone."

Assume that the first sentence of each speaker is true and that the speed limit is at most 75 miles per hour along this stretch of highway. Use the Mean Value Theorem to settle this argument.

Janet's Solution

The motorist must have exceeded 80 miles per hour at some instant during the past hour. Because the position and velocity of the car are continuous functions of time, the Mean Value Theorem states that at some instant of time the instantaneous velocity must equal the average velocity, which is greater than $\frac{80-0}{1} = 80$ because the elapsed time is less than one hour.

 Use Scientific Notebook to record your thoughts.

Activities

1. Define f by the equation $f(x) = x^4 - 6x^3 + 4x - 1$, and define $a = 0$, $b = 1$. Let m be the slope of the secant line connecting the points $(a, f(a))$ and $(b, f(b))$. Solve the equation $f'(c) = m$ for c. Plot the graph of f together with the tangent line at $(c, f(c))$and the secant line connecting the points $(a, f(a))$ and $(b, f(b))$. Explain how your plot illustrates the Mean Value Theorem.

2. Define $f(x) = \dfrac{x+1}{x-1}$. Plot a graph of $y = f(x)$ using the viewing rectangle $0 \leq x \leq 2$ by $-10 \leq y \leq 10$. Add the secant line joining $(0, f(0))$ and $(2, f(2))$ to your plot. Show that there is no value of c between 0 and 2 such that
$$f'(c) = \frac{f(2) - f(0)}{2 - 0}$$
Explain why this does not contradict the Mean Value Theorem.

4.3 Monotonic Functions and the First Derivative Test

A function is monotonic if it is always increasing or always decreasing.

Controlling the Ups and Downs

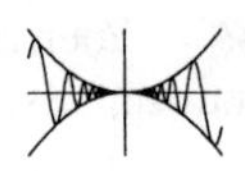

Find a polynomial that is increasing on the interval $(-\infty, -3)$, decreasing on $(-3, 1)$, increasing on $(1, 4)$, decreasing on $(4, \infty)$, and whose graph goes through the origin and the point $(2, 1)$.

Kyle's Solution

Define f by the equation $f(x) = ax^4 + bx^3 + cx^2 + dx + g$, where a, b, c, d, and g are undetermined coefficients. The conditions imply that $f'(-3) = f'(1) = f'(4) = f(0) = 0$ and $f(2) = 1$. Solving the system

$$\begin{aligned} f'(-3) &= 0 \\ f'(1) &= 0 \\ f'(4) &= 0 \\ f(0) &= 0 \\ f(2) &= 1 \end{aligned}$$

produces the polynomial $f(x) = \frac{3}{8}x^4 - x^3 - \frac{33}{4}x^2 + 18x$.

With the insertion point in the equation $f(x) = ax^4 + bx^3 + cx^2 + dx + g$, choose Define + New Definition. Choose Insert + Display and type the conditions in the input boxes, pressing ENTER to create additional display lines as needed. With the insertion point in the system of equations, choose Solve + Exact.

We can verify visually that this polynomial has the desired properties by plotting its graph and marking the points $(-3, f(-3))$, $(0, 0)$, $(1, f(1))$, $(2, 1)$, and $(4, f(4))$ with small circles.

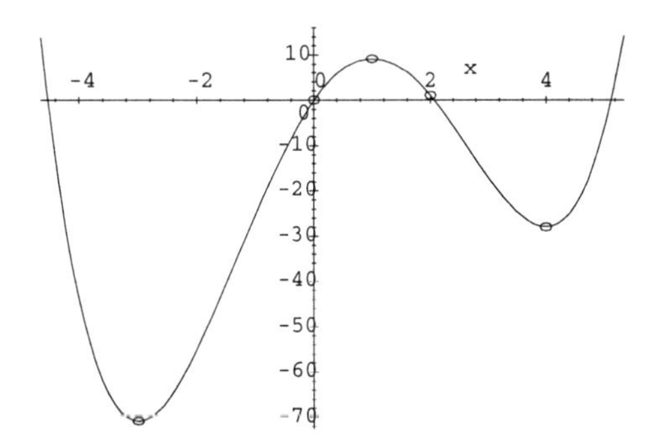

Choose Define + Clear Definitions. With the insertion point in the equation $f(x) = \frac{3}{8}x^4 - x^3 - \frac{33}{4}x^2 + 18x$, choose Define + New Definition. With the insertion point in the expression $\frac{3}{8}x^4 - x^3 - \frac{33}{4}x^2 + 18x$, choose Plot 2D + Rectangular. Select and drag to the plot each of $(-3, f(-3))$, $(0, 0)$, $(1, f(1))$, $(2, 1)$, and $(4, f(4))$. Choose Edit + Properties, Plot Components page. Select each of Item Numbers 2, 3, 4, 5, 6 and change Plot Style to Point, and change Point Symbol to Circle.

Activities

1. Find a polynomial of degree 3 that is decreasing on the interval $-\infty < x < 1$, increasing on $1 < x < 4$, decreasing on $4 < x < \infty$, and whose graph goes through the origin and the point $(4, 1)$. Visually verify your solution by plotting the graph of the polynomial.

2. Find a polynomial of degree 5 that is increasing on the intervals $-\infty < x < -4$, $-1 < x < 2$, and $4 < x < \infty$, decreasing on the intervals $-4 < x < -1$ and $2 < x < 4$, and whose graph goes through the points $(0, 0)$ and $(1, 1)$.

4.4 Concavity and Points of Inflection

An interval of a graph is *concave upward* if it lies above all of its tangents, and *concave downward* if it lies below all of its tangents. A *point of inflection* is a point on a curve where the concavity changes from upward to downward or from downward to upward.

Understanding the Shape of a Function

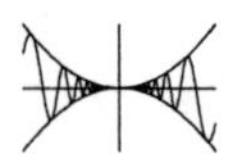

Describe the function $f(x) = 3x^5 - 40x^3 + 30x^2$. Use the graph of f to give a rough estimate of the intervals where the graph is concave upward, the intervals where the graph is concave downward, and the coordinates of the points of inflection. Use the graph of f'' to give better estimates. Use Solve + Numeric on $f''(x) = 0$ to give precise estimates. Explain the connections between the shape of the graph of f and the graph of its second derivative.

Heidi's Solution

From the plot of $f(x) = 3x^5 - 40x^3 + 30x^2$, it appears that the graph is concave downward on the interval $(-\infty, -2)$, concave upward on $(-2, 0)$, concave downward on $(0, 2)$, and concave upward on $(2, \infty)$.

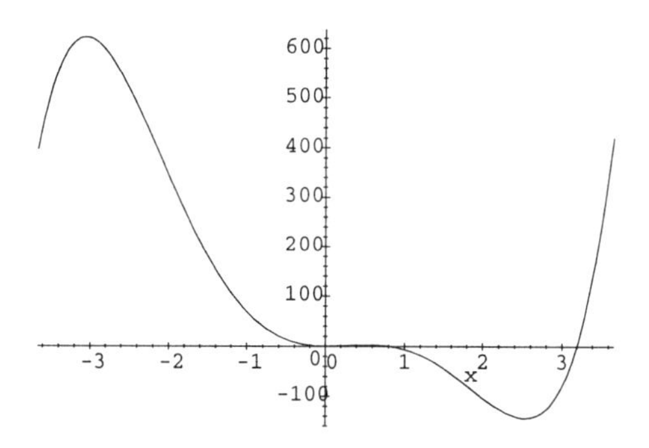

$$f(x) = 3x^5 - 40x^3 + 30x^2$$

With the insertion point in the expression $3x^5 - 40x^3 + 30x^2$, choose Plot 2D + Rectangular. Zoom in just a little to get a better look.

Better estimates can be found by looking at the x-intercepts of $y = f''(x)$. The following graph of $y = f''(x)$ shows x-intercepts at approximately $x = -2.1$, $x = 0.2$ and $x = 1.9$.

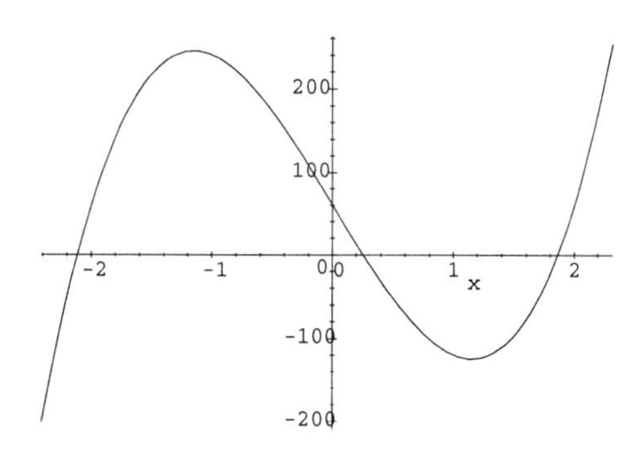

$$y = f''(x)$$

Choose Define + Clear Definitions. With the insertion point in the equation $f(x) = 3x^5 - 40x^3 + 30x^2$, choose Define + New Definition. With the insertion point in the expression $f''(x)$, choose Plot 2D + Rectangular. Zoom in to get a better look at the x-intercepts.

Solving $f''(x) = 0$ numerically we get the improved estimates $x = -2.1149$, $x = .2541$, and $x = 1.8608$.

Here is a graph of f with the inflection points marked.

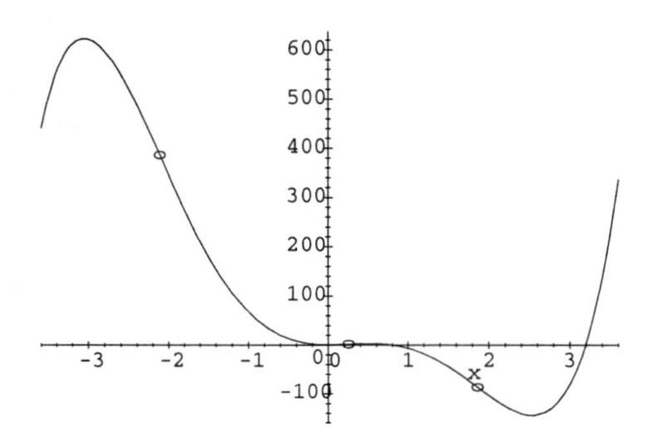

$$f(x) = 3x^5 - 40x^3 + 30x^2$$

With the insertion point in the equation $f''(x) = 0$, choose Solve + Numeric. With the insertion point in the expression $f(x)$, choose Plot 2D + Rectangular. Select and drag to the plot the three inflection points

$$(-2.1149, f(-2.1149), .2541, f(.2541), 1.8608, f(1.8608))$$

Choose Edit + Properties, click Plot Components tab, select Item Number 2, and change Plot Style to Point and Point Symbol to Circle. Change Domain Intervals to $-2.5 < x < 3$. Choose OK.

More Control over the Shape of a Function

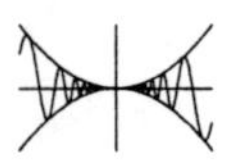

Determine the equation of a fifth-degree polynomial that is concave upward on the intervals $-\infty < x < -3$ and $3 < x < 6$ and concave downward elsewhere, that has a local maximum at $(-5, 5)$, and whose graph goes through the point $(2, -5)$. Verify your polynomial visually by plotting the graph of the polynomial, marking the points $(-5, 5)$ and $(2, -5)$ with small circles, and marking the inflection points with small crosses.

Alice's Solution

Define $f(x) = kx^5 + mx^4 + nx^3 + px^2 + qx + r$. Because the graph of f passes through the points $(2, -5)$ and $(-5, 5)$, f must satisfy $f(2) = -5$ and $f(-5) = 5$. With a local maximum at $(-5, 5)$, we must have $f'(-5) = 0$. Also, because the graph changes concavity at $x = -3$, 3, and 6, we need $f''(-3) = f''(3) = f''(6) = 0$. Thus we solve the system of equations

$$\begin{aligned} f''(-3) &= 0 \\ f''(3) &= 0 \\ f''(6) &= 0 \\ f'(-5) &= 0 \end{aligned}$$

$$\begin{aligned} f(-5) &= 5 \\ f(2) &= -5 \end{aligned}$$

to get the polynomial

$$f(x) = \frac{5}{4704}x^5 - \frac{25}{2352}x^4 - \frac{25}{784}x^3 + \frac{225}{392}x^2 - \frac{2375}{4704}x - \frac{13865}{2352}$$

With the insertion point in the equation $f(x) = kx^5 + mx^4 + nx^3 + px^2 + qx + r$, choose Define + New Definition. Enter the system of equations as shown in the solution, place the insertion point in the system of equations, and choose Solve + Exact. Select each of $k = \frac{5}{4704}$, $m = -\frac{25}{2352}$, $n = -\frac{25}{784}$, $p = \frac{225}{392}$, $q = -\frac{2375}{4704}$, $r = -\frac{13865}{2352}$ with the mouse, and choose Define + New Definition. With the insertion point in the expression $kx^5 + mx^4 + nx^3 + px^2 + qx + r$, choose Evaluate.

To verify visually that $f(x)$ has all the required properties, we mark the points $(-5, 5)$ and $(2, -5)$ with small circles and the inflection points with small crosses. Close inspection confirms that the graph is consistent with the stated properties.

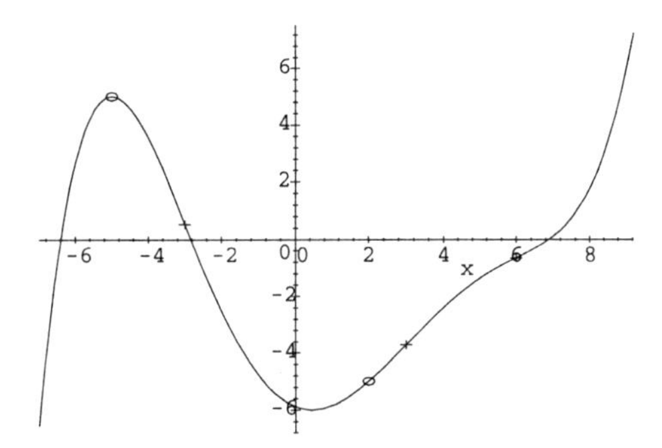

With the insertion point in the expression $\frac{5}{4704}x^5 - \frac{25}{2352}x^4 - \frac{25}{784}x^3 + \frac{225}{392}x^2 - \frac{2375}{4704}x - \frac{13865}{2352}$, choose Plot 2D + Rectangular. Select and drag to the frame $(-5, 5, 2, -5)$. Select and drag to the frame $(-3, f(-3), 3, f(3), 6, f(6))$. Choose Edit + Properties, Plot Components page. For Item Number 2, change Plot Type to Point and Point Style to Circle. For Item Number 3, change Plot Type to Point and Point Style to Cross. Choose OK.

Activities

1. Explore the concavity of $f(x) = \sin x + \frac{1}{100}\cos(100x)$ near the point $(1.0, f(1.0)) = (1.0, .85009)$. Zoom in on the graph of f at $(1.0, f(1.0)) = (1.0, .85009)$ until the graph is clearly of the type $\cap$ or $\cup$. Is it concave upward? Is it concave downward?

Evaluate $f''(1.0)$. Explain the connection between this value and your answer to the previous questions.

2. Construct a polynomial $p(x)$ of degree 5 that has the following properties:

   a. The function p has inflection points at $(-2, p(-2))$, $(1, p(1))$, and $(4, p(4))$.
   b. The function p has a local extremum at -3.
   c. The value of p at 0 is given by $p(0) = 0$.
   d. The value of p at 3 is given by $p(3) = 5$.

   Hint Choose **Define + Clear Definitions** to undefine f; then define $p(x) = ax^5 + bx^4 + cx^3 + dx^2 + fx + g$ and solve a system of six equations for the six unknowns a, b, c, d, f, g.

   Verify your solution by constructing a plot of $y = p(x)$ with tangent lines attached at each of the inflection points. Explain how you know that your function indeed satisfies all the stated conditions.

3. Construct a polynomial $p(x)$ of degree 5 whose second derivative is $h(x) = x^3 + 3x^2 - x - 3$ that satisfies $p(0) = 0$ and $p(1) = -1$. Verify your solution by plotting the graphs of $y = p(x)$ and $y = h(x)$. Explain how you know that your function indeed satisfies all the stated conditions.

   Hint Clear all the definitions from the previous activity, and then define $h(x) = x^3 + 3x^2 - x - 3$. Evaluate h at four different x-coordinates, and then solve a system of six equations and six unknowns, as in the previous activity.

4.5 Curve Sketching

In this activity you will find polynomials with given properties by solving systems of equations and verify the properties by curve sketching. Given a function of the form $f(x) = g(x)(x-a)^n$, where $g(a) \neq 0$, the graph of $f(x)$ looks very similar to the graph of $g(a)(x-a)^n$ for x near a. For example, here is the graph of $f(x) = (x+4)^2(x-3)^3$ together with the graph of $y = (3+4)^2(x-3)^3 = 49\,(x-3)^3$ for x near 3.

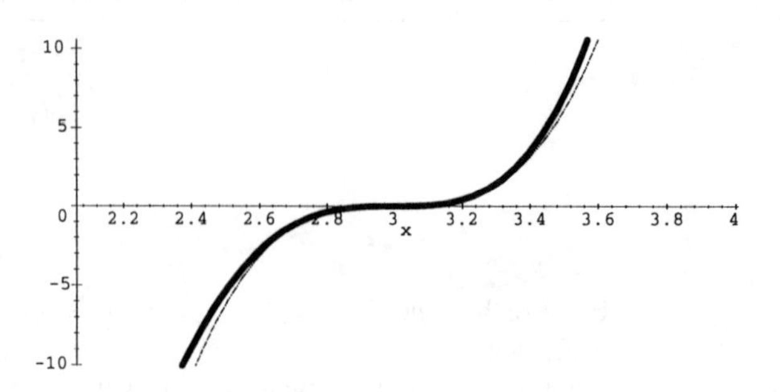

Local and Global Shapes

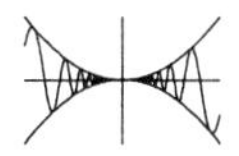

Consider the function $f(x) = x^2(x-2)^3$. Construct functions $g(x)$ and $h(x)$ such that $f(x) = g(x)x^2 = h(x)(x-2)^3$. Plot the graph of f together with the graphs of the expressions $y = g(0)x^2$ and $y = h(2)(x-2)^3$. Describe how the shape of the graph of f is related to the shape of the other two curves. Zoom out and plot the graph of $y = x^5$. Describe how the global shapes are related.

Bryn's Solution

Let $g(x) = (x-2)^3$ and $h(x) = x^2$. The graph of $y = x^2(x-2)^3$ with the curves $y = g(0)x^2$ and $y = h(2)(x-2)^3$ attached is shown below. Notice how the shape of the curve $y = x^2(x-2)^3$ is similar to the shape of the parabola $y = g(0)x^2$ near $x = 0$ and similar to the shape of the cubic $y = h(2)(x-2)^3$ near $x = 2$.

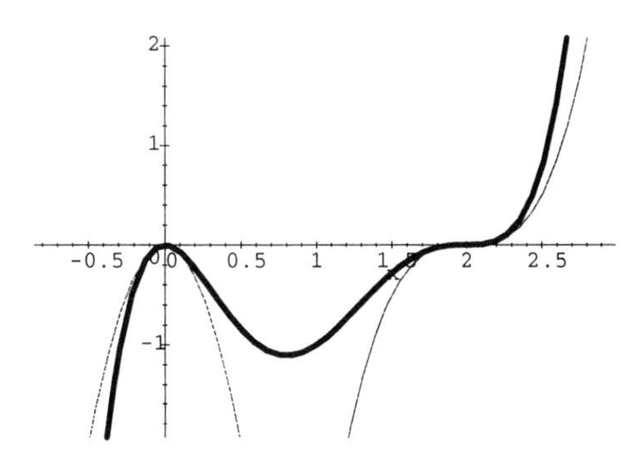

With the insertion point in the equation $g(x) = (x-2)^3$, choose Define + New Definition. With the insertion point in the equation $h(x) = x^2$, choose Define + New Definition. With the insertion point in the expression $x^2(x-2)^3$, choose Plot 2D + Rectangular. Select and drag to the frame $g(0)x^2$ and $h(2)(x-2)^3$. Choose Edit + Properties. On the Plot Components page, select different colors and thicknesses for the graphs. On the View page, turn off Default under View Intervals. Choose OK. Zoom in so that the View Intervals are roughly $-\frac{1}{2} \leq x \leq \frac{5}{2}$ and $-2 \leq y \leq 2$.

To visualize the global shape of the graph of f, we zoom out and attach the graph of $y = x^5$

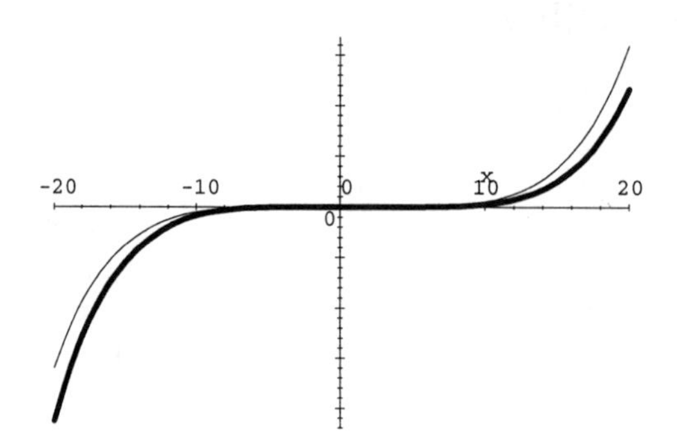

To distinguish between the graphs, it is convenient to draw the graph of f with a fat pen. Notice that the shapes are very similar; the graph of f appears to be translated slightly to the right.

With the insertion point in the expression $x^2(x-2)^3$, choose Plot 2D + Rectangular. Select and drag to the frame x^5. Zoom out so that the Domain Interval is $-20 \le x \le 20$. Choose Edit + Properties, Plot Components page. For Item Number 1, change Thickness to Medium.

Activities

1. Study the shape of the graph of the function $f(x) = (x-2)^2(x+1)^3(x+4)^2$ near $x = 2$, $x = -1$, and $x = -3$. Construct function $g(x)$, $h(x)$, and $k(x)$ such that $f(x) = g(x)(x-2)^2 = h(x)(x+1)^3 = k(x)(x+4)^2$. Plot the graph of f together with the graphs of the expressions $y = g(2)(x-2)^2$, $y = h(x)(x+1)^3$, and $y = k(x)(x+4)^2$. Describe how the shape of the graph of f is related to the shape of the other three curves. Determine a power x^n so that globally the shape of the graph of f matches the shape of the graph of $y = x^n$.

2. Discuss the shape of the graph of the function $f(x) = \sqrt{x}\sqrt{1-x} = g(x)\sqrt{x} = h(x)\sqrt{1-x}$. Use the curves $y = g(0)\sqrt{x}$ and $y = h(1)\sqrt{1-x}$ in your discussion.

3. Study the shape of $f(x) = x^2 \sin x \cos^2 x$ near $x = 0$ and $x = \dfrac{\pi}{2}$. Verify graphically that $y = x^3$ provides a reasonable approximation near $x = 0$. Verify graphically that $y = \dfrac{\pi^2}{4}\left(x - \dfrac{\pi}{2}\right)^2$ provides a reasonable approximation near $x = \dfrac{\pi}{2}$.

4.6 Graphing with Calculus and Calculators

Graphing a function, even with the aid of technology, requires a thoughtful approach.

The default viewing rectangle is rarely the view that provides the most information. With practice, you will quickly adjust the horizontal and vertical components of the viewing rectangle to generate a viewing rectangle that clearly shows important aspects of the function.

What are the important aspects? Here are a few you might wish to consider:

- *Domain:* Especially if the domain is restricted because of such things as square roots or vertical asymptotes.
- *Intercepts:* Does the graph cross the y-axis? If so, where? Does the graph cross the x-axis? If so, where?
- *Symmetry:* Is the function even? Odd? Periodic?
- *Asymptotes:* Are there any horizontal asymptotes? Vertical asymptotes? Slant asymptotes?
- *Intervals of Increase or Decrease:* On which interval(s) is the function increasing? Decreasing?
- *Local Maximum and Minimum Values:* Are there any extreme values? If so, where?
- *Concavity and Points of Inflection:* On which interval(s) is the graph concave upward? Downward? Where are the points of inflection?

If the domain is unrestricted, so that the function is defined for all real numbers, it is not practical to use a viewing window that actually shows $-\infty < x < \infty$. However, you should attempt to find a viewing rectangle that includes at least part of the infinite intervals $-\infty < x < a$ or $b < x < \infty$ where the graph is monotone. Of course, for particular functions, such as periodic functions, this will not be possible.

Occasionally, a single viewing rectangle may not show all of the important aspects. Zooming in too far may lose large-scale features of the graph, and zooming out may obscure small-scale features. Two or more views may be required to adequately describe some functions. The following activities are designed to give you practice at selecting appropriate viewing rectangles and gathering information about functions by looking closely at their graphs.

Getting Into Polynomial Shape

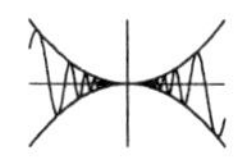

Plot a graph of $f(x) = 4x^4 - 7x^2 + 4x + 6$ using a viewing rectangle that shows the important aspects. Define the function $f(x) = 4x^4 - 7x^2 + 4x + 6$ and locate the extreme values precisely by solving exactly or numerically the equation $f'(x) = 0$. Locate the inflection points precisely by solving exactly or numerically the equation $f''(x) = 0$. Mark each local high point and low point on your plot using a diamond $\diamond$. Mark each inflection point with a circle $\circ$. State the interval(s) where f is increasing. State the interval(s) where the graph is concave upward.

Tane's Solution

Let $f(x) = 4x^4 - 7x^2 + 4x + 6$. Solve the equation $f'(x) = 0$ to get the critical numbers -1.0545, $.32492$, and $.72963$. Solve the equation $f''(x) = 0$ to locate the possible inflection points $(-.54006, f(-.54006))$ and $(.54006, f(.54006))$. In the following plot of f, the extreme values are located with small diamonds and the inflection points with small circles:

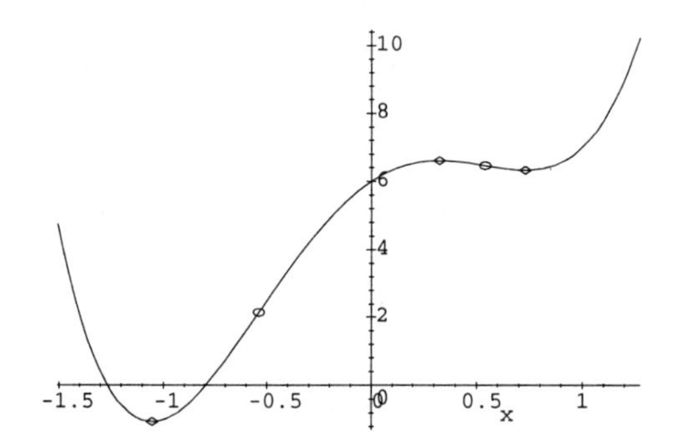

With the insertion point in the equation $f(x) = 4x^4 - 7x^2 + 4x + 6$, choose Define + New Definition. With the insertion point in the expression $f(x)$, choose Plot 2D + Rectangular. With the insertion point in the equation $f'(x) = 0$, choose Solve + Numeric. Drag to the frame

$$[-1.0545, f(-1.0545), .32492, f(.32492), .72963, f(.72963)]$$

With the insertion point in the equation $f''(x) = 0$, choose Solve + Numeric. Drag to the frame

$$[-.54006, f(-.54006), .54006, f(.54006)]$$

Choose Edit + Properties, click the Plot Components tab, choose Item Number 2, and set Plot Style to Point and Point Symbol to Diamond. Choose Item Number 3 and set Plot Style to Point and Point Symbol to Circle. Choose OK.

The function f is increasing on $-1.0545 \le x \le .32492$ and on $.72963 \le x < \infty$. The graph is concave upward on $-\infty < x < -.54006$ and on $.54006 < x < \infty$.

Getting Into Rational Shape

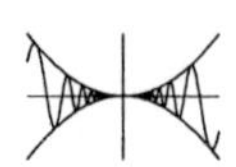

Plot a graph of $f(x) = \dfrac{x^4 + x^3 - 2x^2 + 2}{x^2 + x - 2}$ using a viewing rectangle that shows the important aspects. Locate the vertical asymptotes precisely by solving exactly or numerically the equation $x^2 + x - 2 = 0$. Locate the extreme values precisely by solving exactly or numerically the equation $f'(x) = 0$. Locate the inflection points precisely by solving exactly or numerically the equation $f''(x) = 0$. Mark each local high point and low point on your plot using a diamond $\diamond$. Mark each inflection point with a circle $\circ$. State the interval(s) where f is increasing. State the interval(s) where the graph is concave upward.

Noel's Solution

Consider the function

$$f(x) = \frac{x^4 + x^3 - 2x^2 + 2}{x^2 + x - 2}$$

We can locate the possible vertical asymptotes by solving the equation $x^2 + x - 2 = 0$, which has the roots $x = -2$ and $x = 1$.

 With the insertion point in the equation $x^2 + x - 2 = 0$, choose Solve + Exact.

The critical numbers are located by finding where the derivative is zero. To find approximate values for zeros of the derivative, we look at a plot of $f'(x)$.

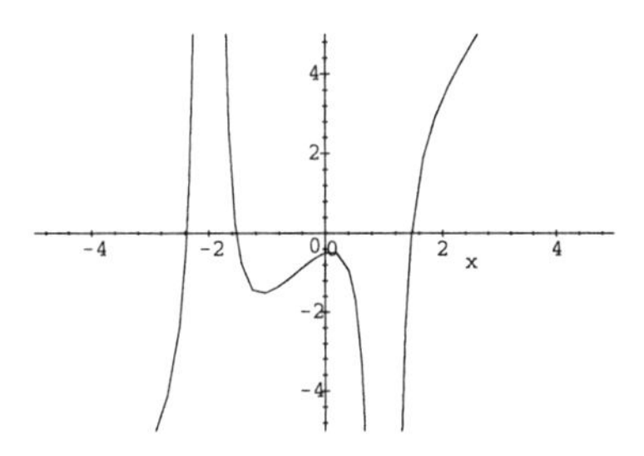

$$y = f'(x)$$

and then find numerical solutions. The three critical numbers are $x = -2.3725$, $x = -1.5428$, and $x = 1.4715$.

 Choose Define + Clear Definitions. With the insertion point in the equation

$$f(x) = \frac{x^4 + x^3 - 2x^2 + 2}{x^2 + x - 2}$$

choose Define + New Definition. With the insertion point in the expression $f'(x)$, choose Plot 2D + Rectangular. On the View page of the Plot Properties dialog, turn off Default under View Intervals and change to $-5 < y < 5$. Choose OK. Choose Insert + Display and type $f'(x) = 0$. Press ENTER and type $x \in (-3, -2)$. (The symbol $\in$ denotes membership, so $x \in (-3, -2)$ means $-3 < x < -2$.) With the insertion point in the system

$$\begin{aligned} f'(x) &= 0 \\ x &\in (-3, -2) \end{aligned}$$

choose Solve + Numeric. Similarly, enter the system

$$\begin{aligned} f'(x) &= 0 \\ x &\in (-2, -1) \end{aligned}$$

and choose Solve + Numeric; and enter the system

$$\begin{aligned} f'(x) &= 0 \\ x &\in (1, 2) \end{aligned}$$

and choose Solve + Numeric.

The inflection points are found by solving $f''(x) = 0$. The graph of

$$f''(x) = 2\frac{-2 + 18x + 12x^2 - 11x^3 - 3x^4 + x^6 + 3x^5}{(x^2 + x - 2)^3}$$

shows approximate values.

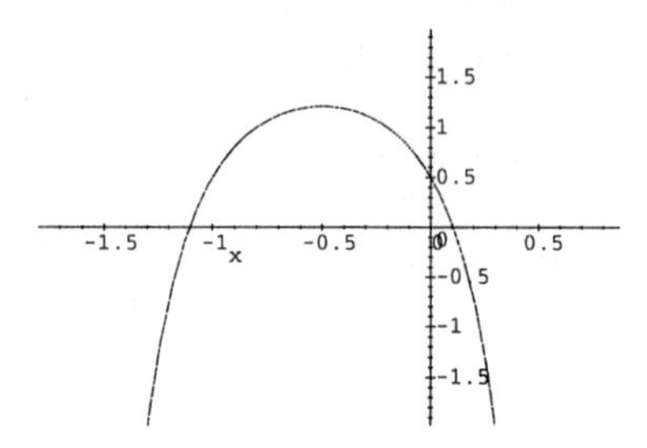

$y = f''(x)$

The inflection points are $(-1.1045, .15868)$ and $(.10454, -1.0503)$.

 With the insertion point in the system

$$\begin{aligned} f''(x) &= 0 \\ x &\in (-1.5, -1) \end{aligned}$$

choose Solve + Numeric. With the insertion point in the system

$$\begin{aligned} f''(x) &= 0 \\ x &\in (-1, 1) \end{aligned}$$

choose Solve + Numeric.

In the following plot, the extreme points are marked with small diamonds and the inflection points are marked with small circles:

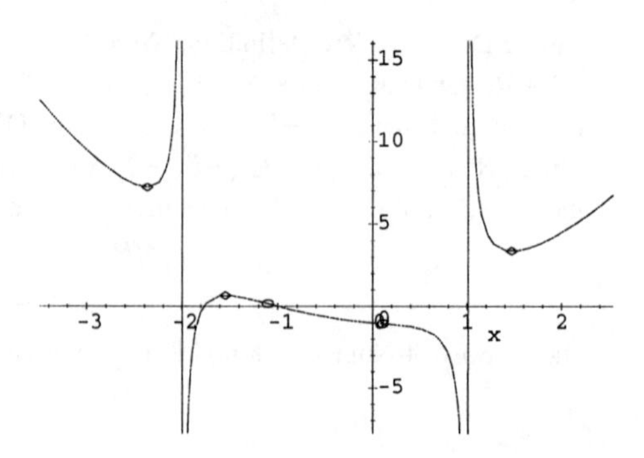

$f(x) = \frac{x^4+x^3-2x^2+2}{x^2+x-2}$

The function is increasing on the intervals $-2.3725 \leq x < -2$, $-2 < x \leq -1.5428$, and $1.4715 \leq x < \infty$. The graph is concave upward on $-\infty < x < 2$, $-1.1045 < x < .10454$, and $1 < x < \infty$. Vertical asymptotes are located at $x = -2$ and $x = 1$.

(These lines are shown in the graph although they are not actually part of the graph of the function $f(x)$.)

With the insertion point in the expression $f(x)$, choose Plot 2D + Rectangular. Drag to the frame $(-2.3725, f(-2.3725), -1.5428, f(-1.5428), 1.4715, f(1.4715))$. Drag to the frame $(-1.1045, f(-1.1045), .10454, f(.10454))$. From the Plot Components page of the Plot Properties dialog, choose Item Number 2 and set Plot Style to Point and Point Symbol to Diamond; choose Item Number 3 and set Plot Style to Point and Point Symbol to Circle. Choose OK.

Remark Because $\dfrac{x^4+x^3-2x^2+2}{x^2+x-2} = x^2 + \dfrac{2}{x^2+x-2}$, it follows that the large-scale behavior of f is determined by the curve $y = x^2$. The following plot is a graph of f together with the curve $y = x^2$:

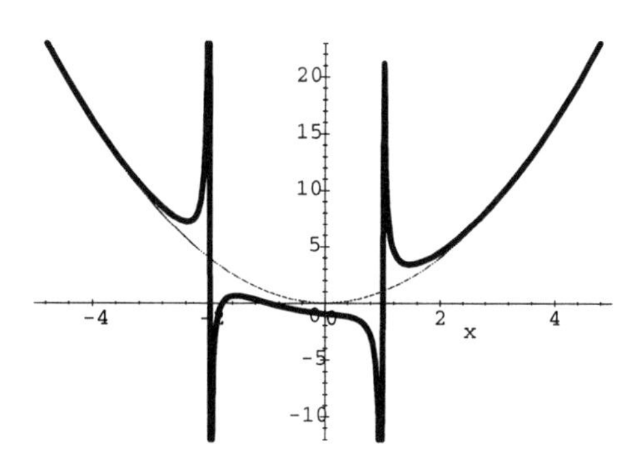

$$y = x^2 + \frac{2}{x^2+x-2}, \; y = x^2$$

Activities

1. Find equations for the following graphs of polynomials. Verify your solution by using a viewing rectangle that displays a graph that looks like the given graph. Solutions may not be unique. Describe the techniques you used to obtain your equations.

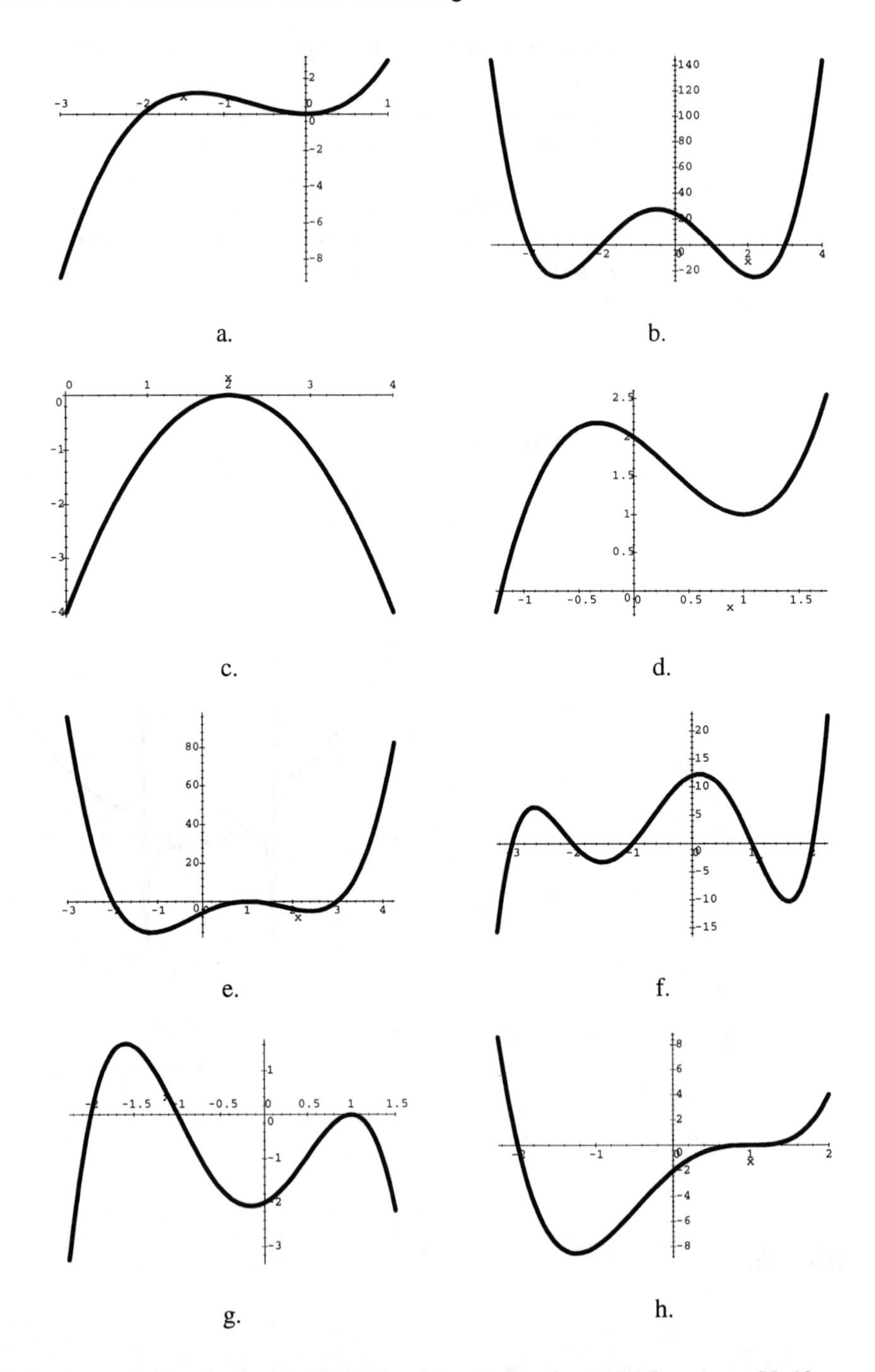

2. Find an equation for each of the following graphs of rational functions. Verify your solution by using a viewing rectangle that displays a graph that looks like the given graph. Solutions may not be unique. Describe the techniques you used to obtain your equations.

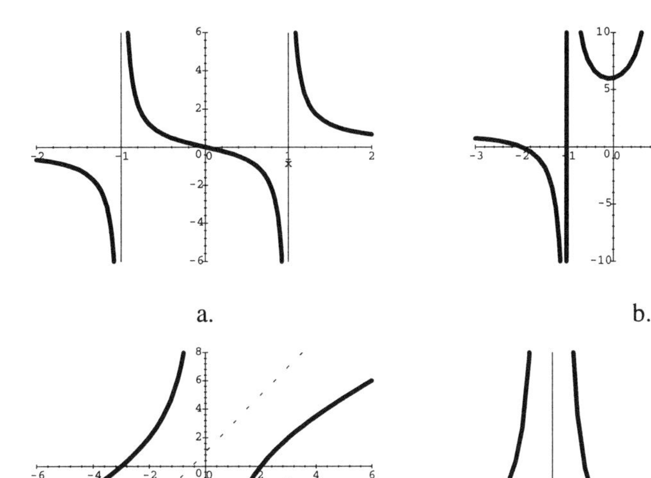

a.

b.

c.

d.

3. Sketch the graph of
$$f(x) = \frac{(x+4)(x-3)^2}{(x+2)(x-5)}$$
by hand, using asymptotes and intercepts but not derivatives. Then sketch the graph with *Scientific Notebook*, using your sketch as a guide for determining an appropriate viewing window. Use Polynomials + Divide to rewrite
$$\frac{(x+4)(x-3)^2}{(x+2)(x-5)}$$
as a polynomial plus a rational function. Drag the polynomial to the frame. Locate the extreme values and mark the high points and low points on the graph using a cross $+$. Locate the inflection points and mark them on the graph with a circle $\circ$. Label each vertical asymptote with its equation. Label each slant asymptote with its equation. Explain how you know that you have correctly identified and labeled the important aspects of f.

4. Sketch a graph of $f(x) = \sqrt{x} + \sqrt{1-x}$ and discuss the important aspects of f. In particular, discuss the domain. Drag the expressions $\sqrt{x}$ and $\sqrt{1-x}$ to the frame and explain how the curves $y = \sqrt{x}$ and $y = \sqrt{1-x}$ are related to the graph of $y = \sqrt{x} + \sqrt{1-x}$.

5. Investigate the graph of $f(x) = x^{\sin x}$. What is the domain of f? In what ways is the graph similar to the graph of $y = x^x$? In what ways is the graph similar to the graph of $y = \sin x$? What is an appropriate value for $f(0)$? Locate three critical numbers. Locate three inflection points.

4.7 Applied Maximum and Minimum Problems

Calculus provides useful tools for solving problems that ask for the "best" way to perform a task. Such problems may be stated

- in economic terms: What is the cheapest or most profitable?
- in physical terms: What is the easiest?
- in geometric terms: What is the largest or smallest? Longest or shortest?
- in terms of time: What is the fastest?

Usually several steps are required to solve such a problem.

1. Read and understand the problem. What is known? What is the unknown?
2. Draw a diagram and label the parts with appropriate symbols.
3. Write an equation that expresses the unknown quantity in terms of the known quantities.
4. Plot a graph to locate approximately the extreme values.
5. Use calculus to locate precisely the extreme values.

Making a Big Box

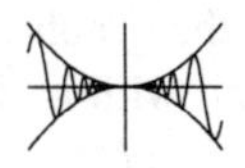

A box with an open top is to be constructed from a rectangular piece of cardboard of dimensions 8 inches by 10 inches by cutting out a square from each of the four corners and bending up the sides. Draw a diagram of the box and label the parts. Determine a formula for the volume of the box in terms of the dimension of an edge of a corner that is cut out. Explain why the interval $[0, 4]$ is a reasonable domain for the volume function. Plot a graph of the volume versus corner dimension. Find the largest volume that such a box can have.

Carla's Solution

Let x denote the length of the edge of a corner that is cut out. Fold the cardboard along the dotted lines and fold to make an open-top box, as indicated to the right.

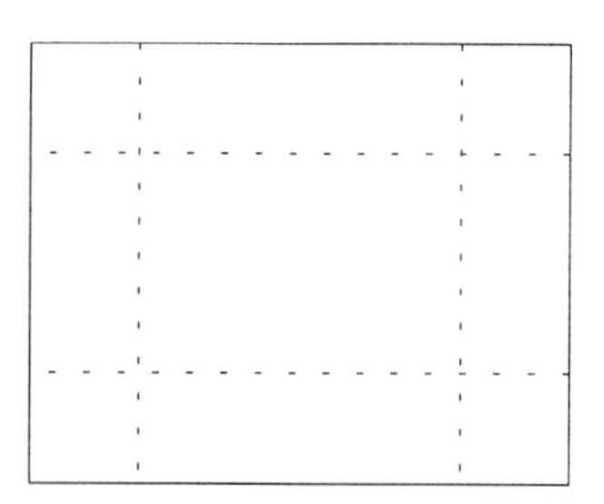

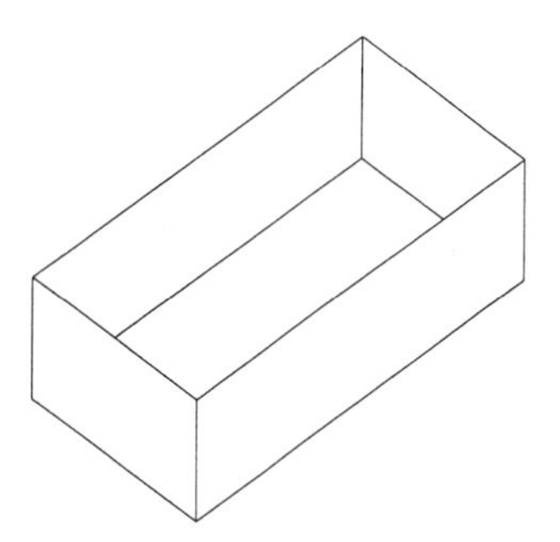

With the insertion point in the list $(-5, -4, 5, -4, 5, 4, -5, 4, -5, -4)$, choose Plot 2D + Rectangular. Select with the mouse and drag each of the following to the plot

$$(-5, -2, 5, -2)$$
$$(-5, 2, 5, 2)$$
$$(-3, -4, -3, 4)$$
$$(3, -4, 3, 4)$$

Choose Edit + Properties, Plot Components page. Set Line Style to Dots for Item Numbers 2, 3, 4, 5. Choose Axes page, set Axes Type to None, and check Equal Scaling Along Each Axis. Choose OK. With the insertion point in the list $(0, 0, 4, 4, 4, 6, 2, 8, -2, 4, -2, 2, 0, 0, 0, 2)$, choose Plot 2D + Rectangular. Select with the mouse and drag each of the following to the plot: $(-2, 4, 0, 2, 4, 6)$, $(3, 5, 2, 6, 2, 8, 2, 6, -1, 3)$. Choose Edit + Properties, Axes page; set Axes Type to None, and check Equal Scaling Along Each Axis. Choose OK.

The open-top box will have dimensions x by $8-2x$ by $10-2x$, and hence the volume is given by $V = x(8-2x)(10-2x) = 80x - 36x^2 + 4x^3$. The domain of the volume function is $[0, 4]$ because for $x > 4$ there is nothing left to fold up. The following is a graph of the volume function:

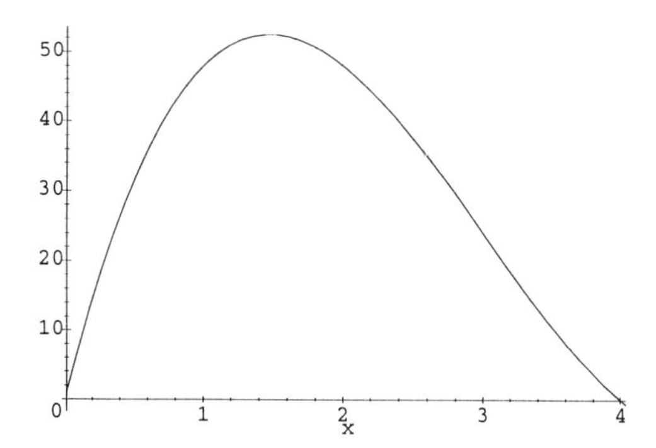

With the insertion point in the expression $x(8-2x)(10-2x)$, choose Expand. With the insertion point in the expression $80x - 36x^2 + 4x^3$, choose Plot 2D + Rectangular. Change Domain Interval to $0 < x < 4$.

There appears to be a critical number at approximately 1.5. Solving

$$\frac{d}{dx}\left[80x - 36x^2 + 4x^3\right] = 0$$

we get the two potential solutions $x = 1.4725$ and $x = 4.5275$. The second potential solution must be thrown out because it is not in the domain. The maximum volume is given by

$$\left[80x - 36x^2 + 4x^3\right]_{x=1.4725} = 52.514 \text{ cubic inches}$$

With the insertion point in the equation $\frac{d}{dx}\left[80x - 36x^2 + 4x^3\right] = 0$, choose Solve + Numeric. With the insertion point in the expression $\left[80x - 36x^2 + 4x^3\right]_{x=1.4725}$, choose Evaluate.

More Braking Distance Models

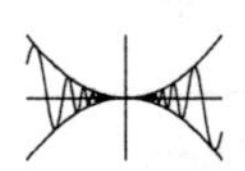

One aspect of building mathematical models is the problem of fitting a curve to data. This can be viewed as the problem of finding a curve that minimizes (in some sense) the distance between the data and the curve. The data we consider in this problem comes from measurements of the total stopping distance of a car in a panic stop. This data is given in the following table, where the numbers in the top row represent velocity in miles per hour and the numbers immediately below give the stopping distance in feet measured at that velocity:

20	25	30	35	40	45	50	55	60	65	70	75	80
42	56	74	92	116	142	173	210	248	292	343	401	464

In later courses you will learn how to fit curves to data using methods such as the "least squares" method, which minimizes the sum of the squares of the distances from the curve to the data points. Now, however, you can use an item called Fit Curve to Data on the Statistics submenu. This command produces polynomials that give a good fit to the data. Use this technique to find polynomials of degrees 1, 2, and 3. Plot the data and these three curves on the same graph. (If you worked the similar problem, "Give me a Brake," at the end of Chapter 1, add that curve to the graph as well.) Look at the graph for the domain $20 < x < 80$. Look at the graph for the domain $0 < x < 300$. Discuss what you observe in these two graphs. Based on these observations, which type of curve do you think is a better model for this problem?

Roger's Solution

The following table gives measurements of total stopping distance for a car traveling at different speeds between 20 and 80 miles per hour:

20	25	30	35	40	45	50	55	60	65	70	75	80
42	56	74	92	116	142	173	210	248	292	343	401	464

A plot of this data looks as follows.

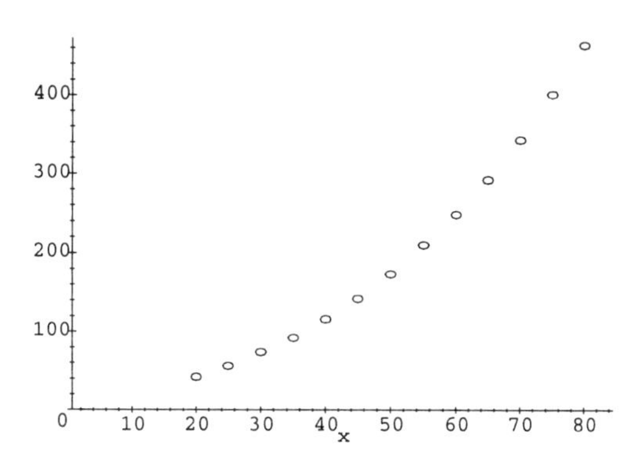

To make the table, use Insert + Matrix, select 2 Rows and 13 Columns, and choose OK. Type the data in the input boxes. To make the graph, first transpose the matrix: With the insertion point in the matrix, choose Matrix + Transpose. With the insertion point in the transposed matrix

20	42
25	56
30	74
35	92
40	116
45	142
50	173
55	210
60	248
65	292
70	343
75	401
80	464

choose Plot 2D + Rectangular. On the Plot Components page of the Plot Properties dialog, change Plot Style to Point and Point Symbol to Circle. Set Domain Intervals at $0 < x < 90$. Choose OK.

The "Fit Curve to Data" command on the statistics submenu produces the following polynomials

$$y = -\frac{339}{1001} + \frac{10459}{6006}x + \frac{15}{2002}x^2 + \frac{29}{53625}x^3$$
$$y = \frac{50411}{1001} - \frac{3957}{2002}x + \frac{887}{10010}x^2$$

$$y = -\frac{1822}{13} + \frac{179}{26}x$$

With the insertion point in the transposed matrix, choose Statistics + Fit Curve to Data. Check Polynomial of Degree (select 1). Choose OK. Repeat for degrees 2 and 3.

These three polynomials, together with the data points, give the following graph:

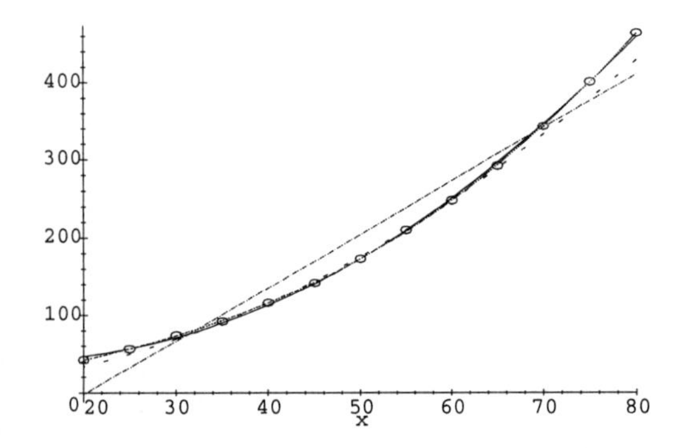

With the insertion point in the transposed matrix, choose Plot 2D + Rectangular. Select and drag to the frame each of the three polynomial expressions: $-\frac{339}{1001} + \frac{10459}{6006}x + \frac{15}{2002}x^2 + \frac{29}{53625}x^3$, $\frac{50411}{1001} - \frac{3957}{2002}x + \frac{887}{10010}x^2$, and $-\frac{1822}{13} + \frac{179}{26}x$. Choose Insert + Properties, and click the Plot Components tab. Change Plot Style to Point and Point Symbol to Circle. Select Item Numbers 2, 3, and 4, and change colors as you wish. Set Domain Interval at $20 < x < 80$. Choose OK.

Changing the view to $0 < x < 300$ gives the following graph:

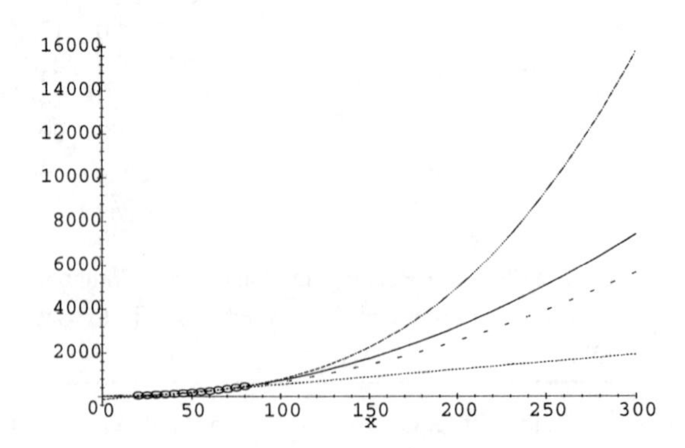

Click the plot to select it. Choose Edit + Properties, click the Plot Components tab, and change the Domain Interval to $0 < x < 300$. Choose OK.

The straight line is not a good model even for the range of the data. (I have added my curve from Chapter 1 as a dotted curve. It is a better model than the line but not as good as either of the other curves in the range of the data.) Both the (new) quadratic and cubic curves give a very good model for the range of the data, and they are barely distinguishable on that portion of the graph. They give quite different predictions for higher speeds, however. The cubic model predicts that at 300 miles per hour it would take over 3 miles to brake a car; the quadratic model predicts about half that distance. It is difficult to say which is the better model based purely on the data given. Possibly neither is valid for speeds much higher than the data used to create the model. (And a panic stop at 300 miles per hour sounds like a really bad idea to me anyway!) However, if the physical description of the problem given in the Chapter 1 activity, "Give Me a Brake," is valid, I would have to go with the quadratic model.

Activities

1. A local building code requires that a rectangular-shaped manufacturing plant be bordered by 100 feet of landscaping on the front and rear and by 80 feet of landscaping on each side. If 800,000 square feet of land are necessary in order to have room for parking lots, buildings, and so forth, what is the minimum number of square feet that must be purchased for the manufacturing site? Label the inner rectangle with sides of length x and y, and use the stated conditions to write y in terms of x. Write a formula for the area of the large rectangle in terms of x. What is an appropriate domain? Plot a graph of the area of the large rectangle versus x. Use calculus to find precise estimates for x and y that minimize the area of the large rectangle.

Manufacturing plant site

2. A new power plant is needed to satisfy the needs of three cities. Bakerville is 100 miles due north of Allentown, and Cinderberg is 100 miles due east of Allentown. Where should the power plant be built in order to minimize the total length of the power lines needed to connect the plant to the three cities? Locate a coordinate system

so that Allentown is at the origin, Bakerville is on the positive y-axis, and Cinderberg is on the positive x-axis. Explain why it is reasonable to assume that the power plant should be located at (x, x) for some positive x, as indicated by the circle in the following plot. Derive an equation that gives the total length of the power lines in terms of x. Plot this equation. Give a precise estimate for the value of x that minimizes the total length of the power lines. Use some trigonometry to calculate the size of the central angles about the power plant.

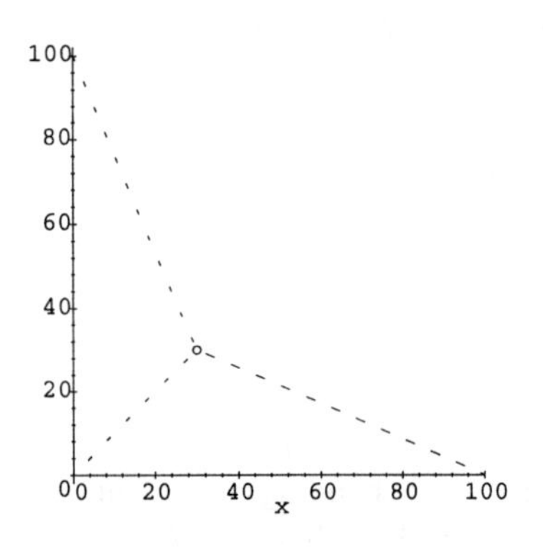

Power plant

3. A rain gutter is to be constructed from a metal sheet of width 25 cm by bending 10 cm on each side so that one side is vertical for attaching to a vertical wall, the opposite side is bent at an angle θ, and the middle 5 cm is horizontal. How should θ be chosen so that the gutter will carry the maximum amount of water? Plot the cross-sectional area versus angle θ. What is a reasonable domain for the angles θ?

Cross section of gutter   Cross section of gutter

4.8 Applications to Economics

If you put \$100 into a savings account that pays 5% per year, after one year your new balance in dollars will be $100 + 100 \times 0.05 \times 1 = 105.00$. If we let P denote principal (in this case $P = 100$), r denote the rate ($r = 5\% = 0.05$), t denote the number of years ($t = 1$), and B the new balance, then we can write $B(P) = P(1 + rt)$.

If interest is added to your account annually and the interest itself earns interest, then you earn *compound interest.*

Compound Interest Is Not Simple

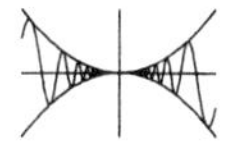

Define $B(P) = P(1 + rt)$, and set $r = 0.05$, $P = 100$, and $t = 1$. You can calculate your balance after two, three, or four years by evaluating $B(B(P))$, $B(B(B(P)))$, or $B(B(B(B(P))))$. Calculate your new balance for each year for ten years. Find an expression that calculates the new balance after n compounding periods. Replace n by the greatest integer function $\lfloor n \rfloor$, and plot a graph of the balance over a 20-year period. Use the graph to estimate the doubling time for the account balance. Solve an equation for n to get a better estimate.

Felecia's Solution

I am earning compound interest on my account. After one year, the new balance is described by the equation $B(P) = P(1 + rt)$, where P denotes principal, r denotes the interest rate, and t denotes the length of the compounding period. Setting $r = .05$, $P = 100$, and $t = 1$, we see that after four years, the \$100 has grown to \$121.55:

$$\begin{aligned} B(100) &= 105.00 \\ B(B(100)) &= 110.25 \\ B(B(B(100))) &= 115.76 \\ B(B(B(B(100)))) &= 121.55 \end{aligned}$$

N Define + New Definition $B(P) = P(1 + rt)$. Select each of $r = 0.05$, $P = 100$, $t = 1$ with the mouse and, for each, choose Define + New Definition. With the insertion point in the display

$$\begin{aligned} &B(B(P)) \\ &B(B(B(P))) \\ &B(B(B(B(P)))) \end{aligned}$$

choose Evaluate. Use "cut and paste" or "drag and drop" to move the values into the display.

Over a 10-year period we have the following yearly balances:

Years	Balance
0	100
1	105.0
2	110.25
3	115.76
4	121.55
5	127.63
6	134.01
7	140.71
8	147.75
9	155.13
10	162.89

Choose Calculus + Iterate (Function B, Starting Value 100, Number of Iterations 10). Choose OK.

Note that after n compound periods the new balance is given by $P(1+rt)^{\lfloor n \rfloor}$, where $\lfloor n \rfloor$ denotes the greatest integer (or floor) function. The balance over the first 20 years can be visualized by looking at the following graph:

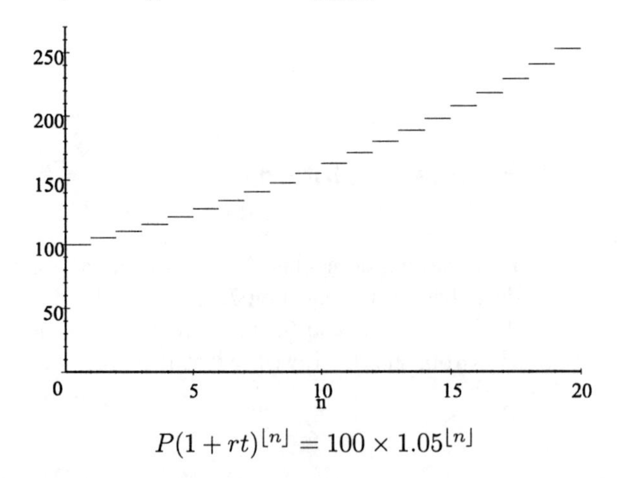

$$P(1+rt)^{\lfloor n \rfloor} = 100 \times 1.05^{\lfloor n \rfloor}$$

To write the greatest integer function, choose Insert + Brackets, select the shapes $\lfloor$ and $\rfloor$, and choose OK. Type n in the input box. With the insertion point in the expression $P(1+rt)^{\lfloor n \rfloor}$, choose Plot 2D + Rectangular. Choose Edit + Properties, Plot Components page. Change Domain Interval to $0 < x < 20$. Choose View page, turn off Default for View Intervals, and change View Intervals to $0 < x < 20, 0 < y < 270$. Choose OK.

According to the graph, it takes about 15 years for the balance to double. A more precise estimate can be found by solving the equation

$$P(1+rt)^n = 2P$$

for n, which produces $n = 14.207$.

With the insertion point in the equation $P(1+rt)^n = 2P$, choose Solve + Exact (Variable to Solve For: n). Then choose Evaluate Numerically.

Activities

1. Every year your grandmother gives you \$100 in December, which you faithfully deposit into your savings account on January 1, the same day the annual interest is credited to your account. Assume the initial deposit (on January 1) is \$100, that the interest rate is $r = 0.05$, and that the compounding period is $t = 1$. Revise the balance function to reflect the additional \$100 deposited annually. Make a table of your account balance for a 10-year period. Plot a graph of your account balance over a 20-year period. How long does it take for your account to double? How long does it take to double again?

2. You have a choice of Bank A, which compounds savings accounts annually and pays 5% interest, or Bank B, which compounds daily but pays an annual interest rate of only 4%. Which bank should you choose? Why?

4.9 Antiderivatives

A function F is called an *antiderivative* of f on an interval I if $F'(x) = f(x)$ for all x in I. The equation $F'(x) = f(x)$ is called a *differential equation*, because it involves an unknown function F and its derivative. If F is an antiderivative of f, then for any number C, $F(x) + C$ is also an antiderivative of f because

$$\frac{d}{dx}[F(x) + C] = F'(x) = f(x)$$

Geometrically, the graph of $y = F(x) + C$ is a translation of the graph of $y = F(x)$. Hence the shape of the two antiderivatives is identical. Given a graph of one antiderivative, other antiderivatives can be plotted by translating the original curve up or down.

The Sky Is the Limit

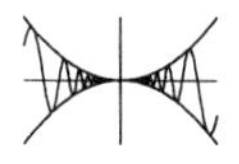

If a baseball is thrown straight up, how high will it go? In order to answer this, we need to know how fast it was thrown and then find some antiderivatives. To keep things simple, we will also neglect air friction. Assume the ball is thrown upward with an initial velocity of 64 feet per second from the player's hand, which is 6 feet above the ground at the instant of release. Assume that acceleration is given by $a(t) = -32$ feet per second per second (ft/sec^2). How high will the ball go? Plot a graph of position versus time.

René's Solution

A ball is thrown upward with an initial velocity of 64 feet/second from a player's hand, which is 6 feet above the ground at the instant of release. We will assume that acceleration is given by $a(t) = -32$ feet/second^2. Our problem is to determine how high the ball will go and to plot a graph of position versus time. Because the velocity $v(t)$ is an antiderivative of the acceleration $a(t)$, we solve the differential equation $\frac{dv}{dt} = a(t)$ to get $v(t) = -32t + C_1$. Evaluation at $t = 0$ gives C_1. Evaluation at $t = 0$ also gives the initial velocity, so $C_1 = 64$. Thus the velocity of the ball at time t is

$$v(t) = -32t + 64$$

With the insertion point in the equation $\frac{dv}{dt} = -32$, choose Solve ODE + Exact. With the insertion point in the equation $v(t) = -32t + C_1$, choose Define + New Definition. With the insertion point in the equation $v(0) = 64$, choose Solve + Exact.

Because the height $y(t)$ is the antiderivative of the velocity $v(t)$, we solve the differential equation $\frac{dy}{dt} = -32t + 64$ to get $y(t) = -16t^2 + 64t + C_1$. Evaluation at $t = 0$ implies $C_1 = 6$. Thus the height of the ball at time t is given by

$$y(t) = -16t^2 + 64t + 6$$

The ball reaches its highest point when the velocity is zero. Solving the equation $v(t) = 0$, we get $t = 2$ seconds. Thus $y(2) = 70$ feet is the highest the ball will go.

With the insertion point in the equation $\frac{dy}{dt} = v(t)$, choose Solve ODE + Exact. With the insertion point in the equation $y(t) = -16t^2 + 64t + 6$, choose Define + New Definition. With the insertion point in the equation $v(t) = 0$, choose Solve + Exact. With the insertion point in the expression $y(2)$, choose Evaluate.

The following graph shows the height of the ball versus time:

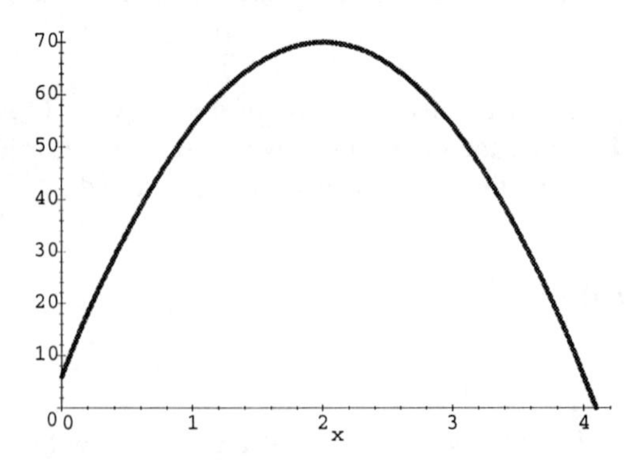

Flight of a baseball

With the insertion point in the expression $y(t)$, choose Plot 2D + Rectangular. Choose Edit + Properties, Plot Components page; change Domain Interval to $0 < x < 4.2$. Choose View page.

Under View Intervals, turn Default off and change to $0 < x < 4.2$ and $0 < y < 72$. Choose the Labeling page. In the Caption Text box, type "Flight of a baseball." Choose OK.

All Antiderivatives Have the Same Shape

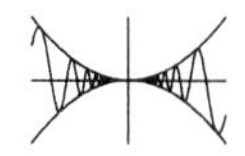

Let $f(x) = x^2 + x + 1$. Show that $F(x) = \frac{1}{3}x^3 + \frac{1}{2}x^2 + x$ is an antiderivative of $f(x)$ by calculating $F'(x)$. Define $G(x) = F(x) + 5$ and $H(x) = F(x) - 3$ and verify that $G'(x) = f(x)$ and $H'(x) = f(x)$. Plot a graph of $y = F(x)$ and drag the expressions $G(x)$ and $H(x)$ to the frame using a viewing rectangle that clearly distinguishes among the three curves. Describe how the three curves are related to each other. Check to make sure that the three curves have the same slope at $x = 1$ by attaching tangent lines to the three curves at $x = 1$. Describe how the three tangent lines are related.

Jeff's Solution

Let $f(x) = x^2 + x + 1$ and $F(x) = \frac{1}{3}x^3 + \frac{1}{2}x^2 + x$. Calculating the derivative of $F(x)$ gives $F'(x) = x^2 + x + 1 = f(x)$, which means that F is an antiderivative of f. Let $G(x) = F(x) + 5$ and $H(x) = F(x) - 3$. Note also that $G'(x) = x^2 + x + 1$ and $H'(x) = x^2 + x + 1$, so that G and H are also antiderivatives of f.

Select each of the equations $f(x) = x^2 + x + 1$, $F(x) = \frac{1}{3}x^3 + \frac{1}{2}x^2 + x$, $G(x) = F(x) + 5$, $H(x) = F(x) - 3$, and, for each, choose Define + New Definition. Select each of the expressions $F'(x)$, $G'(x)$, and $H'(x)$ and, for each, choose Evaluate.

The graphs of F, G, and H are given below.

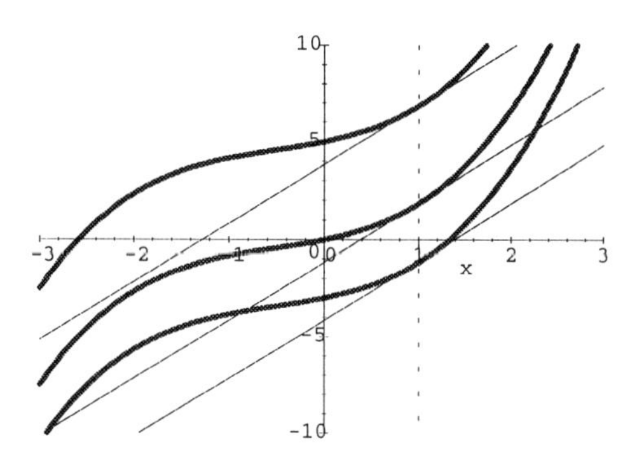

Antiderivatives of $x^2 + x + 1$
Tangent lines at $x = 1$

The graphs appear to be closer together where the curves are steeper. To visually test to see whether the shapes are actually the same, we attached tangent lines to each at $x = 1$.

The lines appear to have the same slope. By repeating this experiment, we could verify visually that the slopes are the same at each of these sampling points.

N With the insertion point in the expression $F(x)$, choose Plot 2D + Rectangular. Select and drag to the frame $G(x)$, $H(x)$. Choose Edit + Properties, click the View tab, turn off Default, and set View Intervals to $-3 \le x \le 3$ and $-10 \le y \le 10$. Choose OK. Select and drag $F(1) + F'(1)(x-1)$, $G(1) + G'(1)(x-1)$, and $H(1) + H'(1)(x-1)$ to the frame. Choose Edit + Properties, and click the Plot Components tab. Set the Line Color to Red for Item Numbers 1, 2, and 3; set the Line Color to Blue for Item Numbers 4, 5, and 6. Choose OK.

Activities

1. Let $f(x) = ax^2 + bx + c$. Differentiation lowers the degree of a polynomial by 1. Explain why it is reasonable to expect that an antiderivative of a polynomial has degree 1 higher. Assume $F(x) = px^3 + qx^2 + rx + s$ and apply Solve + Exact to the system

$$\begin{aligned} F'(0) &= f(0) \\ F'(1) &= f(1) \\ F'(2) &= f(2) \\ F'(3) &= f(3) \end{aligned}$$

of equations to solve for p, q, r, and s. (The equation $s = s$ in the solution indicates that s can be assigned an arbitrary value.) Explain why four equations are required. Describe what would happen if different numbers were used instead of 1, 2, 3, and 4 to generate four equations. Verify your solution by applying Evaluate to $\frac{d}{dx}\left[\frac{1}{3}ax^3 + \frac{1}{2}bx^2 + cx + p\right]$.

2. One way to visualize a family of antiderivatives is by creating a direction field. Create a direction field for $f(x) = \cos\frac{x}{2}$ by applying Plot 2D + Vector Field to the expression $\left[1, \cos\frac{x}{2}\right]$. In the Plot Properties + Plot Components tabbed dialog, change the Domain Intervals to $-10 \le x \le 10$ and $-5 \le y \le 5$, and then increase the Sample Size to 20 for both x and y. Draw by hand a graph of an antiderivative of $\cos\frac{x}{2}$ by starting at $(0,0)$ and following the direction of the arrows, so that the arrows remain parallel to the tangent lines at each x-coordinate. Use your antiderivative to guess a formula for the antiderivative of $\cos\frac{x}{2}$ whose graph goes through the origin. Verify your formula by differentiating it and comparing the derivative with $\cos\frac{x}{2}$.

5 INTEGRALS

The definite integral of a function arose as a solution to the problem of computing areas. It has many other applications as well.

5.1 Sigma Notation

The notation $\sum_{k=1}^{n} a_k$ (read as "the summation from k equals 1 to n of a_k") is an abbreviation for the sum $a_1 + a_2 + \cdots + a_n$. For example, $\sum_{k=1}^{3} k = 1 + 2 + 3$ and $\sum_{k=1}^{5} \frac{k}{k+1} = \frac{1}{2} + \frac{2}{3} + \frac{3}{4} + \frac{4}{5} + \frac{5}{6}$.

Sums of Squares Are Cubic

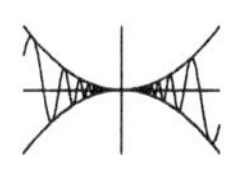

Generate a formula for $1^2 + 2^2 + 3^2 + \cdots + n^2$ by assuming a formula of the form

$$1^2 + 2^2 + 3^2 + \cdots + n^2 = an^3 + bn^2 + cn + d$$

and solving for a, b, c, and d. Verify your formula by using *Scientific Notebook* to evaluate the expression

$$\sum_{i=1}^{n} i^2$$

Lisa's Solution

Define f by the equation $f(n) = an^3 + bn^2 + cn + d$. Note that

$$\begin{aligned} 1^2 &= 1 \\ 1^2 + 2^2 &= 5 \\ 1^2 + 2^2 + 3^2 &= 14 \\ 1^2 + 2^2 + 3^2 + 4^2 &= 30 \end{aligned}$$

We wish to find coefficients a, b, c, and d such that

$$\begin{aligned} f(1) &= 1 \\ f(2) &= 5 \\ f(3) &= 14 \\ f(4) &= 30 \end{aligned}$$

The solution is $d=0$, $c=\frac{1}{6}$, $b=\frac{1}{2}$, $a=\frac{1}{3}$, so that

$$f(n)=\frac{1}{3}n^3+\frac{1}{2}n^2+\frac{1}{6}n$$

This suggests that

$$1^2+2^2+3^2+\cdots+n^2=\frac{1}{3}n^3+\frac{1}{2}n^2+\frac{1}{6}n$$

Choose Define + Clear Definitions. With the insertion point in the equation $f(n)=an^3+bn^2+cn+d$, choose Define + New Definition. With the insertion point in each of 1^2, 1^2+2^2, $1^2+2^2+3^2$, and $1^2+2^2+3^2+4^2$, choose Evaluate. With the insertion point in the system of equations

$$\begin{aligned} f(1) &= 1 \\ f(2) &= 5 \\ f(3) &= 14 \\ f(4) &= 30 \end{aligned}$$

choose Solve + Exact.

To verify this formula we evaluate $\sum_{i=1}^{n} i^2$ directly. This produces

$$\begin{aligned} \sum_{i=1}^{n} i^2 &= \frac{1}{3}(n+1)^3-\frac{1}{2}(n+1)^2+\frac{1}{6}n+\frac{1}{6} \\ &= \frac{1}{3}n^3+\frac{1}{2}n^2+\frac{1}{6}n \\ &= \frac{1}{6}n(n+1)(2n+1) \end{aligned}$$

which agrees with the derived formula.

With the insertion point in the expression $\sum_{i=1}^{n} i^2$, choose Evaluate. With the insertion point in the expression $\frac{1}{3}(n+1)^3-\frac{1}{2}(n+1)^2+\frac{1}{6}n+\frac{1}{6}$, choose Simplify. With the insertion point in the expression $\frac{1}{3}n^3+\frac{1}{2}n^2+\frac{1}{6}n$, choose Factor.

Activities

1. Generate a formula for the sum $1+2+3+\cdots+n$ by following these steps.

   a. Explain how the numbers in the third equation are generated.

$$\begin{array}{ccccccccccc} 1 & + & 2 & + & 3 & + & \cdots & + & n & = & x \\ n & + & (n-1) & + & (n-2) & + & \cdots & + & 1 & = & x \\ (n+1) & + & (n+1) & + & (n+1) & + & \cdots & + & (n+1) & = & 2x \end{array}$$

   b. Explain why $2x=n(n+1)$.

c. Explain why

$$1 + 2 + 3 + \cdots + n = \frac{n(n+1)}{2}$$

2. Explain why

$$1 + 2 + 3 + \cdots + n = \frac{n(n+1)}{2}$$

by counting the number of symbols in the following rectangular array, which has n rows and $n + 1$ columns:

a. Explain how you know (without counting) that the number of circles is the same as the number of stars.

b. Explain how you know that the number of circles is equal to

$$\frac{n(n+1)}{2}$$

c. Explain how you know that the number of circles is equal to

$$1 + 2 + 3 + \cdots + n$$

3. Verify the formula

$$1 + 2 + 3 + \cdots + n = \frac{n(n+1)}{2}$$

by evaluating the sum

$$\sum_{i=1}^{n} i$$

in *Scientific Notebook.*

4. Explain how your formula for $1^2 + 2^2 + 3^2 + \cdots + n^2$ can be used to count the number of oranges in a pyramid in a grocery store, assuming there is one orange on top, four in the next-to-the-top layer, and so forth. If an orange pyramid contains 1240 oranges, how many layers does it contain?

5. Explain how the formula $1 + 3 + 5 + 7 + \cdots + (2n - 1) = n^2$ is related to the

following array:

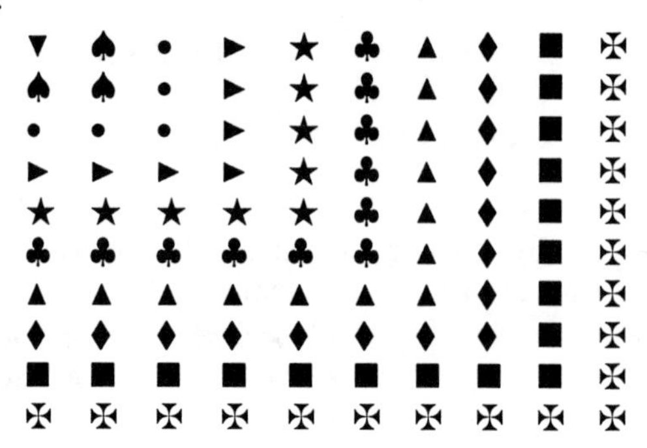

6. Verify the formula

$$1 + 3 + 5 + 7 + \cdots + (2n - 1) = n^2$$

by evaluating the sum

$$\sum_{k=1}^{n} (2k - 1)$$

5.2 Area

The simple formula *Area* = *Length* × *Width* for the area of a rectangle, together with the not-so-simple idea of a limit, can be used to compute the area of any figure whose boundary we can describe with a function.

Area Equals Length Times Width Is a Good Place to Start

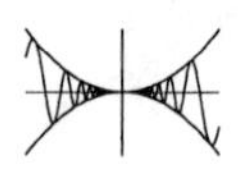

Experiment with Middle Boxes and the Middle formula for $f(x) = x^3 - 5x^2 + 2x + 8$ on the interval $-1 \leq x \leq 4$. What is a geometric interpretation for what the Middle formula estimates?

Fred's Solution

The Midpoint Rule applied to the integral $\int_{-1}^{4} (x^3 - 5x^2 + 2x + 8)\, dx$ can be interpreted visually as the sum of the areas of the rectangles that lie above the x-axis minus the sum of the areas of the rectangles that lie below the x-axis in the following figure, which is based on ten rectangles:

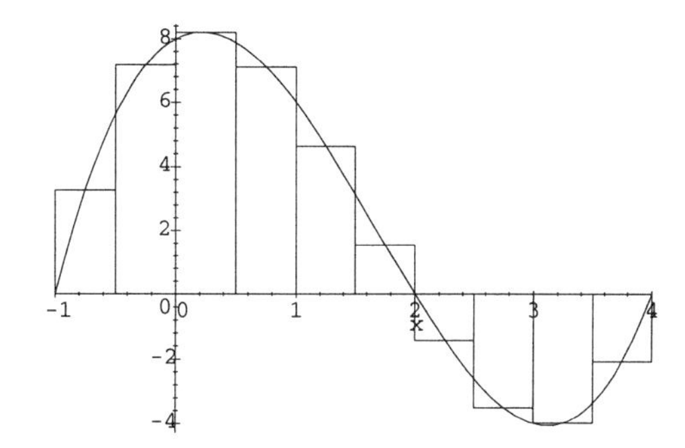

With the insertion point in the expression $x^3 - 5x^2 + 2x + 8$, choose Calculus + Plot Approximate Integral. Select the frame and choose Edit + Properties, Plot Components page. Set Ranges to -1 to 4. Choose OK.

This sum is given numerically as

$$\frac{1}{2}\sum_{i=0}^{9}\left(\left(-\frac{3}{4}+\frac{1}{2}i\right)^3-5\left(-\frac{3}{4}+\frac{1}{2}i\right)^2+\frac{13}{2}+i\right)=10.469$$

which is reasonably close to the approximation

$$\int_{-1}^{4}\left(x^3-5x^2+2x+8\right)dx=10.417$$

The integral represents the area under the graph and above the x-axis between -1 and 2 minus the area above the curve and below the x-axis between 2 and 4.

With the insertion point in the definite integral $\int_{-1}^{4}\left(x^3-5x^2+2x+8\right)dx$ choose Calculus + Approximate Integral (Midpoint formula, Subintervals 10). With the insertion point in the expression $\frac{1}{2}\sum_{i=0}^{9}\left(\left(-\frac{3}{4}+\frac{1}{2}i\right)^3-5\left(-\frac{3}{4}+\frac{1}{2}i\right)^2+\frac{13}{2}+i\right)$, choose Evaluate Numerically. With the insertion point in the integral $\int_{-1}^{4}\left(x^3-5x^2+2x+8\right)dx$, choose Evaluate Numerically.

Activities

1. Consider the problem of estimating the area of the region bounded by $y = x\sin x$, $x = 0$, $x = \pi$, and $y = 0$. A picture of this region follows.

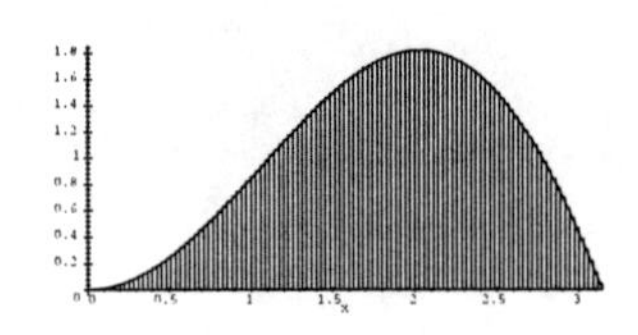

With the insertion point within the expression $x \sin x$, choose Calculus + Plot Approximate Integral. By making changes in the Plot Properties + Plot Components tabbed dialog, change the interval to $0 \leq x \leq 3.1416$ and experiment with the Number of Boxes and the choice of Left Boxes, Right Boxes, and Middle Boxes to estimate the area of the given region.

a. Discuss which collections of rectangles you would expect to give the best results, including appropriate pictures in your discussion.
b. What do you expect to happen when the number of rectangles becomes very large?

2. With the insertion point within the expression $x \sin x$, choose Calculus + Approximate Integral and experiment with the Midpoint formula, Trapezoid formula, and Simpson formula with various numbers of subintervals.

   a. Discuss which formulas you would expect to give the best results.
   b. Verify your predictions by giving numerical evidence. Explain how these numbers support your claims.

5.3 The Definite Integral

A definite integral is a number that arises in a special way and has many useful interpretations.

Areas Are the Limit

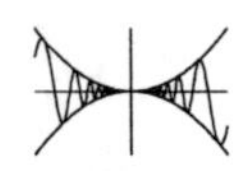

Experiment with the integral $\int_1^7 x^{\sin x}\,dx$. Visualize the integral as an area by applying Calculus + Plot Approx. Integral using Middle Boxes with Number of Boxes set at 10. Define $a = 1$, $b = 7$, and $n = 10$ and evaluate the Midpoint approximation. Create a second Middle Boxes plot with Number of Boxes set at 50, and recalculate the Midpoint approximation with 50 subintervals. Repeat with $n = 100$. Describe the trends you observe in these sums. Explain why this is reasonable. Evaluate Numerically the integral $\int_1^7 x^{\sin x}\,dx$ and explain how this relates to your previous approximations.

Cathy's Solution

The Midpoint Rule with $n = 10$ applied to the integral $\int_1^7 x^{\sin x}\,dx$ can be interpreted visually as the sum of the areas of the rectangles in the following figure:

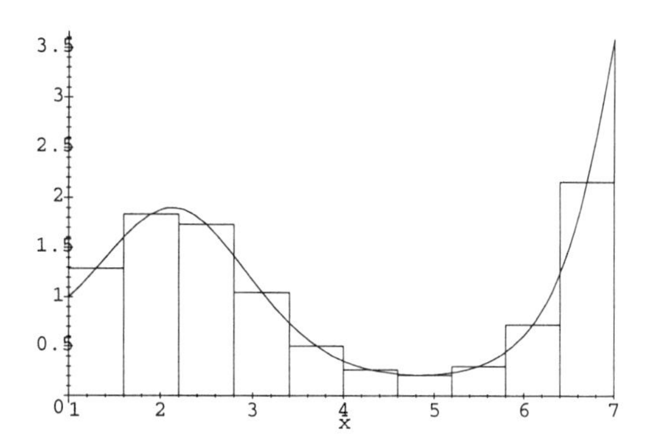

The sum of the areas of these rectangles is given by

$$\frac{3}{5}\sum_{i=0}^{9}\left(\frac{13}{10}+\frac{3}{5}i\right)^{\sin\left(\frac{13}{10}+\frac{3}{5}i\right)} = 6.0325$$

With the insertion point in the expression $x^{\sin x}$, choose Calculus + Plot Approx. Integral (Ranges 1 to 7, Number of Boxes 10, Middle Boxes). With the insertion point in the integral $\int_1^7 x^{\sin x}\,dx$, choose Calculus + Approximate Integral (Midpoint, Subintervals 10). With the insertion point in the expression $\frac{3}{5}\sum_{i=0}^{9}\left(\frac{13}{10}+\frac{3}{5}i\right)^{\sin\left(\frac{13}{10}+\frac{3}{5}i\right)}$, choose Evaluate Numerically.

Repeating this experiment with $n = 50$, we get the figure

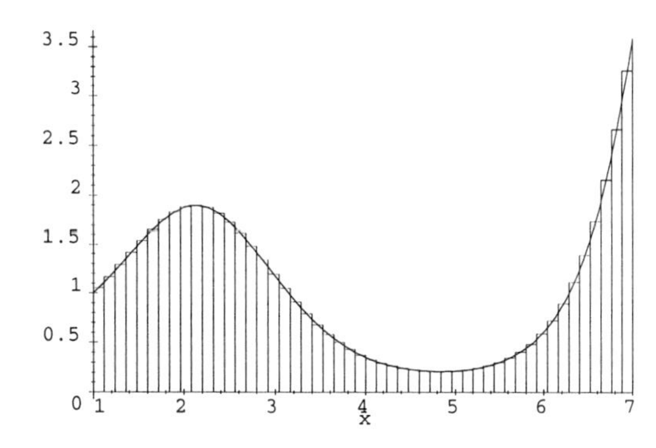

The sum of the areas of these 50 rectangles is given by

$$\frac{3}{25}\sum_{i=0}^{49}\left(\frac{53}{50}+\frac{3}{25}i\right)^{\sin\left(\frac{53}{50}+\frac{3}{25}i\right)} = 6.1025$$

With the insertion point in the expression $x^{\sin x}$, choose Calculus + Plot Approx. Integral (Ranges 1 to 7, Number of Boxes 50, Middle Boxes). With the insertion point in the integral $\int_1^7 x^{\sin x}\,dx$, choose Calculus + Approximate Integral (choose Midpoint and set Subintervals to 50). With the insertion point in the expression $\frac{3}{25}\sum_{i=0}^{49}\left(\frac{53}{50}+\frac{3}{25}i\right)^{\sin\left(\frac{53}{50}+\frac{3}{25}i\right)}$, choose Evaluate Numerically.

With $n = 100$, the figure is given by

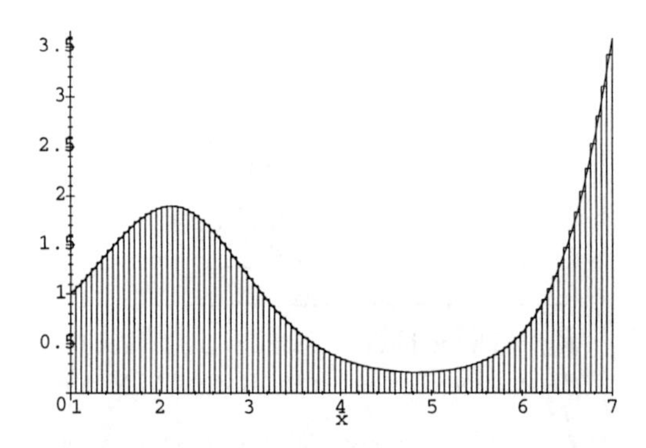

and the sum of the areas of the rectangles is

$$\frac{3}{50}\sum_{i=0}^{99}\left(\frac{103}{100}+\frac{3}{50}i\right)^{\sin\left(\frac{103}{100}+\frac{3}{50}i\right)} = 6.1046$$

With the insertion point in the expression $x^{\sin x}$, choose Calculus + Plot Approx. Integral (Ranges 1 to 7, Number of Boxes 100, Middle Boxes). With the insertion point in the integral $\int_1^7 x^{\sin x}\,dx$, choose Calculus + Approximate Integral (choose Midpoint and set Subintervals to 100). With the insertion point in the expression $\frac{3}{50}\sum_{i=0}^{99}\left(\frac{103}{100}+\frac{3}{50}i\right)^{\sin\left(\frac{103}{100}+\frac{3}{50}i\right)}$, choose Evaluate Numerically.

This compares closely with the numerical approximation

$$\int_1^7 x^{\sin x}\,dx = 6.1053$$

returned by the symbolic computation system.

With the insertion point in the integral $\int_1^7 x^{\sin x}\,dx$, choose Evaluate Numerically.

Activities

1. Consider the definite integral $\int_a^b f(x)dx$. Explain how you know that **Left Boxes** corresponds to the Riemann sum

$$\frac{b-a}{n}\sum_{i=0}^{n-1} f\left(a+i\frac{b-a}{n}\right)$$

Apply **Calculus + Plot Approximate Integral** to the function $f(x)=\sin(x+\cos x)$ on the interval $-5 \le x \le 5$ with $n=10$, and choose **Left Boxes**. Describe what each piece of the **Left Boxes** formula represents in terms of your figure.

2. Explain how you know that **Right Boxes** for the integral $\int_a^b f(x)dx$ corresponds to the Riemann sum

$$\frac{b-a}{n}\sum_{i=1}^{n} f\left(a+i\frac{b-a}{n}\right)$$

Compute the integral $\int_1^3 \left(x^3+2x^2+3\right)dx$ by doing the following steps:

   a. Form a right sum.
   b. Rewrite the right sum in terms of the expressions
   $$\sum_{i=1}^{n} 1 \quad \sum_{i=1}^{n} i \quad \sum_{i=1}^{n} i^2 \quad \text{and} \quad \sum_{i=1}^{n} i^3$$
   c. **Evaluate** each of the expressions
   $$\sum_{i=1}^{n} 1 \quad \sum_{i=1}^{n} i \quad \sum_{i=1}^{n} i^2 \quad \text{and} \quad \sum_{i=1}^{n} i^3$$
   d. **Simplify** your approximation.
   e. **Evaluate** a limit of your approximation as n tends to ∞.
   f. Verify your limit by evaluating the integral $\int_1^3 \left(x^3+2x^2+3\right)dx$.

5.4 The Fundamental Theorem of Calculus

The Fundamental Theorem of Calculus describes an important and surprising connection between integrals and derivatives.

It Is Fundamental

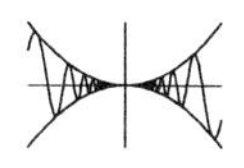

Define $f(x)$ to be a (continuous) generic function and define g by $g(x)=\int_0^x f(t)dt$. Evaluate $g'(x)$. Explain why you get $f(x)$ and not $f(t)$. Justify each of the following steps:

$$\begin{aligned}\frac{g(x+h)-g(x)}{h} &= \frac{\int_0^{x+h} f(t)dt - \int_0^x f(t)dt}{h} \\ &= \frac{1}{h}\int_x^{x+h} f(t)dt \\ g'(x) &= \lim_{h\to 0}\frac{1}{h}\int_x^{x+h} f(t)dt\end{aligned}$$

Use pictures to explain why it is reasonable that

$$\lim_{h\to 0}\frac{1}{h}\int_x^{x+h} f(t)dt = f(x)$$

Jim's Solution

If $f(x)$ is an arbitrary continuous function and $g(x) = \int_0^x f(t)dt$, then Evaluate produces $g'(x) = f(x)$.

With the insertion point in the expression $f(x)$, choose Define + New Definition. With the insertion point in the equation $g(x) = \int_0^x f(t)dt$, choose Define + New Definition. With the insertion point in the expression $g'(x)$, choose Evaluate.

The expression $\int_0^x f(t)dt$ depends on x, not t, so the derivative with respect to x must be in terms of x, not t. The equation

$$\frac{g(x+h)-g(x)}{h} = \frac{\int_0^{x+h} f(t)dt - \int_0^x f(t)dt}{h}$$

involves replacing x by $x+h$ in the equation $g(x) = \int_0^x f(t)dt$. If we interpret $\int_0^x f(t)dt$ as the area under the graph of f between 0 and x, then

$$\int_0^{x+h} f(t)dt - \int_0^x f(t)dt$$

represents the area between 0 and $x+h$ minus the area between 0 and x, which is simply the area between x and $x+h$. Thus the difference may be replaced by the single integral

$$\int_x^{x+h} f(t)dt$$

The next equation

$$g'(x) = \lim_{h\to 0}\frac{1}{h}\int_x^{x+h} f(t)dt$$

is the result of taking the limit on both sides of the equation

$$\frac{g(x+h)-g(x)}{h} = \frac{1}{h}\int_x^{x+h} f(t)dt$$

as h approaches zero. To see why $\lim_{h\to 0}\frac{1}{h}\int_x^{x+h} f(t)dt = f(x)$, take a specific function—say, $f(x) = 2 + \sin(x\cos x)$—and look closely at its graph.

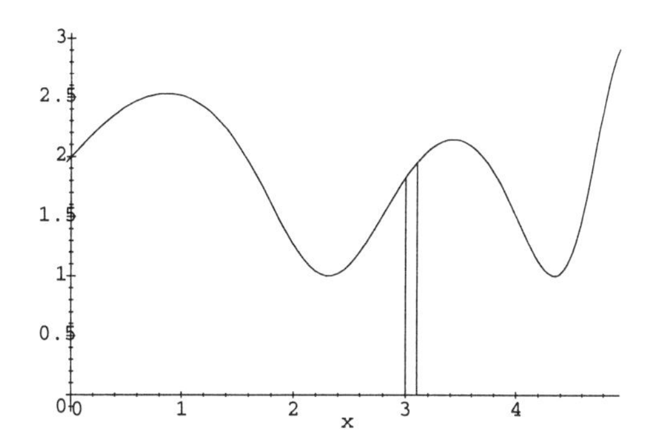

With the insertion point in the equation $f(x) = 2 + \sin(x \cos x)$, choose Define + New Definition. With the insertion point in the expression $f(x)$, choose Plot 2D + Rectangular. Select and drag to the frame each of $(3, 0, 3, f(3))$ and $(3.1, 0, 3.1, f(3.1))$.

The integral $\int_x^{x+h} f(t)dt$ may be interpreted as the area under the graph in the narrow vertical strip. The area inside the vertical strip is roughly $hf(x_h)$, where x_h is some number between x and $x + h$. Thus $\frac{1}{h}\int_x^{x+h} f(t)dt \approx \frac{1}{h}hf(x_h) = f(x_h) \approx f(x)$. Imagining even narrower strips, this approximation must get better and better. It is thus reasonable to expect that

$$\lim_{h \to 0} \frac{1}{h} \int_x^{x+h} f(t)dt = f(x)$$

If $f(x) < 0$ for some x, this argument needs to be revised slightly to consider the negative of the area of a region that lies below the x-axis.

Activities

1. Define F by the equation $F(x) = \int_1^x t \sin t\, dt$. Plot the curve $y = F(x)$ on the interval $-5 \leq x \leq 5$. Explain the significance of the fact that $(1, 0)$ is a point on this curve. Verify that $F'(x) = x \sin x$. Verify that $\int_2^5 t \sin t\, dt = F(5) - F(2)$.

2. Experiment with the Fresnel function $S(x) = \int_0^x \sin(\pi t^2/2)dt$. Plot S and its derivative on the interval $-5 \leq x \leq 5$. Use your plot to estimate a value for
$$\lim_{x \to \infty} S(x)$$
Apply **Evaluate Numerically** to each of the expressions $S(10)$, $S(20)$, $S(30)$, $S(40)$, and $S(50)$. Discuss how these values relate to your estimate for the limit.

5.5 The Substitution Rule

The Substitution Rule for Integrals states that

$$\int_a^b f(g(x))g'(x)\,dx = \int_{g(a)}^{g(b)} f(u)\,du$$

where $u = g(x)$.

When a Variable Goes Flat, Change It

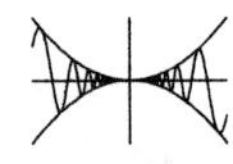

Experiment with the substitution $u = 2x$, which yields

$$\int_0^{\pi/2} \sin 2x\,dx = \frac{1}{2}\int_0^{\pi} \sin u\,du$$

Verify that

$$\int_0^{\pi/2} \sin 2x\,dx = \frac{1}{2}\int_0^{\pi} \sin u\,du$$

by applying Calculus + Change Variable to the integral

$$\int_0^{\pi/2} \sin 2x\,dx$$

with the substitution $u = 2x$. Use Calculus + Plot Approx. Integral with Number of Boxes set at 100 to view each integral as an area, and resize the frames so that the areas can be compared visually. Use the plots to generate a rough approximation to the area of each region. Use the Midpoint Rule to approximate the integral

$$\int_0^{\pi/2} \sin 2x\,dx$$

using $n = 4$. Calculate the area of each of the four rectangles. Use the Midpoint Rule to approximate the integral

$$\frac{1}{2}\int_0^{\pi} \sin u\,du$$

using $n = 4$. Calculate the area of each of the four rectangles. Based on the plots and approximations, explain why you believe that the two integrals

$$\int_0^{\pi/2} \sin 2x\,dx$$

and

$$\frac{1}{2}\int_0^{\pi} \sin u\,du$$

are equal.

Ivy's Solution

The substitution $u = 2x$ yields

$$\int_0^{\pi/2} \sin 2x \, dx = \int_0^{\pi} \frac{1}{2} \sin u \, du$$

With the insertion point inside the integral $\int_0^{\pi/2} \sin 2x \, dx$, choose Calculus + Change Variable (Substitution $u = 2x$).

Each side of this equation can be interpreted as the area. The integral $\int_0^{\pi/2} \sin 2x \, dx$ corresponds to the area below $y = \sin 2x$ between 0 and $\frac{\pi}{2}$, as shown in the following figure:

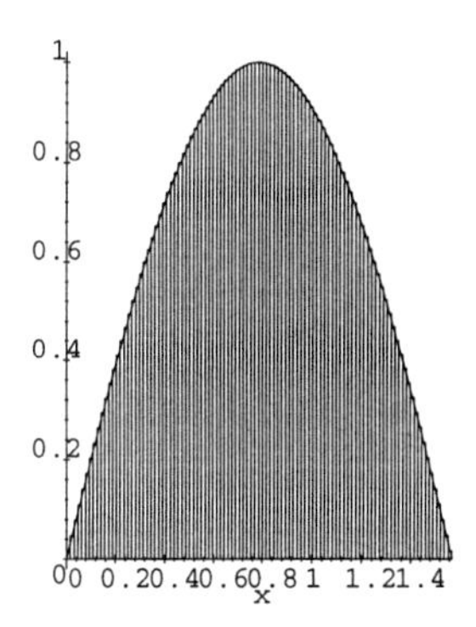

With the insertion point inside the expression $\sin 2x$, choose Calculus + Plot Approximate Integral. Choose Edit + Properties, Plot Components page. Set Number of Boxes to 100 and Ranges to $0 < x < 1.57$. Choose OK. Click inside the frame, and then use the side handles to resize the frame to about half its original width.

The second integral $\int_0^{\pi} \frac{1}{2} \sin u \, du$ corresponds to the area below $\frac{1}{2} \sin u$ between 0 and π, as shown in the second figure.

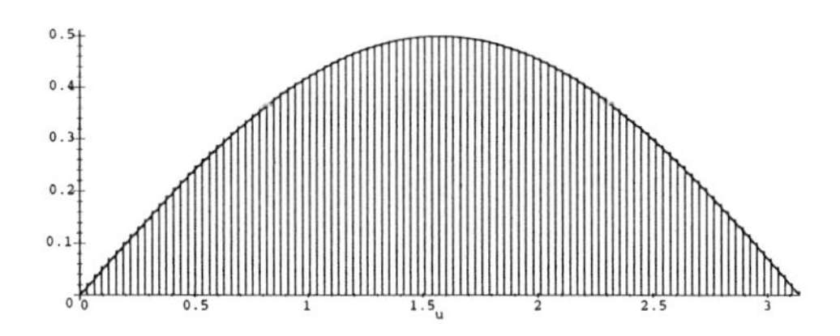

With the insertion point inside the expression $\frac{1}{2} \sin u$, choose Calculus + Plot Approximate Integral. Choose Edit + Properties, Plot Components page. Set Number of Boxes to 100, and Ranges to $0 < x < 3.1416$. Choose OK. Click inside the frame, and then use a handle on the top or bottom to resize the frame to roughly half its original height.

Note that these shapes are similar, with the first twice as tall and the second twice as wide. It is thus reasonable to expect the areas to be approximately equal.

As a check, consider the following Riemann sums, based on the use of midpoints as sampling points. A Riemann sum for the first integral yields

$$\int_0^{\pi/2} \sin 2x \, dx \approx \frac{1}{8}\pi \sum_{i=0}^{3} \sin\left(\frac{\pi}{4}\left(i+\frac{1}{2}\right)\right)$$

With the insertion point in the integral $\int_0^{\pi/2} \sin 2x \, dx$, choose Calculus + Approximate Integral (Formula: Midpoint, Subintervals: 4).

The area of each the four rectangles is

$$\begin{aligned}
\frac{\pi}{8}\sin\frac{\pi}{8} &= .15028 \\
\frac{\pi}{8}\sin\frac{3\pi}{8} &= .36281 \\
\frac{\pi}{8}\sin\frac{5\pi}{8} &= .36281 \\
\frac{\pi}{8}\sin\frac{7\pi}{8} &= .15028
\end{aligned}$$

The second integral has a Riemann sum

$$\int_0^{\pi} \frac{1}{2}\sin u du \approx \frac{\pi}{4}\sum_{i=0}^{3} \frac{1}{2}\sin\left(\frac{\pi}{4}\left(i+\frac{1}{2}\right)\right)$$

With the insertion point in the integral $\int_0^{\pi} \frac{1}{2}\sin u du$, choose Calculus + Approximate Integral (Formula: Midpoint, Subintervals: 4).

The area of each of the four rectangles is

$$\begin{aligned}
\frac{\pi}{4}\frac{1}{2}\sin\frac{\pi}{8} &= .15028 \\
\frac{\pi}{4}\frac{1}{2}\sin\frac{3\pi}{8} &= .36281 \\
\frac{\pi}{4}\frac{1}{2}\sin\frac{5\pi}{8} &= .36281 \\
\frac{\pi}{4}\frac{1}{2}\sin\frac{7\pi}{8} &= .15028
\end{aligned}$$

Because these Riemann sums are equal, I expect the integrals to be equal as well.

Activities

1. Experiment with the substitution $u = x^2$, which yields
$$\int_0^1 x \sin x^2 dx = \int_0^1 \frac{1}{2} \sin u \, du$$
Use **Calculus + Plot Approximate Integral** with **Number of Boxes** set at 100 to view each of the two integrals as an area. Resize the frames so that the areas can be compared visually. Are the areas about the same size? If not, how can you explain the difference? Use **Calculus + Plot Approx. Integral** with **Number of Boxes** set at 10, and choose **Middle Boxes** to generate a plot. Consider a Riemann sum for the integral $\int_0^1 \frac{1}{2} \sin u \, du$, using the partition
$$0 < 0.1^2 < 0.2^2 < 0.3^2 < 0.4^2 < 0.5^2 < 0.6^2 < 0.7^2 < 0.8^2 < 0.9^2 < 1.0^2$$
Use the substitution $u = x^2$ to generate the sampling points. Compare the area of the rectangle sitting on the subinterval $0.4 \le x \le 0.5$, which is used in the approximation for $\int_0^1 x \sin x^2 dx$, with the area of a comparable rectangle sitting on the subinterval $0.4^2 \le x \le 0.5^2$, which is used as a term in the Riemann sum for the integral $\int_0^1 \frac{1}{2} \sin u \, du$ with sampling point 0.45^2.

2. Experiment with the substitution $u^2 = x^3 + 1$, which yields
$$\int_0^2 x^2 \sqrt{x^3 + 1} dx = \int_1^3 \frac{2}{3} u^2 \, du$$
   - Apply **Calculus + Plot Approximate Integral** with Number of Boxes set at 4 to visualize an approximation to the integral
   $$\int_0^2 x^2 \sqrt{x^3 + 1} dx$$
   - Start with the partition $[0, 0.5, 1, 1.5, 2]$ of the interval $0 \le x \le 2$ and use the substitution $u^2 = x^3 + 1$ to generate a partition of the interval $1 \le u \le 3$.
   - Compare a right sum with $n = 4$ for the integral
   $$\int_0^2 x^2 \sqrt{x^3 + 1} dx$$
   with a corresponding Riemann sum for the integral
   $$\int_1^3 \frac{2}{3} u^2 \, du$$
   Do each of the four terms in the first sum agree exactly with the corresponding terms in the second sum? Discuss any discrepancies.

3. Sometimes substitution leads to pictures that look quite different.
   - Apply the substitution $u^2 = x^2 + 1$ to rewrite the integral $\int_{-1}^1 x\sqrt{x^2 + 1} dx$.
   - Generate middle boxes for both integrals.
   - Explain how the integrals are related to the pictures and why the pictures look so different.

5.6 The Logarithm Defined as an Integral

The power rule yields

$$\int x^n dx = \frac{x^{n+1}}{n+1}$$

However, if $n = -1$, then this formula produces a division by zero. We must look elsewhere for an antiderivative of $1/x$. Because $1/x$ is continuous for $x > 0$, it follows from the Fundamental Theorem of Calculus that $\int_1^x \frac{1}{t}dt$ is an antiderivative of $1/x$. How do we get rid of the integration sign? We simply define it as a function.

No Exact Integral? Just Give It a Name

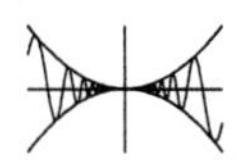

Define $f(x) = \int_1^x \frac{1}{t}dt$ for $x > 0$. Plot graphs of $f(x)$ and $1/x$ and describe features of these graphs that are compatible with the fact that $f(x)$ is an antiderivative of $1/x$. Use the Midpoint Rule with $n = 10$ to discuss the possibility that $f(6) = f(2) + f(3)$. Describe what this means in terms of areas. Use the Midpoint Rule with $n = 10$ to discuss the possibility that $f(8) = 3f(2)$. Describe what this means in terms of areas. State two general rules that these two equations suggest. Test them using additional experiments.

Michael's Solution

Define f by $f(x) = \int_1^x \frac{1}{t}dt$. This definition makes sense only for $x > 0$. The following shows graphs of $f(x)$ and $1/x$, with $f(x)$ drawn with a thick pen and $1/x$ with a regular pen. Note that $f(1) = 0$. For $x > 1$, the graph of f shows the accumulation of area under the graph of $y = 1/x$ starting at 1. For $x < 1$, we have $f(x) < 0$ because $\int_1^x \frac{1}{t}dt = -\int_x^1 \frac{1}{t}dt$ represents the negative of the area under the graph of $y = 1/x$ from x to 1. Because $\lim_{x\to\infty} \frac{1}{x} = 0$, it follows that the graph of $f(x)$ gets flatter and flatter as x gets large.

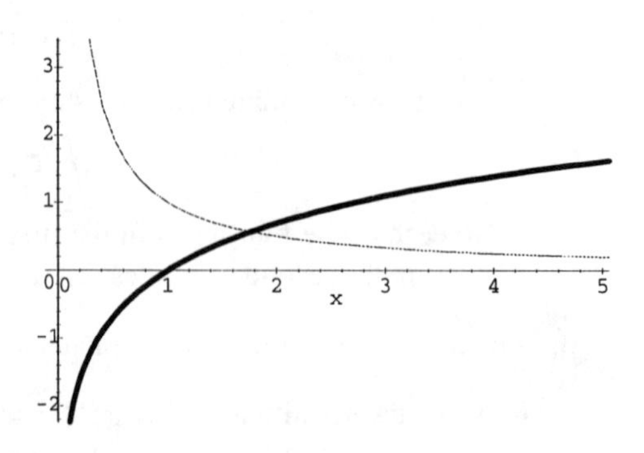

$y = \int_1^x \frac{1}{t}dt$ and $y = \frac{1}{x}$

With the insertion point in the equation $f(x) = \int_1^x \frac{1}{t}dt$, choose Define + New Definition. With the insertion point in the list $f(x), 1/x$, choose Plot 2D + Rectangular.

Note that

$$\begin{aligned} f(6) &= \int_1^6 \frac{1}{t}dt \approx \frac{1}{2}\sum_{i=0}^{9} \frac{1}{\frac{5}{4}+\frac{1}{2}i} = 1.782 \\ f(2) &= \int_1^2 \frac{1}{t}dt \approx \frac{1}{10}\sum_{i=0}^{9} \frac{1}{\frac{21}{20}+\frac{1}{10}i} = .69284 \\ f(3) &= \int_1^3 \frac{1}{t}dt \approx \frac{1}{5}\sum_{i=0}^{9} \frac{1}{\frac{11}{10}+\frac{1}{5}i} = 1.0971 \end{aligned}$$

and $1.0971 + .69284 = 1.7899 \approx 1.782$. This makes the equation $f(6) = f(2) + f(3)$ plausible.

With the insertion point in the expression $f(x)$, choose Calculus + Approximate Integral (Formula: Midpoint, Subintervals: 10, Range: $1 \leq x \leq 6$). Choose OK. Repeat for ranges $1 \leq x \leq 2$ and $1 \leq x \leq 3$.

Also,

$$f(8) = \int_1^6 \frac{1}{t}dt \approx \frac{7}{10}\sum_{i=0}^{9} \frac{1}{\frac{27}{20}+\frac{7}{10}i} = 2.0608$$

and $3f(2) \approx 3(.69284) = 2.0785 \approx 2.0608$ makes the equation $f(8) = 3f(2)$.

With the insertion point in the expression $f(x)$, choose Calculus + Approximate Integral (Formula Midpoint, Subintervals 10 , Range $1 \leq x \leq 8$). Choose OK.

These approximations make the following two identities plausible:

$$\begin{aligned} f(ab) &= f(a) + f(b) \\ f(a^n) &= nf(a) \end{aligned}$$

Testing these with $a = 3$, $b = 4$, and $n = 2$, using the Midpoint Rule with 50 subintervals, yields

$$\begin{aligned} f(3) &\approx \frac{1}{25}\sum_{i=0}^{49} \frac{1}{\frac{51}{50}+\frac{1}{25}i} = 1.0986 \\ f(4) &\approx \frac{3}{50}\sum_{i=0}^{49} \frac{1}{\frac{103}{100}+\frac{3}{50}i} = 1.3862 \\ f(12) &\approx \frac{11}{50}\sum_{i=0}^{49} \frac{1}{\frac{111}{100}+\frac{11}{50}i} = 2.4829 \end{aligned}$$

and hence

$$f(3) + f(4) \approx 1.0986 + 1.3862 = 2.4848 \approx 2.4829$$

Furthermore,

$$f(9) \approx \frac{4}{25} \sum_{i=0}^{49} \frac{1}{\frac{27}{25} + \frac{4}{25} i} = 2.1962$$

and hence

$$2f(3) \approx 2\,(1.0986) = 2.1972 \approx 2.1962$$

These approximations are consistent with the equations $f(ab) = f(a)+f(b)$ and $f\,(a^n) = nf(a)$. They are strongly reminiscent of the equations $\ln ab = \ln a + \ln b$ and $\ln a^n = n \ln a$ for the natural logarithm.

Activities

1. Give evidence to support the theory that the function $f(x) = \int_1^x \frac{1}{t} dt$ is actually the natural logarithm $\ln x$.

2. Define $s(x) = \int_0^x \frac{dt}{\sqrt{1-t^2}}$. Give a reasonable domain of definition for $s(x)$. Use the Midpoint Rule with 50 subintervals to estimate $s(1/2)$. Multiply this number by 6 and compare it with π. Explain why this approximation is so good.

3. Define $t(x) = \int_0^x \frac{du}{1+u^2}$. Give a reasonable domain of definition for $t(x)$. Use the Midpoint Rule with 50 subintervals to estimate $t(1)$. Multiply this number by 4. What number does this approximate? Explain why this approximation is so good.

6 APPLICATIONS OF INTEGRATION

You will use integration to find areas, volumes, lengths of curves, and centers of gravity.

6.1 Areas Between Curves

You are not limited to using the x-axis as one of the boundaries of an area.

Caught in the Middle, Again

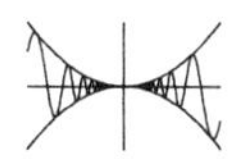

Define $f(x) = 3x^2+4x-3$ and $g(x) = 2x^2+4x+1$. Plot graphs of f and g. Determine the points where the two curves cross. Determine the area between the two curves by interpreting the area as a difference of integrals. Plot the curve $y = f(x) - g(x)$ using the same viewing rectangle as before. Determine the area under the curve and above the x-axis by interpreting this area as an integral. Explain why the two areas are equal. Attach an appropriate rectangle to each plot that approximates the area between $x = -1$ and $x = -0.5$, and use the areas of these rectangles in your discussion.

Josie's Solution

Define $f(x) = 2x^2 + 4x + 1$ and $g(x) = 3x^2 + 4x - 3$. The region bounded by the graphs of f and g is shown in the following figure.

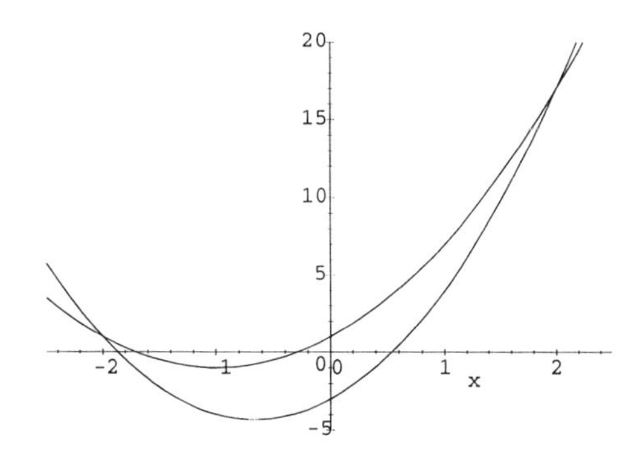

$y = f(x)$ and $y = g(x)$

With the insertion point in the equation $f(x) = 2x^2 + 4x + 1$, choose Define + New Definition. Repeat for $g(x) = 3x^2 + 4x - 3$. With the insertion point in the expression $2x^2 + 4x + 1$, choose Plot 2D + Rectangular. Choose Edit + Properties, Plot Components page; set Domain at $-2.5 \leq x \leq 2.5$. Choose View page, turn off Default and set View Intervals to $-2.5 \leq x \leq 2.5, -5 \leq y \leq 20$. Select and drag to the frame $3x^2 + 4x - 3$.

Setting $f(x) = g(x)$ and solving, we see that the curves intersect at $x = -2$ and again at $x = 2$. The difference of the integrals

$$\int_{-2}^{2} f(x)\,dx - \int_{-2}^{2} g(x)\,dx = \frac{32}{3}$$

represents the area between the curves. We also plot the graph of $y = f(x) - g(x)$.

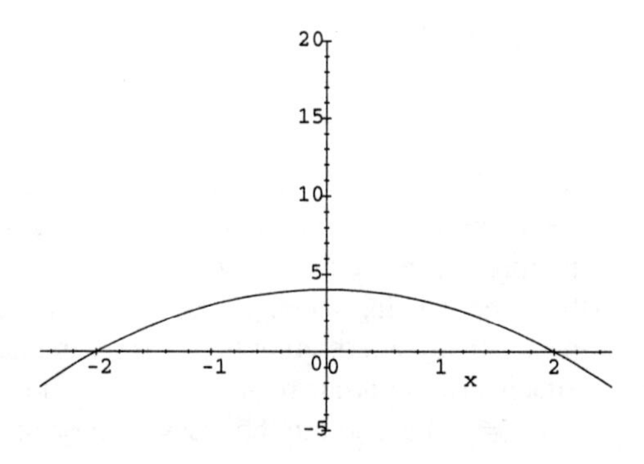

$y = f(x) - g(x)$

With the insertion point in the equation $f(x) = g(x)$, choose Solve + Exact. With the insertion point in the expression $\int_{-2}^{2} f(x)\,dx - \int_{-2}^{2} g(x)\,dx$, choose Evaluate. With the insertion point in the expression $f(x) - g(x)$, choose Plot 2D + Rectangular. Choose Edit + Properties, Plot Components page; set Domain to $-2.5 \leq x \leq 2.5$. Choose View page, turn off Default, and set View Intervals to $-2.5 \leq x \leq 2.5, -5 \leq y \leq 20$.

We observe that the area under this curve between -2 and 2 is given by

$$\int_{-2}^{2} (f(x) - g(x))\,dx = \frac{32}{3}$$

To understand why these two regions have equal areas, it is useful to visualize Riemann sums. Draw a rectangle between $x = -1$ and $x = -.5$ in each figure, with the height determined by left endpoints.

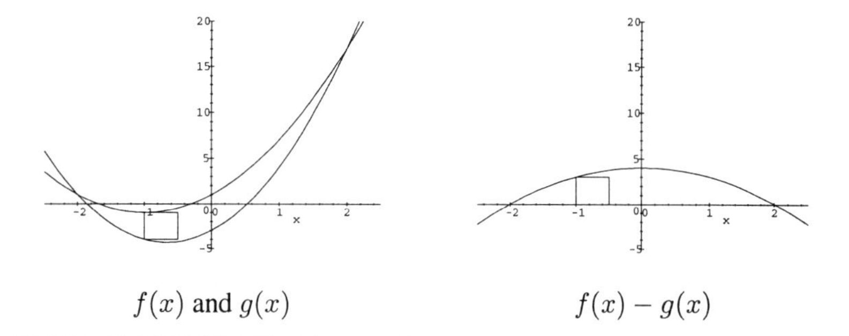

$f(x)$ and $g(x)$ $\qquad$ $f(x) - g(x)$

 Select and drag to the graph of $f(x)$ and $g(x)$ the list of points

$$[-1, f(-1), -1, g(-1), -.5, g(-1), -.5, f(-1), -1, f(-1)]$$

Select and drag to the graph of $f(x) - g(x)$ the list of points

$$[-1, 0, -1, (f(-1) - g(-1)), -.5, (f(-1) - g(-1)), -.5, 0]$$

The area of each of the approximating rectangles is $0.5 \times 3 = 1.5$. Imagine doing this for another pair of approximating rectangles between $x = a$ and $x = b$. The width of such a rectangle would be $b - a$, and the height would be $f(a) - g(a)$. It seems reasonable to conclude that the two areas are equal.

Activities

1. Let f and g be your favorite functions and discuss the equation
$$\int_a^b f(x)dx - \int_a^b g(x)dx = \int_a^b (f(x) - g(x))\, dx$$
Include appropriate graphs in your discussion.

2. Interpret geometrically the integral $\int_{-4}^{4} (f(x) - g(x))\, dx$, where $f(x) = x^2 - x - 1$ and $g(x) = 2x + 3$.

6.2 Volume

This will add another dimension to your work.

A Revolutionary Volume

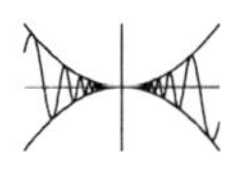

Use Plot 3D + Tube to visualize the solid of revolution generated by rotating the graph of $y = x^2 - 1$, $y = 0$, $x = 0$, and $x = 2$ about the x-axis. Find an integral that represents the volume of this solid of revolution and evaluate the integral. Is one integral enough to represent this volume? Are two integrals required? Justify your reasoning. Show that the value of the integral is reasonable by approximating the volume of the solid with two cone-shaped solids.

Richard's Solution

The surface generated by rotating the graph of $y = x^2 - 1$ about the x-axis between $x = 0$ and $x = 2$ is shown in the following figure. The volume is given by the integral $\int_0^2 \pi \left(x^2 - 1\right)^2 dx$, which evaluates to $\frac{46\pi}{15} = 9.6342$.

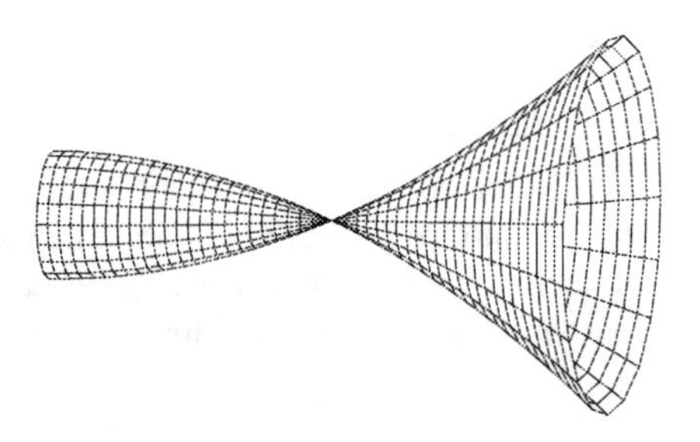

With the insertion point in the triple $[x, 0, 0]$, choose Plot 3D + Tube. Choose Edit + Properties, Plot Components page. In the Radius input box, type $x^2 - 1$. Set the Domain Interval to $0 \leq t \leq 1$ and Plot Style to Hidden Line. Choose OK. Double-click inside the frame to bring up the rotate tool; then rotate the solid so that the positive x-axis extends to the right. With the insertion point inside the integral $\int_0^2 \pi \left(x^2 - 1\right)^2 dx$, choose Evaluate.

To show that this is reasonable, note that the volume between $x = 0$ and $x = 1$ can be approximated using a cone of height 1 and base radius 1 for a volume of $V_1 = \frac{1}{3}\pi 1^2 1 = 1.0472$. The volume between $x = 1$ and $x = 2$ can be approximated using a cone of height 1 and base radius 3 for a volume of $V_2 = \frac{1}{3}\pi 3^2 1 = 9.4248$. That gives a total approximation of $1.0472 + 9.4248 = 10.472$. It is reasonable to expect our estimate to be too large because the shape of the larger cone strictly includes the corresponding volume of revolution, because the graph of $y = x^2 - 1$ is concave upward.

Activities (How to Weigh an Elephant)

1. A boat in the shape of an inverted truncated pyramid was built by a wise prince in order to weigh an elephant by estimating the displacement of water when the elephant stepped onto the boat. Explain how this could work.

2. **Rain gauge.** Design a rain gauge in the shape of an inverted pyramid that has a 2 in. × 2 in. opening at the top (the base of the pyramid) and is 12 in. tall (inside dimensions). The outside should have markings that make it easy to read the amount of rainfall.

   Evaporation from the rain gauge. After a heavy rain, Sam forgot to empty the rain gauge. The next day the humidity dropped and evaporation caused the reading to drop from 1.5 in. to 1.3 in. during a 6-hour period. Assuming the rate of evaporation is proportional to surface area, how long will it take for the reading to drop from 1.3 down to 1.0? From 1.0 down to 0.1? From 0.1 down to 0.01?

3. Your uncle George has acquired a cylindrical tank he wants to use for a gasoline tank. This tank is 4 feet in diameter, 12 feet long, and is lying on its side. He can measure the depth of the liquid through a hole in the top with a dipstick, but he doesn't know how to translate this into gallons. Design a dipstick for him with useful markings.

4. Consider the volume of revolution generated by rotating the region bounded by $y = x$, $y = 0$, $x = 0$, and $x = 5$ about the x-axis. Visualize the region in the plane being rotated by applying Calculus + Plot Approximate Integral to x on the interval $0 \leq x \leq 5$ with Number of Boxes set to 100. Visualize the solid of revolution by applying Plot 3D + Tube to $[x, 0, 0]$ with Radius x and Domain Interval $0 \leq t \leq 5$. Rotate the solid so that the positive x-axis points to the right side of the figure. (You should see a cone of base radius 5 and height 5. The base of the cone should be to the right.) Plot the expression $\lfloor x \rfloor$. (These brackets can be found by clicking the brackets button.) Explain what this expression calculates. Visualize a Riemann sum with $n = 5$ by applying Plot 3D + Tube to $[x, 0, 0]$ with Radius $\lfloor x \rfloor$ and Domain Interval $0 \leq t \leq 4.999$. (Explain why it is appropriate to use 4.999 rather than 5.) Rotate the solid as before. Calculate the volume represented by the Riemann sum and add together the volumes of five right circular cylinders. Explain why this represents a lower bound for the volume of the original solid. Give an integral that represents the volume of the original solid of revolution. Evaluate the integral and compare its value with your Riemann sum. Is the value of the integral reasonable? Explain your reasoning.

5. Find an integral that represents the volume of a right circular cone of height h and base radius r by rotating an appropriate region in the first quadrant about the x-axis. Evaluate the integral and verify that it is compatible with the usual formula for the volume of a cone.

6. Consider a frustum of a right circular cone with height h, lower base radius R, and upper base radius r. Find an equation of a line $y = mx + b$ that goes through the points $(0, R)$ and (h, r). Rotate the region bounded by $y = mx + b$, $y = 0$, $x = 0$, and $x = h$ about the x-axis to generate a frustum of a right circular cone. Find an integral that represents of volume of the frustum. Evaluate the integral. Verify that the volume of a frustum of a right circular cone is equal to the height multiplied by the average of three areas—namely, the two bases and an ellipse whose major semiaxis equals one base radius and whose minor semiaxis equals the other base radius. Verify

that the formula yields the usual formula for the volume of a right circular cone in the case where one base radius is zero. Verify that the formula yields the usual formula for the volume of a right circular cylinder in the case where the two base radii are equal.

7. Consider a frustum of a pyramid with square base of side b, square top of side a, and height h. Find a function $f(x)$ that represents the cross-sectional area of the frustum for any number x between 0 and h, so that in particular $f(0) = b^2$ and $f(h) = a^2$. Express the volume of the frustum as an integral. Evaluate the integral. Verify that the volume of a frustum of a pyramid with square base of side b and square top of side a is equal to the height multiplied by the average of three areas—namely, the area of the top, the area of the bottom, and the area of a rectangle of dimensions a by b. Verify that the formula yields the usual formula for the volume of a pyramid with a square base in the case where $a = 0$. Verify that the formula yields the usual formula for a rectangular solid in the case where $a = b$.

6.3 Volumes by Cylindrical Shells

The method introduced in the previous section for calculating volumes works well for many solids. A second method, based on the use of cylindrical shells, provides a convenient alternative method for many solids. Instead of parallel slices cut by a plane, this method is more like peeling a potato to create potato chips.

Sometimes Peeling Works Better Than Slicing

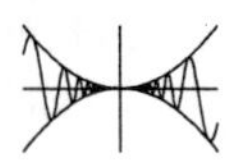

Consider the problem of finding the volume of the solid obtained by rotating the region bounded by $y = x - x^2$ and $y = 0$ about the line $x = 2$. The disk method would require solving for x in terms of y and integrating with respect to y. Create a picture of the region with a typical vertical slice shown; then draw a picture that shows the volume generated by rotating the slice about the line $x = 2$. Create a picture of the solid of revolution. Find an integral that represents the volume and evaluate it. Give some crude estimates that demonstrate that the computed volume is reasonable.

Lee's Solution

The region bounded by $y = x - x^2$ and $y = 0$ is shown in the following figure. A typical vertical strip has dimensions dx by $x - x^2$. (In the picture, $dx = 0.1$ and $x = 0.6$.)

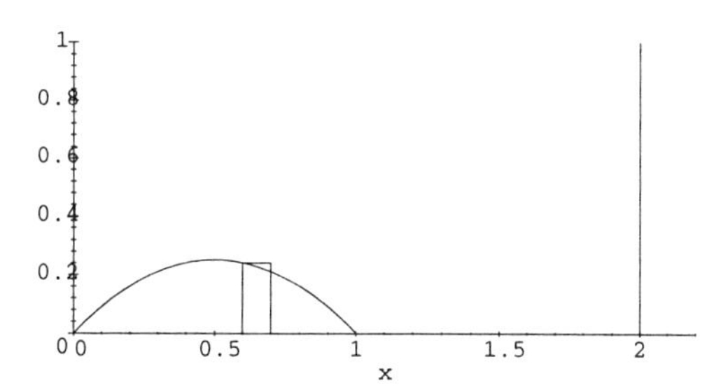

With the insertion point in the expression $x - x^2$, choose Plot 2D + Rectangular. Select and drag $(2, 0, 2, 1)$ to the frame. Select and drag $(.6, 0, .6, .6 - .36, .7, .6 - .36, .7, 0)$ to the frame. Choose Edit + Properties, View page. Turn off the Default, and set the View Intervals to $0 \leq x \leq 2.2, 0 \leq y \leq 1$.

The next figure shows the volume generated by rotating the thin vertical strip about the line $x = 2$. Evidently the volume of this shell is given approximately by $dV = (x - x^2)\, 2\pi(2 - x)dx$

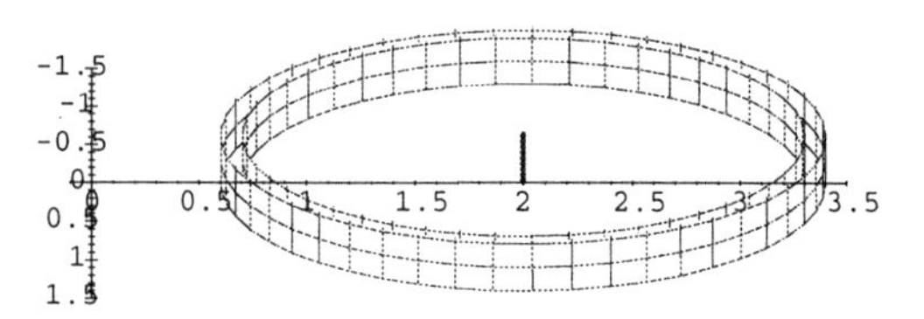

With the insertion point in the expression $(2, y, 0)$, choose Plot 3D + Tube. Choose Edit + Properties, View page. Turn off the Default and set the View Intervals to $-.1 \leq x \leq 3.5$, $0 \leq y \leq .24$, and $-1.5 \leq z \leq 1.5$. Choose Axes page and set Axes Type to Normal. Set the Orientation to Turn -90, Tilt -20. Turn on Equal Scalingon Each Axis. Go to the Plot Components page and set the Radius: 1.4, Sample Size: 10, Number of Tube Points: 50, Plot Style: Hidden Line. Choose Add Item, type $(2, y, 0)$ in the edit box and change the Radius to 1.3. Choose Add Item, type $(2, y, 0)$ in the edit box, and change the Radius to .01.

The third figure shows the solid of revolution. Because the volume of the shell is approximately $\left(x - x^2\right) 2\pi(2 - x)dx$, it following that the volume is given by

$$\int_0^1 \left(x - x^2\right) 2\pi(2 - x)dx = \frac{\pi}{2} \approx 1.5708$$

Is this number reasonable? The cross-sectional area is given by $\int_0^1 \left(x - x^2\right) dx = \frac{1}{6}$ and this region travels through a circle of radius $3/2$, which has a circumference of $2\pi\left(3/2\right) = 3\pi$. If this half-bagel-shaped solid were cream cheese, and the cheese were spread out evenly over a disk of radius 2, it seems as though the thickness should be about $\frac{1}{10}$. Multiplying this by the area of a circle of radius 2 produces $4\pi\left(\frac{1}{10}\right) = \frac{2}{5}\pi \approx 1.2566$. Thus the estimate 1.5708 appears to be reasonable.

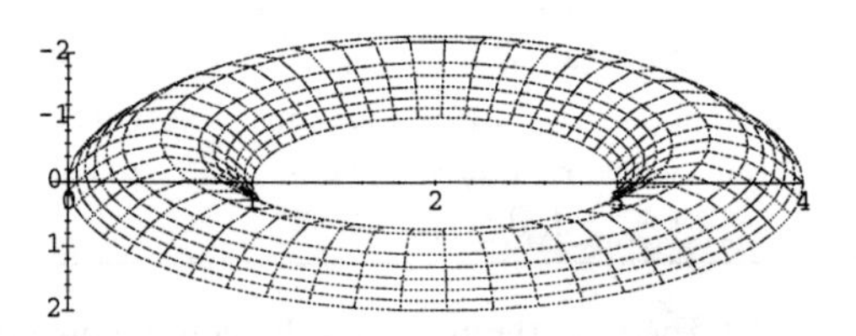

With the insertion point in the expression $y = x - x^2$, choose Solve + Exact, Solve for x. Rotating about $x = 2$, this gives an outside tube radius of $2 - \left(\frac{1}{2} - \frac{1}{2}\sqrt{(1 - 4y)}\right)$ and an inside tube radius of $2 - \left(\frac{1}{2} + \frac{1}{2}\sqrt{(1 - 4y)}\right)$. With the insertion point in the expression $(2, y, 0)$, choose Plot 3D + Tube. Choose Edit + Properties, View page. Turn off the Default and set the View Intervals to $-.1 \leq x \leq 3.5$, $0 \leq y \leq .24$, and $-1.5 \leq z \leq 1.5$. Choose Axes page and set Axes Type to Normal. Set the Orientation to Turn: -90, Tilt: -20. Turn on Equal Scaling. Go to the Plot Components page and set the Radius to $2 - \left(\frac{1}{2} - \frac{1}{2}\sqrt{(1 - 4y)}\right)$, Sample Size: 25, Number of Tube Points: 50, Plot Style: Hidden Line. Choose Add Item, type $(2, y, 0)$ in the edit box, and change the Radius to $2 - \left(\frac{1}{2} + \frac{1}{2}\sqrt{(1 - 4y)}\right)$.

Activities

1. Use the disk method to find the volume of the solid obtained by rotating the region bounded by $y = x^2$ and $y = \sqrt{x}$ about the y-axis. Create a picture of the region bounded by $y = x^2$ and $y = \sqrt{x}$ with a typical horizontal slice shown. Give a formula for the volume of the disk generated by rotating this slice about the y-axis. Create a picture of the solid of revolution. Find an integral based upon the disk

method that represents the volume and evaluate it. Give some crude estimates that demonstrate that the computed volume is reasonable.

2. Use the shell method to find the volume of the solid obtained by rotating the region bounded by $y = x^2$ and $y = \sqrt{x}$ about the y-axis. Create a picture of the region bounded by $y = x^2$ and $y = \sqrt{x}$ with a typical vertical slice shown. Give a formula for the volume of the shell generated by rotating this slice about the y-axis. Find an integral based on the shell method that represents the volume and evaluate it. Compare this evaluated integral with what you found in the first activity. Discuss some general criteria you would use in the future to decide whether to apply the disk method or the shell method to find the volume of a particular solid of revolution.

6.4 Work

For a constant force F and fixed distance d, work is defined as the product $W = Fd$. In calculus, this formula is a starting point for computing work done by a variable force.

Exercise Is Work

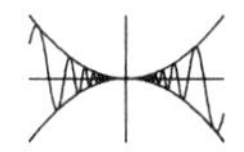

Modern exercise equipment provides a variable resistance designed to optimize muscle development. Explore the force required to lift a fixed weight that is attached to a circular pulley with an off-centered pivot point.

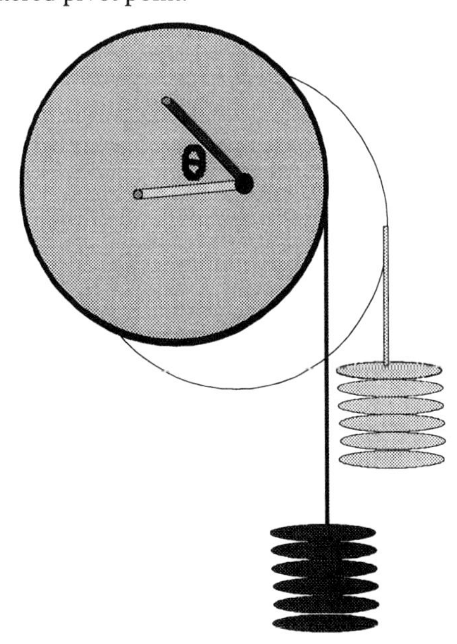

Variable force

Assume that a circular disk of radius 1 foot has a pivot point at the origin and that the equation of the circle is given by $(x - .5\cos(\pi + \theta))^2 + (y - .5\sin(\pi + \theta))^2 = 1$, where θ is an angle between 0 and $\frac{\pi}{2}$ through which the disk has been rotated. Assume that a 100-pound weight is attached to a cable that is partially wrapped around the circumference of the disk. Plot the equation $(x - .5\cos(\pi + \theta))^2 + (y - .5\sin(\pi + \theta))^2 = 1$ for several angles θ between 0 and $\frac{\pi}{2}$. Imagine that the cable hangs straight down from the right edge of each of these disks. The person performing the exercise pushes against a bar attached to a 1-foot arm that rotates about the origin. Calculate the amount of work done in rotating the disk through an angle of 90°.

Greth's Solution

Following is a picture that shows the orientation of the disk at angles between 0 and $\frac{\pi}{2}$. The initial position is drawn with a thick pen. The force applied to the 1-foot arm is equal to the weight (100 pounds) multiplied by horizontal distance between the origin and a vertical tangent line on the right. Two such tangent lines are shown in the picture. It follows that the total amount of work done is given by the integral

$$100\int_0^{\pi/2}\left(1 + \frac{1}{2}\cos(\pi + t)\right)dt = 100\left(\frac{1}{2}\pi - \frac{1}{2}\right) = 107.08 \text{ ft-lbs}$$

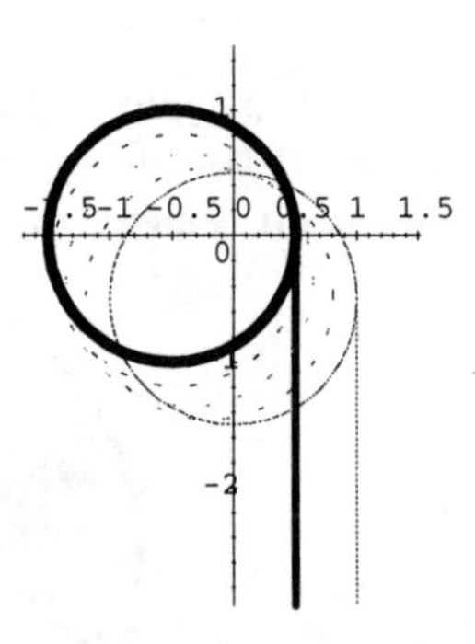

With the insertion point in the expression $(-.5\cos 0 + \cos t, -.5\sin 0 + \sin t)$, choose Plot 2D + Parametric. Select and drag to the frame each of the expressions

$$(-.5\cos(\pi/8) + \cos t, -.5\sin(\pi/8) + \sin t)$$
$$(-.5\cos(\pi/4) + \cos t, -.5\sin(\pi/4) + \sin t)$$
$$(-.5\cos(3\pi/8) + \cos t, -.5\sin(3\pi/8) + \sin t)$$
$$(-.5\cos(\pi/2) + \cos t, -.5\sin(\pi/2) + \sin t)$$
$$(.5, t)$$
$$(1, -.5 - t)$$

Choose Edit + Properties. From the Axes page, turn on Equal Scaling Along Each Axis. From the View page, turn off the Default. From the Plot Components page, change the Thickness to Medium for Item Number 1. Change the Line Style and Color to show the various positions of the disk and cable.

As a check, calculate the total distance through which the 100-pound weight has been lifted, and multiply by 100. New cable is wrapped around one-fourth of the disk for a total length of $\frac{2\pi}{4} = \frac{\pi}{2}$. The location of the vertical tangent at the end of the lift is $(1, -.5)$ for a net distance lifted of $\frac{\pi}{2} - \frac{1}{2}$. Thus, the work is given by

$$W = Fd = 100\left(\frac{\pi}{2} - \frac{1}{2}\right) = 107.08 \text{ ft-lbs}$$

Activities

1. Use Tube Plot to visualize a spring with five coils. Assume that a force of 1000 lb is required to hold the spring stretched 1 foot beyond its natural length. Create a table that shows the force required to hold the spring x feet beyond its natural length, where $x = 0, 0.1, 0.2, 0.3, 0.4, 0.5, 0.6, 0.7, 0.8, 0.9$, and 1. Add a third column that estimates the work done in stretching the spring from x ft to $x + 0.1$ ft. Describe the method you used to fill in the third column in your table. Discuss a variation that you think might provide more precise results. Sum this column and compare it with the value of an integral that represents the total work done in stretching the spring from its natural position to 1 foot beyond its natural length.

   Hint With the insertion point in the expression $(\cos t, \sin t, t)$, choose Plot 3D + Tube. Choose Edit + Properties, Plot Components page. Set the Domain Interval to $0 \leq t \leq 31.416$, Plot Style Hidden Line, Sample Size 99, Radius 0.1.

2. Pick an interesting shape for a water tank. Choose reasonable dimensions, and make a picture that indicates the location of the outlet. Assume the tank is full of water and that water weighs 62.5 lb/ft^3. Divide the depth into 10 equal subdivisions. Create a table with 11 rows that lists on each row the current depth of water and an estimate of the work required to reduce the depth to the next table entry. Sum these estimates and compare the total with an evaluated integral that represents the total work done in emptying the tank.

6.5 Average Value of a Function

Given a finite list $y_1, y_2, \ldots, y_n$ of numbers, the average or mean is given by

$$ave = \frac{y_1 + y_2 + \cdots + y_2}{n}$$

In the following activity, this idea is extended to the average value of a function defined on an interval.

It All Averages Out

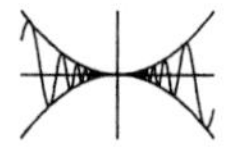

Let $f(x) = x \sin x$. Find the average value of f on the interval $[0, \pi]$, using two different methods. First, use n equally distributed sampling points, find the average of the function at these sampling points, and then calculate a limit as the number of sampling points goes to infinity. Next, find a constant function $g(x) = c$ such that $\int_0^\pi f(x)\,dx = \int_0^\pi g(x)\,dx$. Plot regions that represent these two integrals and decide whether or not the areas appear to be equal. Explain how these two methods are related and why it is not surprising that you get the same "average" value in each case.

Mark's Solution

Let $f(x) = x \sin x$. Given n, choose the sampling points $\frac{\pi}{2n}, \frac{\pi}{2n} + \frac{\pi}{n}, \frac{\pi}{2n} + 2\frac{\pi}{n}, \ldots, \frac{\pi}{2n} + (2(n-1)+1)\frac{\pi}{n}$. These points are equally spaced a distance $\frac{\pi}{n}$ apart. The average of the function values at these points is given by

$$ave = \frac{1}{n}\sum_{i=0}^{n-1} f\left(\frac{(2i+1)\pi}{2n}\right)$$

With $n = 5$, 10, 20, and 50, this average is given in the following table:

n	5	10	20	50
ave	1.0166	1.0041	1.001	1.0002

The limit is given by

$$\lim_{n\to\infty} \frac{1}{n}\sum_{i=0}^{n-1} f\left(\frac{(2i+1)\pi}{2n}\right) = 1$$

With the insertion point in the equation $f(x) = x \sin x$, choose Define + New Definition. With the insertion point in the equation $h(n) = \frac{1}{n}\sum_{i=1}^{n} f\left(\frac{(2i-1)\pi}{2n}\right)$, choose Define + New Definition. With the insertion point in the expression $h(5)$, choose Evaluate Numerically. Repeat this for the arguments 10, 20, and 50. Choose Insert + Table with 2 Rows and 5 Columns and fill in the entries. Select the first row of the table and choose Edit + Properties. In the Alignment page, choose Aligned Center. In the Lines page, choose Single Line, Bottom, and click OK. Select the first column and choose Edit + Properties. In the Lines page, choose Single Line, Right, and click OK. With the insertion point in the expression $\lim_{n\to\infty} h(n)$, choose Evaluate.

The integral of a constant c on the interval $[0, \pi]$ is given by $\int_0^\pi c\,dx = c\pi$. Solving the equation $\int_0^\pi c\,dx = \int_0^\pi f(x)\,dx$ for c, we get $c = 1$. The following figure includes a graph of $y = x \sin x$ and the line $y = 1$ on the interval $[0, \pi]$:

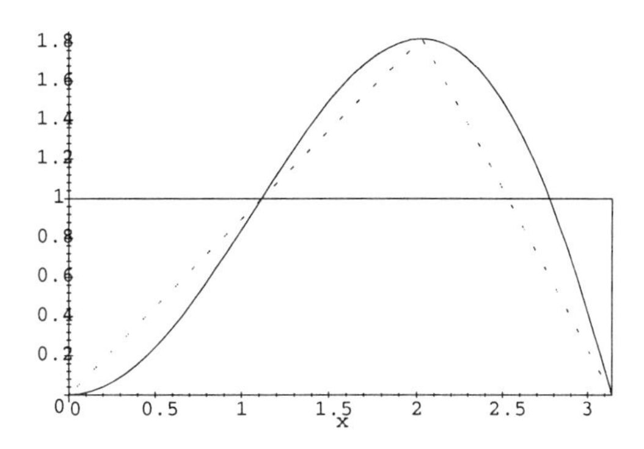

Because the shape of the area is a bit like a triangle, we could make the very rough estimate that the area under the graph of $y = x \sin x$ is approximately the area of a triangle with base π and height 1.8—namely, $\frac{1}{2}\pi\,(1.8) = 2.8274$. The area of the rectangular region is $\pi \times 1 = 3.1416$. The exact value of the integral that gives the area under the curve $x \sin x$ is

$$\int_0^{\pi} x \sin x \, dx = \pi$$

Thus, the rectangular region actually has area exactly the same as the region bounded by the curve $x \sin x$ and the x-axis.

With the insertion point in the expression$=x \sin x$, choose Plot 2D + Rectangular. Select the expression $(0, 1, 3.1416, 1, 3.1416, 0)$ and drag it to the frame. Choose Edit + Properties, and in the Plot Components page set the Domain Interval to $0 \le x \le 3.1416$.

The formula $ave = \frac{1}{n}\sum_{i=0}^{n-1} f\left(\frac{(2i+1)\pi}{2n}\right)$ is very similar to the middle-point formula $\frac{\pi}{n}\sum_{i=0}^{n-1} f\left(\frac{\pi}{2n} + i\frac{\pi}{n}\right)$. In fact, $\frac{(2i+1)\pi}{2n} = \frac{\pi}{2n} + i\frac{\pi}{n}$, so the middle-point formula is π multiplied by the formula for the average of n equally spaced sampling points. Thus, $ave = \frac{1}{\pi}M$, where M is the middle-point approximation to the integral. In general, the average value of a function f defined on an interval $[a, b]$ is given by

$$ave = \frac{1}{b-a}\int_a^b f(x)\, dx$$

Activities

1. Visualize the average value of the function $f(x) = x^3$ on the interval $[0, 2]$ by using Middle Boxes to represent values of the function at equally spaced sampling points. Estimate the average value of f on $[0, 2]$ using 5, 10, 20, and 50 equally spaced sampling points. Calculate the exact average value by evaluating an integral.

2. Visualize the average value of the function $f(x) = \cos(x + \sin x)$ on the interval

$[0, 2\pi]$ by using Middle Boxes to represent values of the function at equally spaced sampling points. Estimate the average value of f on $[0, 2\pi]$ using 5, 10, 20, and 50 equally spaced sampling points.

7 TECHNIQUES OF INTEGRATION

Various techniques of integration can give insight into applications of integrals.

7.1 Integration by Parts

Integration by parts is a powerful tool for evaluating integrals. It allows an integration problem to be replaced by a differentiation problem and a different integration problem. Differentiation is almost always easier than integration, and with any luck (and some foresight) the second integral may be easier than the first.

► **To integrate by parts**

- With the insertion point in an indefinite or definite integral, choose Calculus + Integrate by Parts. In the dialog box, specify a function for Part to be Differentiated.

  This is illustrated in the following two examples.

- Calculus + Integrate by Parts (Specify $\ln x$)

$$\int x \ln x dx = \frac{1}{2} (\ln x) x^2 - \int \frac{1}{2} x \, dx$$

- Calculus + Integrate by Parts (Specify $\sin x$)

$$\int_0^1 e^x \sin x dx = e \sin 1 - \int_0^1 (\cos x) e^x \, dx$$

In the first example, the new integral $\int \frac{1}{2} x dx$ is certainly simpler than the original integral $\int x \ln x dx$. In the second example, it is not immediately clear that any progress has been made. However, a second application of integration by parts followed by solving an equation for the unknown integral finally leads to success.

Integration by Differentiation

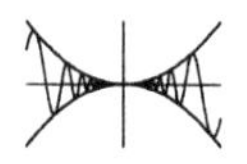

Use integration by parts to evaluate $\int \tan^{-1} x \, dx$ by specifying $\tan^{-1} x$ as the part to be differentiated. Fill in the details in the process by defining

$$u = \tan^{-1} x \qquad dv = dx$$

and from these equations determining du and v. Verify your final answer by direct evaluation of $\int \tan^{-1} x\, dx$.

Virginia's Solution

Integration by parts produces

$$\int \tan^{-1} x\, dx = (\arctan x)\, x - \int \frac{1}{1+x^2} x\, dx$$

With the insertion point in the integral $\int \tan^{-1} x\, dx$, choose Calculus + Integrate by Parts (specify $\tan^{-1} x$).

To verify this, let

$$u = \tan^{-1} x \qquad dv = dx$$

Then

$$du = \frac{dx}{1+x^2} \qquad v = x$$

so integration by parts implies

$$\begin{aligned} \int \tan^{-1} x\, dx &= \int u dv \\ &= uv - \int v du \\ &= x \tan^{-1} x - \int \frac{x}{1+x^2} dx \end{aligned}$$

The integral $\int \frac{x}{1+x^2} dx$ may be evaluated by using the substitution $u = 1 + x^2$ to yield

$$\int \frac{x}{1+x^2} dx = \int \frac{1}{2u}\, du = \frac{1}{2} \ln u = \frac{1}{2} \ln\left(1+x^2\right)$$

It follows that

$$\int \tan^{-1} x\, dx = x \tan^{-1} x - \frac{1}{2} \ln\left(1+x^2\right) + C$$

With the insertion point in the integral $\int \frac{1}{1+x^2} x\, dx$, choose Calculus + Change Variable (Specify $u = 1 + x^2$). With the insertion point in the integral $\int \frac{1}{2u}\, du$, choose Evaluate. With the insertion point in the expression $\left[\frac{1}{2} \ln u\right]_{u=1+x^2}$, choose Evaluate. Combine these steps to evaluate $\int \tan^{-1} x\, dx$, adding the "$+\, C$" at the end.

Direct evaluation produces

$$\int \tan^{-1} x dx = (\arctan x)\, x - \frac{1}{2} \ln\left(1+x^2\right)$$

and good judgment dictates that we add the "$+\, C$"

$$\int \tan^{-1} x dx = (\arctan x)\, x - \frac{1}{2} \ln\left(1+x^2\right) + C$$

With the insertion point in the integral $\int \tan^{-1} x\, dx$, choose Evaluate.

Activities

1. Use integration by parts to evaluate $\int_0^3 x\sqrt{x+1}\,dx$ by specifying x as the part to be differentiated. Use the Midpoint Rule with $n = 3$ to verify that your answer is reasonable. Draw a graph and give an argument why the Midpoint Rule produces an approximation that is too large (or too small).

2. Use integration by parts to evaluate $\int \cos\sqrt{x}\,dx$ by specifying $\sqrt{x}$ as the part to be differentiated. Fill in the details in the process by defining
$$u = \sqrt{x} \qquad dv = \frac{\cos\sqrt{x}}{\sqrt{x}}dx$$
and from these equations determining du and v. Verify your final answer by direct evaluation of $\int \cos\sqrt{x}\,dx$.

3. Use integration by parts to find an antiderivative of $\sin^{-1} x$. Check your answer by differentiation. Use graphs of $y = \sin^{-1} x$ and the antiderivative you found to give an argument that your answer is plausible.

4. Use integration by parts to find an antiderivative of $\sec^{-1} x$. Check your answer by differentiation. Use graphs of $y = \sec^{-1} x$ and the antiderivative you found to give an argument that your answer is plausible.

7.2 Trigonometric Integrals

Integration of trigonometric functions calls for some special techniques.

From Trig to Algebra

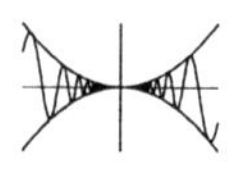

Evaluate the integral $\int \sin^6 x \cos^5 x dx$ by using the substitution $u = \sin x$. Justify the crucial substitution step. Check your answer by differentiation and by direct evaluation.

Debby's Solution

Change of Variable with the substitution $u = \sin x$ produces
$$\int \sin^6 x \cos^5 x dx = \int u^6 \left(1 - u^2\right)^2 du$$

With the insertion point in the integral $\int \sin^6 x \cos^5 x dx$, choose Calculus + Change Variable (Substitution: $u = \sin x$).

This change of variables is based on the replacement

$$\cos^5 x = \cos^4 x \cos x = \left(\cos^2 x\right)^2 \cos x = \left(1 - \sin^2 x\right)^2 \cos x$$

which in turn gives

$$\int \sin^6 x \cos^5 x dx = \int \sin^6 x \left(1 - \sin^2 x\right)^2 \cos x dx = \int u^6 \left(1 - u^2\right)^2 du$$

The last integral can be evaluated by expanding $u^6 \left(1 - u^2\right)^2$ and integrating term-by-term to get

$$\int \left(u^6 - 2u^8 + u^{10}\right) du = \frac{1}{7}u^7 - \frac{2}{9}u^9 + \frac{1}{11}u^{11} + C$$

Replacing u by $\sin x$, we get

$$\left[\frac{1}{7}u^7 - \frac{2}{9}u^9 + \frac{1}{11}u^{11} + C\right]_{u=\sin x} = \frac{1}{7}\sin^7 x - \frac{2}{9}\sin^9 x + \frac{1}{11}\sin^{11} x + C$$

We check our answer by differentiating:
$\frac{d}{dx}\left(\frac{1}{7}\sin^7 x - \frac{2}{9}\sin^9 x + \frac{1}{11}\sin^{11} x + C\right) = \sin^6 x \cos x - 2\sin^8 x \cos x + \sin^{10} x \cos x$

Select $u^6 \left(1 - u^2\right)^2$ with the mouse and, with the CTRL/COMMAND key pressed down, choose Expand (in-place replacement). With the insertion point in $\int \left(u^6 - 2u^8 + u^{10}\right) du$, choose Evaluate. Add $+C$. With the insertion point in each of the two expressions

$$\left[\frac{1}{7}u^7 - \frac{2}{9}u^9 + \frac{1}{11}u^{11} + C\right]_{u=\sin x} \quad \text{and} \quad \frac{d}{dx}\left(\frac{1}{7}\sin^7 x - \frac{2}{9}\sin^9 x + \frac{1}{11}\sin^{11} x + C\right)$$

choose Evaluate.

To verify that this expression is equivalent to the original expression $\sin^6 x \cos^5 x$, we first compare their graphs. (They appear identical.)

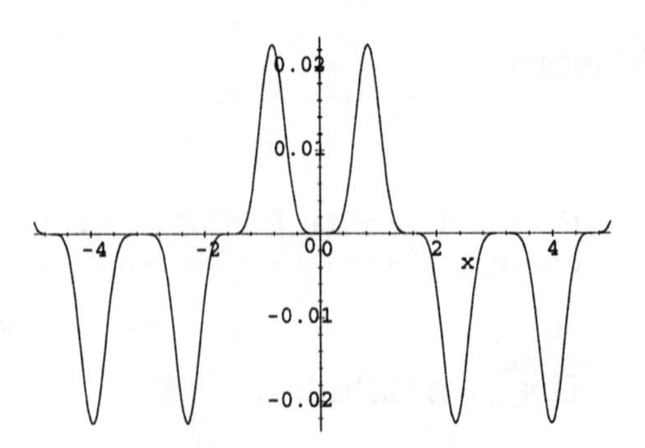

$$y = \sin^6 x \cos^5 x,\ y = \sin^6 x \cos x - 2\sin^8 x \cos x + \sin^{10} x \cos x$$

Then we use the built-in Check Equality to verify that

$$\sin^6 x \cos^5 x = \sin^6 x \cos x - 2\sin^8 x \cos x + \sin^{10} x \cos x$$

With the insertion point in the expression $\sin^6 x \cos^5 x$, choose Plot 2D + Rectangular. Select and drag $\sin^6 x \cos x - 2\sin^8 x \cos x + \sin^{10} x \cos x$ to the frame. With the insertion point in the equation

$$\sin^6 x \cos x - 2\sin^8 x \cos x + \sin^{10} x \cos x = \sin^6 x \cos^5 x$$

choose Check Equality

We can also compare our answer

$$\frac{1}{7}\sin^7 x - \frac{2}{9}\sin^9 x + \frac{1}{11}\sin^{11} x + C$$

with the machine answer

$$\begin{aligned}\int \sin^6 x \cos^5 x dx &= -\frac{1}{11}\sin^5 x \cos^6 x - \frac{5}{99}\sin^3 x \cos^6 x \\ &\quad -\frac{5}{231}\sin x \cos^6 x + \frac{1}{231}\cos^4 x \sin x \\ &\quad +\frac{4}{693}\cos^2 x \sin x + \frac{8}{693}\sin x\end{aligned}$$

With the insertion point in the integral $\int \sin^6 x \cos^5 x dx$, choose Evaluate.

Check Equality says the two answers are equal. The graphs both look like

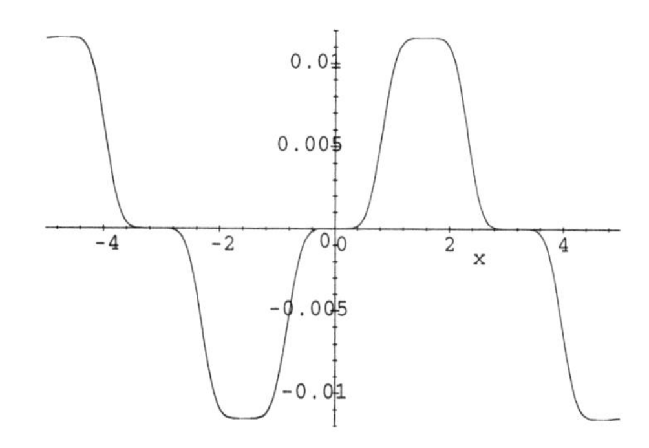

$y = \frac{1}{7}\sin^7 x - \frac{2}{9}\sin^9 x + \frac{1}{11}\sin^{11} x$

With the insertion point in the expression $\frac{1}{7}\sin^7 x - \frac{2}{9}\sin^9 x + \frac{1}{11}\sin^{11} x$, choose Plot 2D + Rectangular. Select and drag $\int \sin^6 x \cos^5 x dx$ to the plot. With the insertion point in the equation

$$\int \sin^6 x \cos^5 x dx = \frac{1}{7}\sin^7 x - \frac{2}{9}\sin^9 x + \frac{1}{11}\sin^{11} x$$

choose Check Equality.

Activities

1. Evaluate $\int \sin^5 x\, dx$ by using the substitution $u = \cos x$. Justify the crucial substitution step. Check your answer by differentiation and by direct evaluation.

2. Evaluate $\int \sin^4 x\, dx$ by using the half-angle formulas for $\sin^2 u$ and $\cos^2 u$. Check your answer by differentiation and by direct evaluation.

3. Discuss a general strategy for evaluating $\int \sin^m x \cos^n x dx$, based on the integers m and n being even or odd.

7.3 Trigonometric Substitution

The identities

$$\begin{aligned} \sin^2\theta + \cos^2\theta &= 1 \\ 1 + \tan^2\theta &= \sec^2\theta \end{aligned}$$

written in the forms

$$\begin{aligned} 1 - \sin^2\theta &= \cos^2\theta \\ 1 + \tan^2\theta &= \sec^2\theta \\ \sec^2\theta - 1 &= \tan^2\theta \end{aligned}$$

can be used to simplify many integrals involving the expressions

$$\begin{aligned} a^2 - x^2 \\ a^2 + x^2 \\ x^2 - a^2 \end{aligned}$$

From Algebra to Trig

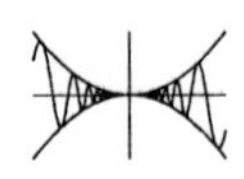

Evaluate $\displaystyle\int \frac{x}{x^4+9}\, dx$ by using the trigonometric substitution $x^2 = 3\tan\theta$.

Jeremy's Solution

Think of $x^4 + 9$ as $\left(x^2\right)^2 + 3^2$, which suggests the substitution $x^2 = 3\tan\theta$. This leads to the sequence of equalities:

$$\begin{aligned} \int \frac{x}{x^4+9}\, dx &= \int \frac{1+\tan^2\theta}{6+6\tan^2\theta}\, d\theta \\ &= \frac{1}{6}\int d\theta \end{aligned}$$

$$= \frac{1}{6}\theta + C$$
$$= \frac{1}{6}\arctan\frac{x^2}{3} + C$$

With the insertion point in the integral $\int \frac{x}{x^4+9}dx$, choose Calculus + Change Variable (Substitution: $x^2 = 3\tan\theta$). Type = and copy the result to the right of the = sign. Select the expression $\frac{1+\tan^2\theta}{6+6\tan^2\theta}$ and, with the CTRL/COMMAND key pressed down, choose Simplify (in-place replacement). With the insertion point in the expression $\frac{1}{6}\int d\theta$, choose Evaluate, and add $+C$. With the insertion point in the expression $\left[\frac{1}{6}\theta + C\right]_{\theta=\arctan(x^2/3)}$, choose Evaluate.

Checking, we get

$$\frac{d}{dx}\left(\frac{1}{6}\arctan\frac{x^2}{3} + C\right) = \frac{1}{9}\frac{x}{1+\frac{1}{9}x^4} = \frac{x}{x^4+9}$$

As an additional check, direct evaluation produces

$$\int \frac{x}{x^4+9}dx = \frac{1}{6}\arctan\frac{1}{3}x^2$$

With the insertion point in the expression $\frac{d}{dx}\left(\frac{1}{6}\arctan\frac{x^2}{3} + C\right)$, choose Evaluate. With the insertion point in the expression $\frac{1}{9}\frac{x}{1+\frac{1}{9}x^4}$, choose Simplify. With the insertion point in the integral $\int \frac{x}{x^4+9}dx$, choose Evaluate.

Activities

1. Evaluate the integral $\int \frac{dx}{x^2\sqrt{16-x^2}}$ using an appropriate trigonometric substitution. Verify your solution by differentiation and by direct evaluation of the integral.

2. Evaluate the integral $\int \frac{\sqrt{x^2-9}}{x}dx$ using an appropriate trigonometric substitution. Verify your solution by differentiation and by direct evaluation of the integral.

3. Evaluate the integral $\int \frac{x^3}{\sqrt{9x^2+49}}dx$ using an appropriate trigonometric substitution. Verify your solution by differentiation and by direct evaluation of the integral.

4. Discuss a general strategy for using trigonometric substitution, based on the identities $\sin^2\theta + \cos^2\theta = 1$ and $\tan^2\theta + 1 = \sec^2\theta$.

7.4 Integration of Rational Functions by Partial Fractions

A *rational function* is a function of the form $\frac{f(x)}{g(x)}$, where f and g are polynomial functions. Rational functions can be rewritten as "partial fractions." This is the name of a technique for writing a rational function as a sum of simpler rational functions.

▶ **To express a rational function $\frac{f(x)}{g(x)}$ as partial fractions**

- With the insertion point in a rational expression $\frac{f(x)}{g(x)}$, choose Polynomials + Partial Fractions.

  This is illustrated in the following example.

▶ Polynomials + Partial Fractions

$$\frac{5x^6 - 3x^2 + 5x - 3}{(x^2+1)(2x-1)^2(x+3)} = \frac{5}{4}x - \frac{5}{2} - \frac{15}{56(2x-1)^2} + \frac{943}{1960(2x-1)} + \frac{360}{49(x+3)} - \frac{1}{5}\frac{1+2x}{x^2+1}$$

A polynomial is *irreducible* if it cannot be written as a product of polynomials of smaller degree. The following theorem is useful for determining the possible types of irreducible factors of a polynomial.

> **Theorem** *Every polynomial with real coefficients can be factored as a product of linear and irreducible quadratic polynomials with real coefficients.*

This is an example of an "existence theorem" in mathematics. It provides no clue about how to actually factor polynomials. Indeed, this can be a very difficult problem, even for a computer algebra system.

However, we can recognize an irreducible factor when we find one. Every linear polynomial $ax + b$ is necessarily irreducible. (Why?) An irreducible quadratic polynomial can be characterized by the fact that its graph lies entirely above, or entirely below, the x-axis. (Why?)

Writing Rational Functions as Sums

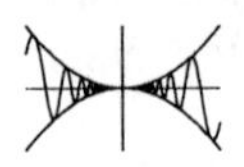

Explore the partial fraction format by converting rational functions such as the following to partial fractions. Give a general rule that describes what a partial fraction term can be.

$$\frac{x^5}{(x+2)^4}$$

$$\frac{x^7 + x + 2}{(x^2+3)^3}$$

$$\frac{4x^5 - 3x^2 + x - 4}{(x^2-1)^2}$$

The following steps provide a method for finding a partial fraction decomposition for a rational function. Explain the rationale for each step. Illustrate the method by using it to find the partial fraction decomposition for

$$\frac{2x^8 + 11x^7 + x^6 - x^5 - 13x^4 - 13x^3 - 4x^2 + 7x - 22}{2x^5 + x^4 - 5x^2 - 4x - 3}$$

1. Choose Polynomials + Divide to rewrite the partial fraction as a polynomial plus a rational function with a numerator that has smaller degree than the denominator.

2. Do an in-place replacement by selecting the denominator and, with the CTRL/COMMAND key held down, choosing Factor.

3. Form a sum of all possible terms of the forms
$$\frac{m}{(ax+b)^k}$$
and
$$\frac{mx+n}{(ax^2+bx+c)^k}$$
using a set of different unassigned letters in place of the m and n in the preceding expressions.

4. Choose Simplify to rewrite the sum as a rational function.

5. Rewrite the numerator as a polynomial in x by applying Polynomials + Collect (Variable: x).

6. Choose Insert + Display and enter into the input boxes a collection of equations determined by equating coefficients.

7. Choose Solve + Exact to determine the unassigned letters.

Verify your answer by applying Polynomials + Partial Fractions to the original rational function.

Joan's Solution

Expanding as partial fractions, we get

$$\begin{aligned}
\frac{x^5}{(x+2)^4} &= x - 8 - \frac{32}{(x+2)^4} + \frac{80}{(x+2)^3} - \frac{80}{(x+2)^2} + \frac{40}{x+2} \\
\frac{x^7+x+2}{(x^2+3)^3} &= x + \frac{-26x+2}{(x^2+3)^3} + 27\frac{x}{(x^2+3)^2} - 9\frac{x}{x^2+3} \\
\frac{4x^5 - 3x^2 + x - 4}{(x^2-1)^2} &= 4x - \frac{1}{2(x-1)^2} + \frac{17}{4(x-1)} - \frac{3}{(x+1)^2} + \frac{15}{4(x+1)}
\end{aligned}$$

With the insertion point in the expression $\frac{x^5}{(x+2)^4}$, choose Polynomials + Partial Fractions. Repeat for each of the expressions $\frac{x^7+x+2}{(x^2+3)^3}$ and $\frac{4x^5-3x^2+x-4}{(x^2-1)^2}$.

A general rule appears to be that if $(ax+b)^n$ appears as a factor in the denominator, there will be a possible term of the form

$$\frac{m}{(ax+b)^k}$$

for each integer k in the range $1 \le k \le n$. If a power $\left(ax^2+bx+c\right)^n$ of an irreducible quadratic ax^2+bx+c appears as a factor in the denominator, there will be a possible term of the form

$$\frac{mx+n}{(ax^2+bx+c)^k}$$

for each integer k in the range $1 \le k \le n$. If the degree of the numerator is greater than or equal to the degree of the denominator, then a polynomial will also appear as a term in the partial fraction decomposition.

Long division gives

$$\frac{2x^8+11x^7+x^6-x^5-13x^4-13x^3-4x^2+7x-22}{2x^5+x^4-5x^2-4x-3} =$$
$$x^3+5x^2-2x+3+\frac{-13+13x+18x^2+13x^4}{2x^5+x^4-5x^2-4x-3}$$

The denominator factors as

$$2x^5+x^4-5x^2-4x-3=(2x-3)\left(x^2+x+1\right)^2$$

and hence the general partial fraction expansion for

$$\frac{-13+13x+18x^2+13x^4}{2x^5+x^4-5x^2-4x-3}$$

is of the form

$$\frac{a}{2x-3}+\frac{mx+n}{x^2+x+1}+\frac{rx+s}{\left(x^2+x+1\right)^2}$$

for some unknown coefficients a, m, n, r, and s.

With the insertion point in the expression $\frac{2x^8+11x^7+x^6-x^5-13x^4-13x^3-4x^2+7x-22}{2x^5+x^4-5x^2-4x-3}$, choose Polynomials + Divide. Copy the denominator $2x^5+x^4-5x^2-4x-3$, and with the insertion point in this expression, choose Factor.

Adding the partial fractions and equating coefficients in the numerator, we get the system

$$\begin{aligned}
2m+a &= 13\\
2n-m+2a &= 0\\
-n-m+3a+2r &= 18\\
-3r+2s-n-3m+2a &= 13\\
a-3s-3n &= -13
\end{aligned}$$

The solution is given by

$$m=4,\quad r=2,\quad s=9,\quad n=-3,\quad a=5$$

which indicates that

$$\frac{-13+13x+18x^2+13x^4}{2x^5+x^4-5x^2-4x-3}=\frac{5}{(2x-3)}+\frac{4x-3}{(x^2+x+1)}+\frac{2x+9}{\left(x^2+x+1\right)^2}$$

 With the insertion point in the expression $\frac{a}{2x-3}+\frac{mx+n}{x^2+x+1}+\frac{rx+s}{(x^2+x+1)^2}$, choose Simplify. Select the numerator

$$ax^4+2ax^3+3ax^2+2ax+a+2mx^4-mx^3-mx^2-3mx$$
$$+2nx^3-nx^2-nx-3n+2rx^2-3rx+2sx-3s$$

with the mouse and choose Polynomials + Collect (Variable: x). Enter the system of equations obtained from equating coefficients and, with the insertion point in the system of equations, choose Solve + Exact.

As a check, applying Polynomials + Partial Fractions directly, we get the result

$$\frac{2x^8+11x^7+x^6-x^5-13x^4-13x^3-4x^2+7x-22}{2x^5+x^4-5x^2-4x-3}=$$
$$x^3+5x^2-2x+3+\frac{5}{2x-3}+\frac{4x-3}{x^2+x+1}+\frac{2x+9}{(x^2+x+1)^2}$$

With the insertion point in the rational expression

$$\frac{2x^8+11x^7+x^6-x^5-13x^4-13x^3-4x^2+7x-22}{2x^5+x^4-5x^2-4x-3}$$

choose Polynomials + Partial Fractions.

Activities

1. Apply the techniques you developed in the introductory activity to determine a partial fraction expansion of
$$\frac{3+2x-12x^2-315x^4+88x^3+741x^5-1152x^6+1168x^7-704x^8+192x^9}{64x^6-192x^5+240x^4-160x^3+60x^2-12x+1}$$
Verify your answer by direction application of Polynomials + Partial Fractions.

2. Calculate the integral
$$\int_1^2 \frac{x^3-2x^2-x+3}{81x^4-216x^3+216x^2-96x+16}dx$$
by finding the partial fraction expansion of
$$\frac{x^3-2x^2-x+3}{81x^4-216x^3+216x^2-96x+16}$$
and integrating term by term. Verify your answer by evaluating the integral using Evaluate.

7.5 Rationalizing Substitutions

The use of substitutions to change functions into rational functions is based on the premise that polynomials and rational functions are easier to deal with than expressions that include radicals.

There are basically two ways to eliminate a radical such as $\sqrt{x}$. The first is to use the substitution $u = \sqrt{x}$. The second is to let $u^2 = x$. On the surface these appear to be the same. However, they sometimes lead to different expressions.

Eliminate the Radicals!

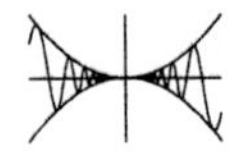

Evaluate the integral

$$\int_0^8 \frac{1}{1+\sqrt[3]{x}}\,dx$$

by using a rationalizing substitution. Interpret the original and rewritten integrals as areas of regions and compare the regions visually. Use the graphs to give rough estimates for the integrals. Explain why it is reasonable that the two areas are equal.

Elbert's Solution

The substitution $u^3 = x$ yields

$$\int_0^8 \frac{1}{1+\sqrt[3]{x}}\,dx = \int_0^2 \frac{3u^2}{1+u}\,du$$

With the insertion point in the integral $\int_0^8 \frac{1}{1+\sqrt[3]{x}}dx$, choose Calculus + Change Variable (Substitution: $u^3 = x$).

Because both integrands are positive, we can interpret each integral as the area of a region. The region on the left-hand side is shown in the following figure:

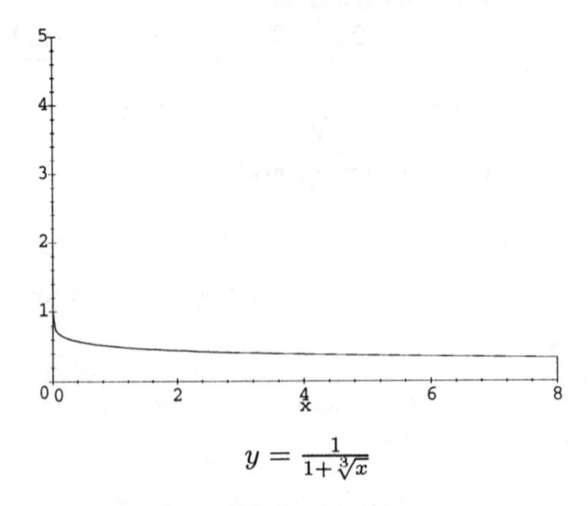

$y = \frac{1}{1+\sqrt[3]{x}}$

With the insertion point in the expression $\frac{1}{1+\sqrt[3]{x}}$, choose Plot 2D + Rectangular. Choose Edit + Properties, Plot Components page. Change Domain Interval to $0 \le x \le 8$. Choose View page, and change View Intervals to $0 \le x \le 8$ and $0 \le y \le 5$. Choose OK.

From the appearance of the graph, we see that the area can be approximated by a trapezoid of height 8 with bases .6 and .3 to yield an approximate area of $.5 \times 8(.6 + .3) = 3.6$.

We plot the second region using the same view rectangle.

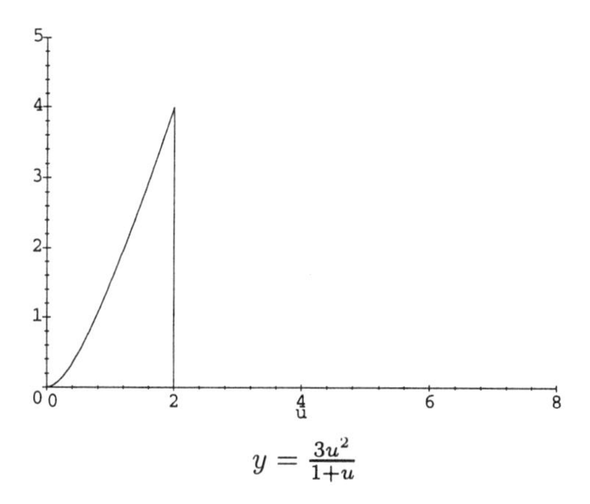

$y = \frac{3u^2}{1+u}$

With the insertion point in the expression $\frac{3}{1+u}u^2$, choose Plot 2D + Rectangular. Choose Edit + Properties, Plot Components page. Change Domain Interval to $0 \le x \le 2$. Choose View page. Change View Intervals to $0 \le x \le 8$ and $0 \le y \le 5$. Choose OK. To add the vertical line, select the list of points $(2, 0, 2, 4)$ and drag it to the plot.

This region is roughly triangular, with base 2 and height 4, which area equals $1/2 \times 2 \times 4 = 4$. This somewhat overestimates the area because the curve appears to be concave upward.

Direct evaluation produces the two definite integrals

$$\int_0^8 \frac{1}{1+\sqrt[3]{x}}dx = 3\ln 3 = 3.2958$$

$$\int_0^2 \frac{3u^2}{1+u}du = 3\ln 3 = 3.2958$$

With the insertion point in the integral $\int_0^8 \frac{1}{1+\sqrt[3]{x}}dx$, choose Evaluate and then choose Evaluate Numerically. Repeat for the integral $\int_0^2 \frac{3u^2}{1+u}du$.

Our visual estimates were a bit high but at least reasonable. The approximating figures are shown here:

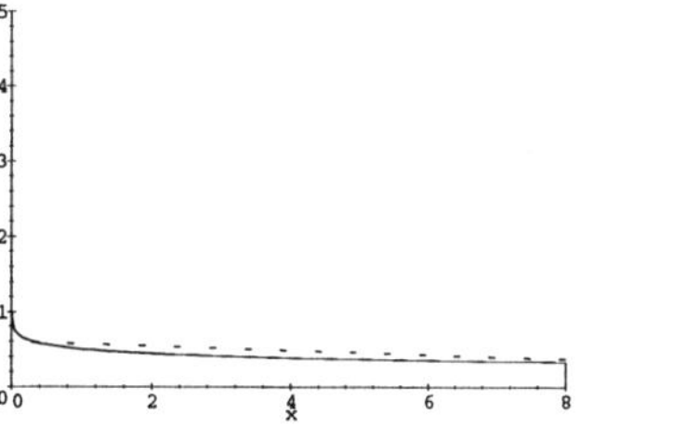

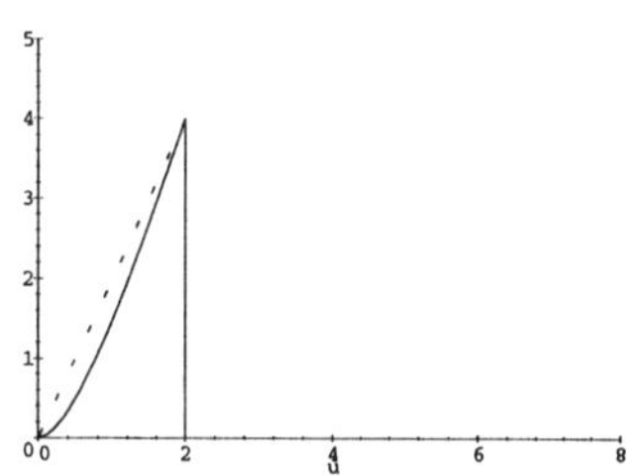

Activities

1. Evaluate the integral
$$\int_0^7 x\sqrt[3]{x+1}dx$$
by using a rationalizing substitution. Interpret the original and rewritten integrals as areas of regions and compare the regions visually. Use the graphs to give rough estimates for the integrals. Explain why it is reasonable that the two areas are equal. Compare your estimates with what you get by direct evaluation of the integral.

2. Evaluate the integral
$$\int_0^{25} \frac{dx}{\sqrt{4+\sqrt{x}}}$$
by using the substitution $u^2 = x$, followed by the substitution $v^2 = 4 + u$. Interpret the three integrals as areas of regions and compare the regions visually. Use the graphs to give rough estimates for the integrals. Explain why it is reasonable that the three areas are equal. Compare your estimates with what you get by direct evaluation of one of these integrals.

3. Use the substitution $u^6 = x$ to rewrite the indefinite integral
$$\int \frac{dx}{\sqrt{x}+\sqrt[3]{x}}$$
Evaluate the rewritten integral, then use $x = u^6$ to find an antiderivative of
$$\frac{1}{\sqrt{x}+\sqrt[3]{x}}$$
Check your answer by differentiation.

4. Discuss a strategy for evaluating an integral that includes several expressions of the form $\sqrt[n]{x}$ for various choices of n. Test your strategy on the integral
$$\int \frac{dx}{\sqrt[3]{x}+\sqrt[4]{x}+\sqrt[6]{x}}$$

7.6 Strategy for Integration

To develop a strategy for integration, you should first memorize basic formulas.

Activity

Compile the shortest list of integration formulas that you think would be useful for you to remember. Give examples showing how you would use these formulas and the techniques that have been introduced in the previous sections to do some interesting integrals.

7.7 Using Tables of Integrals and Computer Algebra Systems

There are a few things to keep in mind when using a symbolic computation system to find integrals. First, the constant of integration is omitted when an indefinite integral is evaluated. One way around this is to solve the differential equation $\frac{dy}{dx} = f(x)$ rather than to evaluate the integral $\int f(x)dx$. Solution of the differential equation will provide a constant, usually "$+ C_1$." A second solution is simply to type "$+ C$" after the computer-supplied answer. A third alternative is simply to ignore the constant but remember that such a constant of integration can and must be added when needed.

A computer algebra system can evaluate an amazing number of integrals, but often in a very messy form. For example, consider the integral $\int x\left(x^2+1\right)^{14} dx$. A computer-supplied answer will probably have 15 terms, all even powers of x. The substitution $u = x^2 + 1$ allows the integral to be rewritten as

$$\int x\left(x^2+1\right)^{14} dx = \int \frac{1}{2}u^{14}\,du$$

which leads to the solution

$$\int x\left(x^2+1\right)^{14} dx = \int \frac{1}{2}u^{14}\,du = \frac{1}{30}u^{15} = \frac{1}{30}\left(x^2+1\right)^{15} + C$$

(with a user-supplied "$+ C$").

When a computer algebra system cannot evaluate a certain integral, it simply returns the same integral. (If only students could get by with this on tests. . . .) For example, Evaluate produces

$$\int \frac{x^4 dx}{\sqrt{x^{10}-2}} = \int \frac{x^4}{\sqrt{(x^{10}-2)}}\,dx$$

People are often better than computers at pattern recognition. In this case, the substitution $u^2 = x^{10} - 2$ leads to

$$\begin{aligned}
\int \frac{x^4 dx}{\sqrt{x^{10}-2}} &= \int \frac{1}{5\sqrt{(u^2+2)}}\,du \\
&= \frac{1}{5}\operatorname{arcsinh}\left(\frac{1}{2}\sqrt{2}u\right) \\
&= \frac{1}{5}\operatorname{arcsinh}\left(\frac{1}{2}\sqrt{2}\sqrt{x^{10}-2}\right) + C
\end{aligned}$$

This can be rather easily checked using

$$\begin{aligned}
\frac{d}{dx}\left(\frac{1}{5}\operatorname{arcsinh}\left(\frac{1}{2}\sqrt{2}\sqrt{x^{10}-2}\right) + C\right) &= \frac{1}{\sqrt{(x^{10}-2)}}\frac{x^9}{\sqrt{(x^{10})}} \\
&= \frac{x^4}{\sqrt{(x^{10}-2)}}
\end{aligned}$$

(A friendly Simplify provided the last result.)

The bottom line is that people working alone can do a lot of mathematics and are good at pattern recognition, but they tend to be slow and make lots of errors. Computers working alone are fairly useless. However, the human/machine combination is much more powerful than the sum of its parts. With each partner doing what he or she or it does best, the combination can tackle mathematical problems that are subtle, difficult, complex, interesting, and important.

Activities

Here are a few integrals that foiled a computer, or for which the answer was returned in an unnecessarily complicated form. Work with your computer to provide reasonable answers. Check your answer by differentiating.

1. $\int_0^{25} \frac{dx}{\sqrt{4+\sqrt{x}}}$

2. $\int \sqrt{2+3\cos x}\tan x dx$

3. $\int e^{\sin x} \sin 2x dx$

4. $\int \left(4x^3+7\right)\left(x^4+7x\right)^{10} dx$

7.8 Approximate Integration

When definite integration was first introduced, you used Riemann sums based on the left rule, right rule, and middle rule to estimate the values of these definite integrals. Then a miracle occurred. The Fundamental Theorem of Calculus gave you a tool that enabled you to evaluate integrals without pain.

Well, maybe some pain. Several methods of integration were introduced that enable you to evaluate many types of integrals. However, some integrals cannot be evaluated symbolically using any of the techniques of integration. For these integrals, some type of numerical integration must be used. In these activities, you will learn additional numerical techniques and will look closely at the associated errors.

Streams and Virtual Streams

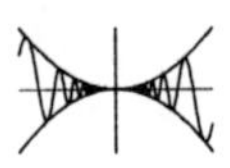

Water is a precious and limited resource in Colorado. To track the quantity and source of spring runoff, it is necessary to make frequent estimates of stream flow at many sites. Consider the flow in a particular stream that is roughly 10 feet wide and 2 feet deep. Surface velocity is estimated at 1-foot intervals across the stream by dropping a twig into the stream and timing how long it takes to go a fixed distance. Field measurements provided the following data. Here x denotes the location across the stream in 1-foot increments, b indicates the depth in feet at x, and a lists the surface velocity in feet per second at x.

x	0	1	2	3	4	5	6	7	8	9	10
b	0	.3	.5	.9	1.6	2.1	2.2	1.9	1.3	.4	0
a	0	.5	.7	1.1	1.7	2.6	2.7	2.4	1.8	.6	0

Plot a picture of a cross section of the stream at this location. Estimate the flow through a virtual stream of width 1 foot by assuming that velocity is a linear function of depth, and that velocity is maximal at the surface and zero at the bottom of the stream. Use Simpson's rule to estimate the total flow of the stream in cubic feet per second.

Linda's Solution

Here is a picture of the cross section of the stream at this particular location. This is based on the following measurements of depth:

x	0	1	2	3	4	5	6	7	8	9	10
b	0	$-.3$	$-.5$	$-.9$	-1.6	-2.1	-2.2	-1.9	-1.3	$-.4$	0

A virtual stream is indicated by the dotted lines for the part of the stream between $x = 3.5$ and $x = 4.5$.

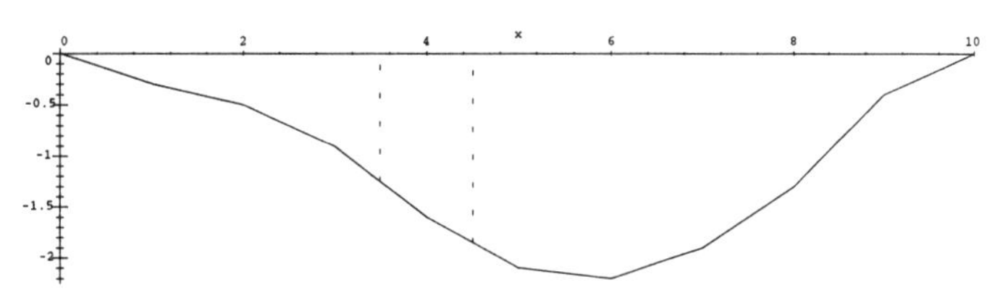

With the insertion point in the list of points

$$(0, 0, 1, -.3, 2, -.5, 3, -.9, 4, -1.6, 5, -2.1, 6, -2.2, 7, -1.9, 8, -1.3, 9, -.4, 10, 0)$$

choose Plot 2D + Rectangular. Select and drag to the plot the list of points $\left(4.5, 0, 4.5, \frac{-1.6-2.1}{2}\right)$ and the list of points $\left(3.5, 0, 3.5, \frac{-.9-1.6}{2}\right)$. Choose Edit + Properties, Plot Components page. Change Domain to $0 < x < 10$. For Item Number 1 and Item Number 2, change Line Style to Dots. Choose OK.

Assume that the velocity at depth y $(0 \le y \le b)$ along a vertical line is given by $v(y) = (b - y)a$, where b is the depth of the channel and a is the velocity at the surface. The average velocity is thus $\frac{1}{b}\int_0^b (b - y)a\,dy = \frac{1}{2}ab$. The flow through a virtual stream of width 1 foot (say, between $x - \frac{1}{2}$ and $x + \frac{1}{2}$) is thus given approximately by $b\left(\frac{1}{2}ab\right) = \frac{1}{2}ab^2$.

With the insertion point in the integral $\frac{1}{b}\int_0^b (b-y)a\,dy$, choose Evaluate. With the insertion point in the expression $b\left(\frac{1}{2}ab\right)$, choose Evaluate.

The flows through the virtual streams of width 1 are given in the following table, where x is the location, $f = \frac{1}{2}ab^2$ is the flow, and a is velocity in feet per second:

x	0	1	2	3	4	5
f	0	.0225	.0875	1.408	2.176	5.733
a	0	.5	.7	1.1	1.7	2.6

x	6	7	8	9	10
f	6.534	4.332	1.521	.048	0
a	2.7	2.4	1.8	.6	0

To find the values for the flow, enter the numbers into a display.

$$\frac{1}{2}(0)^2 0$$
$$\frac{1}{2}(.3)^2 .5$$
$$\frac{1}{2}(.5)^2 .7$$
$$\frac{1}{2}(.9)^2 1.1$$
$$\frac{1}{2}(1.6)^2 1.7$$
$$\frac{1}{2}(2.1)^2 2.6$$
$$\frac{1}{2}(2.2)^2 2.7$$
$$\frac{1}{2}(1.9)^2 2.4$$
$$\frac{1}{2}(1.3)^2 1.8$$
$$\frac{1}{2}(.4)^2 .6$$
$$\frac{1}{2}(0)^2 0$$

Then place the insertion point in the display, and choose Evaluate.

Using Simpson's rule, the total flow is given approximately by

$$\tfrac{1}{3}(0 + 4\,(.0225) + 2\,(.0875) + 4\,(1.408) + 2\,(2.176) + 4\,(5.733) + 2\,(6.534) + 4\,(4.332) + 2\,(1.521) + 4\,(.048) + 0) = 22.27 \text{ ft}^3\text{/sec}$$

For this sum (and the next one), place the insertion point in the sum and choose Evaluate. Add the appropriate units.

Is this a reasonable number? As a check, we calculate the cross-sectional area of the stream using Simpson's rule:

$$\tfrac{1}{3}(0 + 4(.3) + 2(.5) + 4(.9) + 2(1.6) + 4(2.1) + 2(2.2) + 4(1.9) + 2(1.3) + 4(.4) + 0) = 11.2 \text{ ft}^2$$

Then we guess an average velocity by looking at the velocities. It would appear that the average velocity is somewhere around 2 feet/second. This gives the rough estimate of $2(11.2) = 22.4$ ft^3/sec. This is much closer to the original estimate than we might expect.

You Call This Normal?

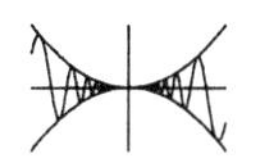

In statistics, the *normal distribution* is given by the equation

$$p(x) = \frac{1}{\sqrt{2\pi}} \exp\left(-\frac{x^2}{2}\right)$$

where we write $\exp(\ldots)$ instead of $e^{(\ldots)}$ in order to avoid writing exponents on top of exponents. Plot a graph of p and discuss its symmetry properties. Evaluate the improper integral

$$\int_{-\infty}^{\infty} \frac{1}{\sqrt{2\pi}} \exp\left(-\frac{x^2}{2}\right) dx$$

What happens if you attempt to evaluate the integral $\int_{-\infty}^{2} p(x)dx$? What about numerical evaluation? Define f by $f(x) = \int_{-\infty}^{x} p(t)dt$. The number $f(x)$ can be interpreted as the probability that a dart thrown at random that lands above the x-axis and below the graph of $y = p(x)$ lies to the left of the number x. Experiment with the simpler function $g(x) = \int_0^x e^{-t^2} dt$. Plot a graph of $y = e^{-x^2}$ and $y = g(x)$ and explain why one graph appears to be the graph of the derivative of the other. Attach horizontal asymptotes by evaluating $\int_0^{\infty} e^{-t^2} dt$ and $\int_0^{-\infty} e^{-t^2} dt$.

Jerry's Solution

The graph of $y = \frac{1}{\sqrt{2\pi}} \exp\left(-\frac{x^2}{2}\right)$ is symmetric with respect to the y-axis.

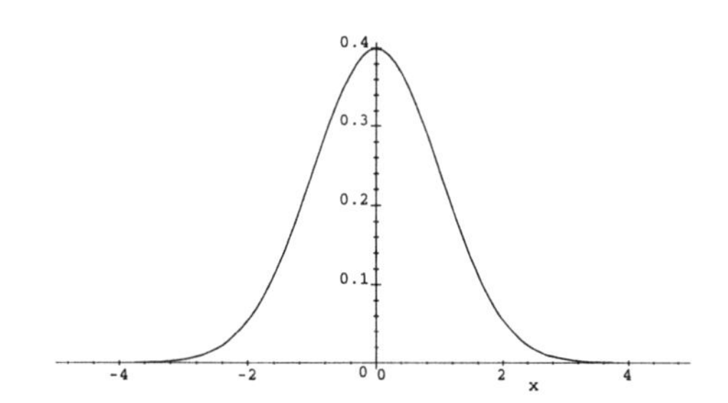

With the insertion point in the expression $\frac{1}{\sqrt{2\pi}} \exp\left(-\frac{x^2}{2}\right)$, choose Plot 2D + Rectangular.

The area under the curve is given by the improper integral

$$\int_{-\infty}^{\infty} \frac{1}{\sqrt{2\pi}} \exp\left(-\frac{x^2}{2}\right) dx$$

which evaluates to 1. However, if we attempt to evaluate the integral

$$\int_{-\infty}^{2} \frac{1}{\sqrt{2\pi}} \exp\left(-\frac{x^2}{2}\right) dx$$

we simply get the same integral back again as an answer. This is the underlying computer algebra system's way of saying that it is unable to further simplify the integral.

With the insertion point in the integral $\int_{-\infty}^{\infty}\frac{1}{\sqrt{2\pi}}\exp\left(-\frac{x^2}{2}\right)dx$, choose Evaluate. With the insertion point in the integral $\int_{-\infty}^{2}\frac{1}{\sqrt{2\pi}}\exp\left(-\frac{x^2}{2}\right)dx$, choose Evaluate.

Numerical evaluation yields

$$\int_{-\infty}^{2}\frac{1}{\sqrt{2\pi}}\exp\left(-\frac{x^2}{2}\right)dx = .97725$$

This means that approximately 97.7% of the area is to the left of the vertical line $x = 2$.

With the insertion point in the integral $\int_{-\infty}^{2}\frac{1}{\sqrt{2\pi}}\exp\left(-\frac{x^2}{2}\right)dx$, choose Evaluate Numerically .

Consider the functions $f(x) = e^{-x^2}$ and $g(x) = \int_0^x e^{-t^2}dt$. Graphs of these two functions follow, with $f(x) = e^{-x^2}$ drawn with a fat pen and $g(x) = \int_0^x e^{-t^2}dt$ drawn with a regular pen. The dotted horizontal lines indicate horizontal asymptotes.

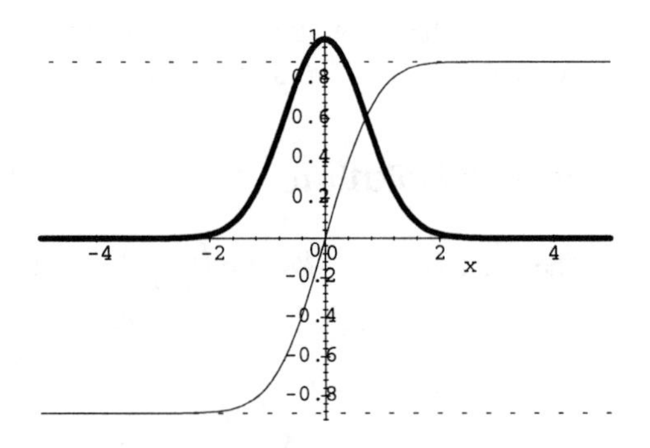

$f(x) = e^{-x^2}$ and $g(x) = \int_0^x e^{-t^2}dt$

Notice that the graph of g is always increasing, which is compatible with its derivative always being positive. The largest value of the derivative is at $x = 0$, which is where the graph of g is steepest. To get the horizontal asymptotes of g we evaluate the improper integrals

$$\int_0^{\infty} e^{-t^2}dt = \frac{1}{2}\sqrt{\pi}$$

$$\int_0^{-\infty} e^{-t^2}dt = -\frac{1}{2}\sqrt{\pi}$$

With the insertion point in the expression $\int_0^x e^{-t^2}dt$, choose Plot 2D + Rectangular. Select and drag to the plot e^{-x^2}. With the insertion point in the integral $\int_0^{\infty} e^{-t^2}dt$, choose Evaluate. Repeat for the integral $\int_0^{-\infty} e^{-t^2}dt$. Select and drag to the plot each of $\frac{1}{2}\sqrt{\pi}$ and $-\frac{1}{2}\sqrt{\pi}$. Choose Edit + Properties, Plot Components page. For Item Number 1, change Thickness to Medium. For Item Numbers 3 and 4, change Line Style to Dots. Choose OK.

Activities

1. Define the function $f(x) = ax^2 + bx + c$.

   a. Solve for a, b, and c so that $f(-h) = y_1$, $f(0) = y_2$, and $f(h) = y_3$.
   b. Define a, b, and c as given in part a.
   c. Evaluate $\int_{-h}^{h} f(x)dx$.
   d. Explain the connection between the formula in part c and Simpson's formula.

2. Let $f(x)$ be a function that has a continuous second derivative on the interval $(-h, h)$, and let $g(x)$ be the quadratic approximation to f at 0, so that $g(x) = f(0) + f'(0)x + \frac{f''(0)}{2}x^2$. Use $\int_{-h}^{h} g(x)dx$ as an approximation to $\int_{-h}^{h} f(x)dx$. Experiment with this method applied to the integral $\int_{-1}^{1} e^{-x^2} dx$. Plot the graphs of f and g on the interval $[-1, 1]$ and use the graphs to explain why the approximation is too large or too small.

7.9 Improper Integrals

A definite integral is *improper* when either the interval of integration is unbounded (Type 1), or the function values are unbounded on the interval of integration (Type 2). Improper integrals are defined in terms of limits and are called convergent or divergent depending on whether or not the limit converges.

A Very Wild Function

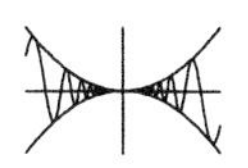

Study the convergence of the improper integral $\int_0^\infty \sin x^2 dx$. Draw graphs of $y = \sin x^2$ and $y = \int_0^x \sin t^2 dt$ together. Based on these graphs, explain why it is plausible that the integral $\int_0^\infty \sin t^2 dt$ converges. Construct a table of values that provides evidence about the value of $\int_0^\infty \sin x^2 dx$. Evaluate $\int_0^\infty \sin x^2 dx$ and relate this value to the figure.

Cameron's Solution

Let $g(x) = \sin x^2$ and define $f(x) = \int_0^x \sin t^2 dt$ so that f is an antiderivative of g. In the following figure the graph of f is drawn with a fat pen and the graph of g with a regular pen. The graph of g oscillates between -1 and 1, but the oscillations of f seem to damp out for large values of x. It appears that $\lim_{x\to\infty} f(x) \approx 0.6$.

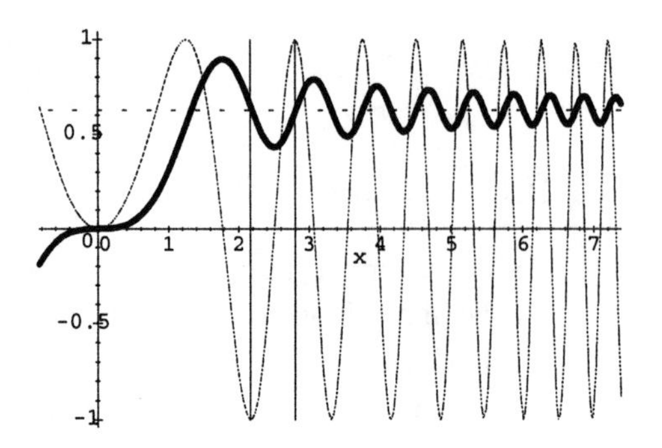

$g(x) = \sin x^2$ and $f(x) = \int_0^x \sin t^2 dt$

The following table provides some additional clues:

n	$x^2 = n\pi$	x	$f(x)$	$x^2 = \left(n + \frac{1}{2}\right)\pi$	x	$f(x)$
1	$x^2 = \pi$	1.7725	.89483	$x^2 = \frac{3}{2}\pi$	2.1708	.64835
2	$x^2 = 2\pi$	2.5066	.43041	$x^2 = \frac{5}{2}\pi$	2.8025	.61587
3	$x^2 = 3\pi$	3.07	.78826	$x^2 = \frac{7}{2}\pi$	3.316	.63328
4	$x^2 = 4\pi$	3.5449	.48625	$x^2 = \frac{9}{2}\pi$	3.7599	.62199
5	$x^2 = 5\pi$	3.9633	.75244	$x^2 = \frac{11}{2}\pi$	4.1568	.63007

The fourth column shows the f values at the high points and low points on the curve $y = f(x)$. The seventh column shows the f values at x-coordinates where the curve $y = g(x)$ has its highs and lows. The seventh column provides good approximations to the improper integral $\int_0^\infty \sin t^2 dt$.

With the insertion point in the expression $\int_0^x \sin t^2 dt$, choose Plot 2D + Rectangular. Select $\sin x^2$ and drag it to the plot. Select each of $(2.1708, -1, 2.1708, 1)$ and $(2.8025, -1, 2.8025, 1)$ and drag them to the plot. Choose Edit + Properties, Plot Components page. Change Domain to $-1 < x < 8$. For Item Number 1, change Thickness to Medium. Choose OK.

As a check, note that $\int_0^\infty \sin t^2 dt = \frac{1}{4}\sqrt{2}\sqrt{\pi} = .62666$.

With the insertion point in the integral $\int_0^\infty \sin t^2 dt$, choose Evaluate and then choose Evaluate Numerically. Select $.62666$ and drag it to the plot. Choose Edit + Properties, Plot Components page. For Item Number 5, change Line Style to Dots. Choose OK.

Activities

1. Show that the integral $\int_0^\infty \frac{dx}{(x^2+1)\sqrt{x}}$ is improper for two reasons (it is both of Type 1 and of Type 2). Write this integral as a sum of two integrals, where the first is improper of Type 1, and the second is improper of Type 2. Show that each of these integrals converges by comparison with known improper integrals. Use a rationalizing substitution to rewrite these integrals without radicals. Evaluate the original improper integral.

2. What is wrong with the following calculations?
$$\int_{-1}^{1} \frac{dx}{x} = [\ln|x|]_{-1}^{1} = \ln 1 - \ln 1 = 0$$

3. Show that $\int_{-1}^{1} \frac{dx}{\sqrt[3]{x}} = 0$. Justify each step.

8 FURTHER APPLICATIONS OF INTEGRATION

Integration is used to solve problems in many fields of study, including geometry, physics, chemistry, biology, and economics.

8.1 Differential Equations

One of the most important applications of integration lies in the solution of differential equations.

The Faulty Furnace

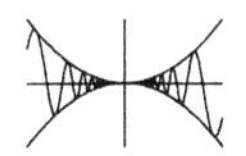

A faulty furnace releases 200 feet^3 per minute of an air/CO mix into air ducts that flow directly into the house. The concentration of CO being released is 25 parts per 100,000. Leaks in the house let 200 feet^3 of air escape per minute. The house has a volume of $12,000$ feet^3. Assuming the furnace runs continuously for one hour, what is the maximal concentration of CO inside the house? A concentration above 0.015% is considered dangerous. How long will it take before the concentration level becomes dangerous?

Elaine's Solution

This is an example of a mixture problem. The rate at which the volume of CO changes is the rate at which CO is being added minus the rate at which CO is being removed. Because the concentration of CO in the air ducts is 25 parts per 100,000 and 200 cubic feet of air/gas mixture leaks into the house each minute, it follows that the rate of change of CO being added to the house is $\frac{25}{100000} \times 200 = .05$ cubic feet per minute. Assuming the air/gas mixture in the house is well mixed, the volume of gas removed each minute is the volume v of CO multiplied by the fraction $\frac{2}{12000}$ of air that is removed from the house each minute. The volume of CO in the house initially is zero, so $v(0) = 0$. A differential equation that measures the rate at which the volume of CO in the house is changing is given by

$$\begin{aligned} \frac{dv}{dt} &= \frac{50}{1000} - \frac{200v}{12000} \\ v(0) &= 0 \end{aligned}$$

The volume of CO in the house after t minutes is given by

$$v(t) = 3 - 3e^{-\frac{1}{60}t}$$

Thus, the concentration of gas in the house is

$$\frac{v(t)}{12000} = \frac{1}{4000} - \frac{1}{4000}e^{-\frac{1}{60}t}$$

Choose Insert + Display. Type $\frac{dv}{dt} = \frac{50}{1000} - \frac{200v}{12000}$ in the input box. Press ENTER. Type $v(0) = 0$ in the input box. With the insertion point in the system of equations

$$\begin{aligned} \frac{dv}{dt} &= \frac{50}{1000} - \frac{200v}{12000} \\ v(0) &= 0 \end{aligned}$$

choose Solve ODE + Exact .

A graph of the concentration of CO is given in the following figure. It appears that a lethal concentration will be reached after about 50 or 60 minutes. To get a more precise estimate, solve the equation

$$\frac{v(t)}{12000} = .00015$$

to get approximately $t = 55$ minutes. Better get everyone out of the house.

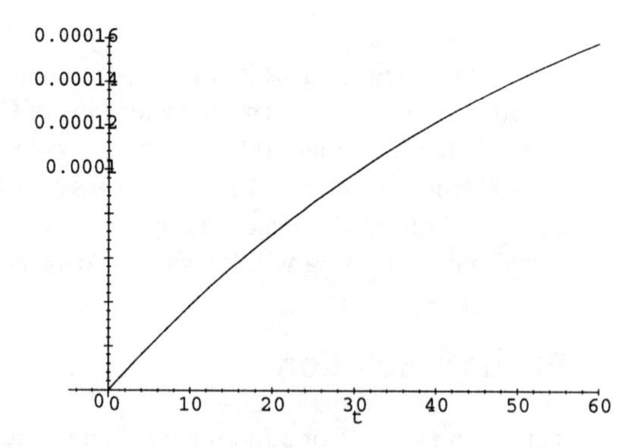

With the insertion point in the expression $\frac{300-300e^{-\frac{1}{6000}t}}{12000}$, choose Plot 2D + Rectangular. Choose Edit + Properties, Plot Components page, and change Domain Interval to $0 < x < 60$. Choose OK.

Activities

1. A lake has been contaminated with pesticides. Local and state laws have stopped an offending supplier from dumping waste into the lake, which holds 110,000,000,000

gallons of water/pesticide mixture. The pollution level is currently 10 parts per million. Clean water flows into (and out of) the lake at a rate of 600,000 gallon/hour. How long will it take for the pesticide level to be down to the official safe level of 1 part per million?

2. Assume that the pesticide in the previous problem has a half-life of 12 years. During each 12-year period, half of the remaining pesticide breaks down chemically into harmless ingredients. How does this affect the amount of time it will take for the lake to reach a safe level of the pesticide?

8.2 Arc Length

Can you find the distance along a curve?

The Length of a Hanging Cable

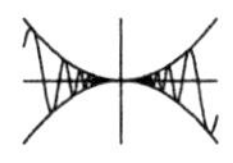

All of the internal forces are in equilibrium when a cable hangs freely. Consideration of these forces leads to the differential equation

$$a\frac{d^2y}{dx^2} = \sqrt{1 + \left(\frac{dy}{dx}\right)^2}$$

which must be satisfied by the equation of the curve formed by a hanging cable whose ends are supported from the same height. Verify that the solution is given by

$$y(x) = \left(\cosh\frac{1}{a}x\right) a\cosh C_1 - \left(\sinh\frac{1}{a}x\right) a\sinh C_1 + C_2$$

Assume a cable is hung between two 30-foot high poles that are 36 feet apart, and the lowest point of the cable is 24 feet off the ground. What is the total length of this section of cable?

Mahlon's Solution

The solution of the differential equation

$$a\frac{d^2y}{dx^2} = \sqrt{1 + \left(\frac{dy}{dx}\right)^2}$$

is given by

$$y(x) = \left(\cosh\frac{1}{a}x\right) a\cosh C_1 - \left(\sinh\frac{1}{a}x\right) a\sinh C_1 + C_2$$

N With the insertion point in the equation

$$a\frac{d^2y}{dx^2} = \sqrt{1 + \left(\frac{dy}{dx}\right)^2}$$

choose Solve ODE + Exact.

Assuming the cable is hung from two 30-foot poles 36 feet apart and that the low point is at $(0, 24)$, we can solve the following system of equations

$$\begin{aligned} y(0) &= 24 \\ y(18) &= 30 \\ y'(0) &= 0 \end{aligned}$$

which yields

$$\begin{aligned} a &= 27.946 \\ C_2 &= -3.9464 \\ C_1 &= 0 \end{aligned}$$

 With the insertion point in the equation

$$y(x) = \left(\cosh \frac{1}{a} x\right) a \cosh C_1 - \left(\sinh \frac{1}{a} x\right) a \sinh C_1 + C_2$$

choose Define + New Definition. Choose Insert + Display and type the equation $y(0) = 24$ in the input box. Press ENTER and type $y(18) = 30$. Press ENTER and type $y'(0) = 0$. With the insertion point in the system of equations

$$\begin{aligned} y(0) &= 24 \\ y(18) &= 30 \\ y'(0) &= 0 \end{aligned}$$

choose Solve + Numeric. Select each of $a = 27.946$, $C_2 = -3.9464$, and $C_1 = 0$ with the mouse, and choose Define + New Definition.

As a visual check, we plot the graph of $y(x)$.

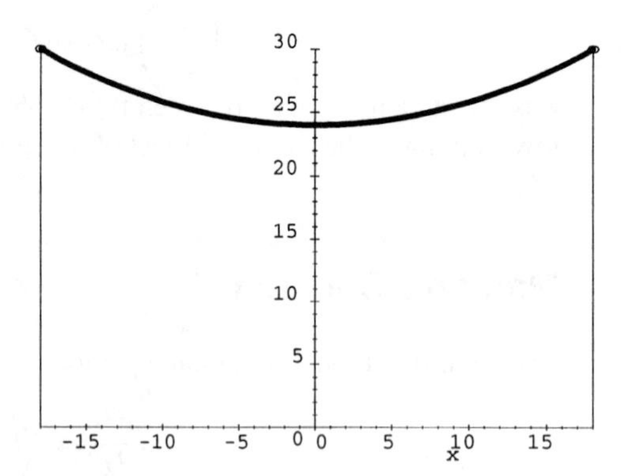

 With the insertion point in the expression $y(x)$, choose Plot 2D + Rectangular.

The length of the cable is given by

$$\int_{-18}^{18} \sqrt{1 + (y'(x))^2} dx = 38.541$$

which is roughly $2\frac{1}{2}$ feet longer than the distance between the two poles.

With the insertion point in the integral $\int_{-18}^{18} \sqrt{1 + (y'(x))^2}\, dx$, choose Evaluate Numerically.

To make sure that this estimate is reasonable, consider twice the distance between the low point $(0, 24)$ and the top of one of the poles $(18, 30)$. This is given by

$$2\sqrt{18^2 + (30 - 24)^2} = 37.947$$

Thus the cable is roughly half a foot longer than the sum of the straight-line distances.

With the insertion point in the expression $2\sqrt{18^2 + (30 - 24)^2}$, choose Evaluate Numerically.

Activity

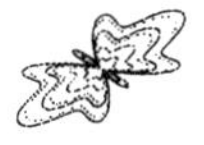

Plot a graph of $y = \sin x$ and change the Domain Interval to $0 \leq x \leq \pi$. Select and drag to the frame the expression $\left(0, 0, \frac{\pi}{4}, \sin\frac{\pi}{4}, \frac{\pi}{2}, \sin\frac{\pi}{2}, \frac{3\pi}{4}, \sin\frac{3\pi}{4}, \pi, \sin\pi\right)$. Evaluate numerically the sum of the lengths of the four line segments.

Define $f(x) = \sin x$ and explain what the sum

$$\sum_{i=0}^{n-1} \sqrt{\left((i+1)\frac{\pi}{n} - i\frac{\pi}{n}\right)^2 + \left(f\left((i+1)\frac{\pi}{n}\right) - f\left(i\frac{\pi}{n}\right)\right)^2}$$

represents. Evaluate this sum numerically for $n = 4$, 6, 10, 20, and 50. (For example, select the equation $n = 4$, choose Define + New Definition, and then evaluate the sum numerically.)

Determine an integral that represents the arc length of the curve $y = f(x)$ for $0 \leq x \leq \pi$ and evaluate the integral numerically. Explain the connection between the integral and the evaluated sums.

8.3 Area of a Surface of Revolution

Parabolic reflectors are used to record the sounds of players colliding on a football field, for receiving TV signals from a satellite, for concentrating solar energy in a solar oven, and for concentrating light in large telescopes. We can visualize a parabolic reflector by rotating a parabola $y = ax^2$ about the y-axis.This following graph was generated as a 3D Tube plot of $(0, 0, t)$ with radius $\sqrt{t}$ and Domain Interval $0 < t < 1$.

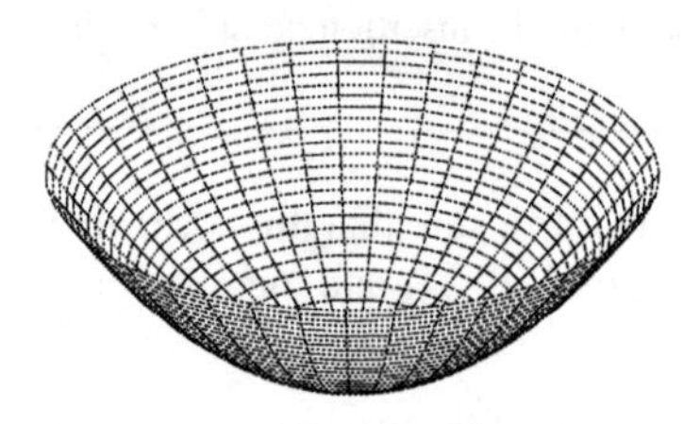

The Area of a Parabolic Reflector

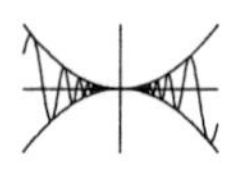

A solar furnace is to be constructed by coating the inside surface of a parabolic reflector with a very expensive reflective material. A parabolic reflector has been designed that is 10 feet in diameter and 2 feet deep. Calculate the area of the inside surface. Show that the computed area is reasonable by using a simple estimate.

Lana's Solution

Define f by $f(x) = ax^2$. To find a, we solve $f(5) = 2$ to get $a = \frac{2}{25}$. The surface is given by the integral

$$\int_0^5 2\pi x\sqrt{1 + (f'(x))^2}\, dx = 90.012 \text{ ft}^2$$

With the insertion point in the equation $f(x) = ax^2$, choose Define + New Definition. With the insertion point in the equation $f(5) = 2$, choose Solve + Exact. Select $a = \frac{2}{25}$ with the mouse, and choose Define + New Definition. With the insertion point in the integral $\int_0^5 2\pi x\sqrt{1 + (f'(x))^2}\, dx$, choose Evaluate Numerically.

This is reasonable because it is slightly larger than the area inside a circle of radius 5—namely, $\pi 5^2 = 78.54$.

Activities

1. Verify that the formula $S = \int_a^b f(x)\sqrt{1 + (f'(x))^2}\, dx$ is consistent with the formula $S = 2\pi r\ell$ for the lateral surface area of a frustum of a cone, where $r = \frac{r_1+r_2}{2}$

is the average of the radii of the two bases and ℓ is the slant height. To do this, let f be a linear function that goes through the two points $(0, r_1)$ and (h, r_2), set $\ell = \sqrt{h^2 + (r_2 - r_1)^2}$ and find a formula for f. Find an integral that represents the surface area generated by rotating the graph of $y = f(x)$ about the x-axis between $x = 0$ and $x = h$ and evaluate the integral. Visualize this surface by setting $r_1 = 2$, $r_2 = 1$, $h = 2$, and with the insertion point in the expression $(x, 0, 0)$ choosing Plot 3D + Tube, Edit + Properties, Plot Components page, Radius $f(x)$, Domain Interval $0 \leq t \leq 1$.

2. Visualize Gabriel's horn by applying Plot 3D + Tube to the expression $(x, 0, 0)$ and setting the Radius to $1/x$, $1 \leq x \leq 10$. Set up an improper integral that represents the surface area of Gabriel's horn on the interval $1 \leq x < \infty$ and verify that the surface area is infinite. Set up an improper integral that represents the volume inside Gabriel's horn, and show that the volume is finite. Discuss the apparent paradox of Gabriel's horn that can be filled with a finite amount of paint, but that would require an infinite amount of paint to cover its interior surface.

8.4 Moments and Centers of Mass

You solved some of these problems when you learned to ride a seesaw.

How to Balance a Sculpture

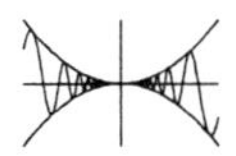

You wish to include as part of an abstract sculpture a piece of quarter-inch steel plate in the shape of a leaf. The leaf is to lie horizontally supported by a steel rod.

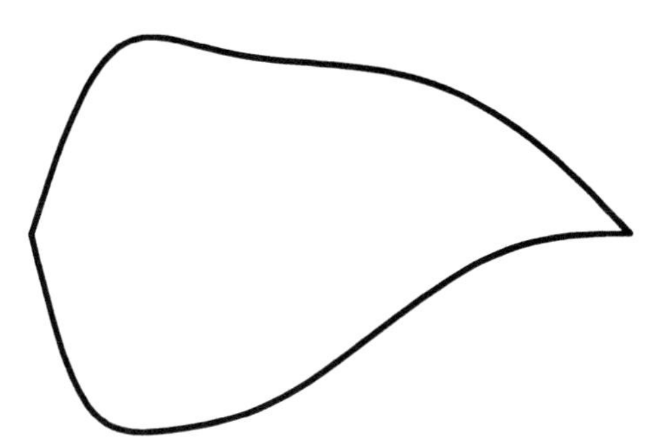

To minimize stress, you wish to balance the leaf on the rod. The leaf is bounded by the curves

$$y = \sin(x + \sin(x + \sin x))$$

and

$$y = -\sin(x + \sin(x + \sin(x + \sin x)))$$

Find the balance point and mark it with a small cross. Decide whether or not you think the leaf might balance on the marked point.

Philip's Solution

Let $f(x) = \sin(x + \sin(x + \sin x))$ and $g(x) = -\sin(x + \sin(x + \sin(x + \sin x)))$. The region bounded by these graphs is given in the following figure.

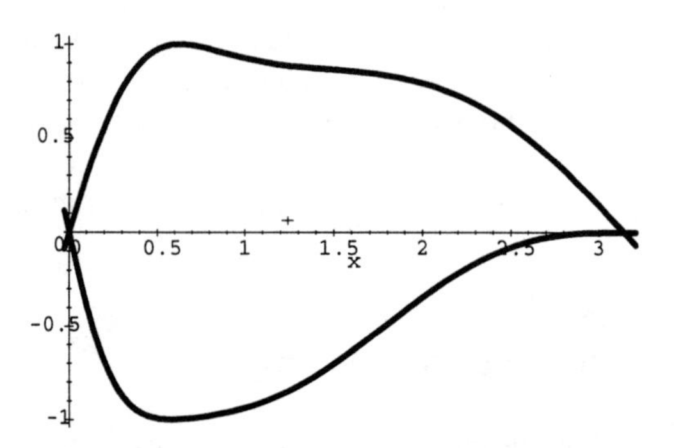

With the insertion point in the equation $f(x) = \sin(x + \sin(x + \sin x))$, choose Define + New Definition. Repeat for $g(x) = -\sin(x + \sin(x + \sin(x + \sin x)))$. With the insertion point in the list $f(x), g(x)$, choose Plot 2D + Rectangular. Double-click on the plot and use the large mountain zoom tool to zoom in on the leaf area.

By observation, the two curves intersect at the origin. There is also an intersection point somewhere between 3 and 4. To determine that intersection point, we solve the system

$$\begin{aligned} \sin(x + \sin(x + \sin x)) &= 0 \\ x &\in (3,4) \end{aligned}$$

and observe that the curves intersect when $x = 3.1416$, which looks suspiciously like π.

 With the insertion point in the system

$$\begin{aligned} f(x) &= g(x) \\ x &\in (3,4) \end{aligned}$$

choose Solve + Numeric.

A quick mental check (evaluating each expression starting with the innermost parentheses) verifies that indeed $f(\pi) = g(\pi) = 0$.

Numerical evaluation shows that the area of the region is given by

$$\int_0^\pi (f(x) - g(x))\,dx = 3.8137$$

 With the insertion point in the integral $\int_0^\pi (f(x) - g(x))\,dx$, choose Evaluate Numerically.

Letting $A = 3.8137$, we then numerically evaluate integrals that calculate $\overline{x}$ and $\overline{y}$.

$$\frac{1}{A}\int_0^\pi x\,(f(x) - g(x))\,dx = 1.241$$

$$\frac{1}{A}\int_0^\pi \frac{1}{2}\left((f(x))^2 - (g(x))^2\right)dx = 6.2757 \times 10^{-2}$$

The centroid $(1.241, .062757)$ is marked in the figure with a small cross.

With the insertion point in the equation $A = 3.8137$, choose Define + New Definition. With the insertion point in the integral $\frac{1}{A}\int_0^\pi x\,(f(x) - g(x))\,dx$, choose Evaluate Numerically. Repeat for the integral $\frac{1}{A}\int_0^\pi \frac{1}{2}\left((f(x))^2 - (g(x))^2\right)dx$. Select and drag to the frame $(1.241, .062757)$. Choose Edit + Properties, Plot Components page. For the Item Number corresponding to the point $(1.241, .062757)$, change Plot Style to Point and Point Symbol to Cross.

Yes, I think it could balance there. The leaf is fatter toward the left, and the centroid is closer to the left side of the region. Also, there appears to be more area above the x-axis than below, so I would expect the centroid to be above the x-axis.

Activities

1. Find the centroid of the triangular region with vertices $(0, 0)$, $(a, 0)$, and $(0, b)$. Use this information to make a conjecture about the centroid of a general triangular region. Test your conjecture on a triangular region with vertices $(0, 0)$, $(a, 0)$, and (b, c).

2. Find the centroid of the region in the first quadrant bounded by the graph of $y = \sqrt{a^2 - x^2}$, the x-axis, and the y-axis.

8.5 Hydrostatic Pressure and Force

Water pressure depends on depth. Finding the total force on a dam requires calculating the pressure at various levels and adding together the resulting forces.

How to Measure the Dam Force

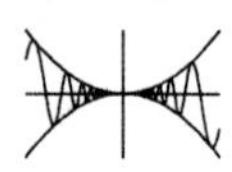

A vertical concrete dam in the shape of a parabola is all that separates the town of Maple Grove from millions of cubic feet of water. Calculate the hydrostatic force against the dam, assuming the parabola has the equation $y = \frac{1}{50}x^2 - 200$.

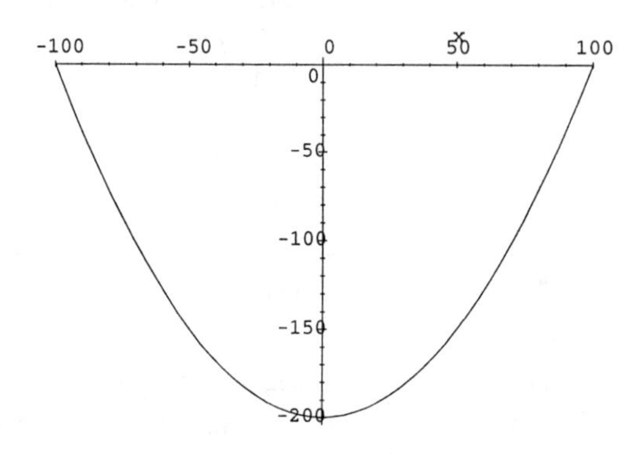

Mary's Solution

To find the hydrostatic force against the dam, it is convenient to integrate with respect to depth. Solving the equation $y = \frac{1}{50}x^2 - 200$ for x in terms of y, we get two potential solutions

$$x = -5\sqrt{(2y+400)}$$
$$x = 5\sqrt{(2y+400)}$$

With the insertion point in the equation $y = \frac{1}{50}x^2 - 200$, choose Solve + Exact (Variable to Solve for: x).

We choose the positive solution, and note that at depth y the total width of the dam is given by

$$w(y) = 10\sqrt{2y+400}$$

The water pressure at depth y ($-200 \le y \le 0$) is given by

$$p(y) = -62.5y$$

The hydrostatic force against the dam is given by the integral

$$\int_{-200}^{0} p(y)w(y)dy = 1.3333 \times 10^8$$

or roughly 133,000,000 lb.

With the insertion point in the equation $w(y) = 10\sqrt{2y+400}$, choose Define + New Definition. Repeat for the equation $p(y) = -62.5y$. With the insertion point in the integral $\int_{-200}^{0} p(y)w(y)dy$, choose Evaluate.

Activity

Livestock ponds are often limited to a depth of 10 feet. A valve at the bottom of such a pond is circular with diameter 1 ft., and has a vertical orientation. Imagine the valve subdivided into horizontal strips, as in the following figure:

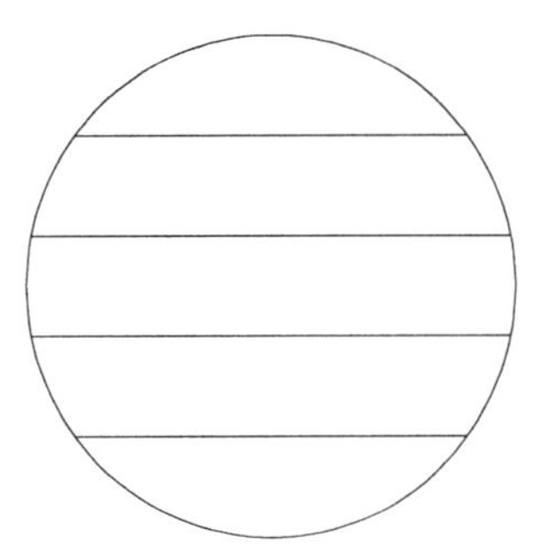

If the bottom of the circle rests on the bottom of the pond (10 feet below the surface), make a table that shows the approximate force applied to each of the five horizontal strips, assuming pressure is $62.5y$ lb where y is the depth in feet. Describe the methods used to create the approximations. Calculate the approximate total force against the circular valve. Set up an integral that represents the force against the circular valve. Compare the evaluated integral with your estimate. Explain any discrepancy.

8.6 Applications to Economics and Biology

The flow of blood through a vessel depends on viscosity, length, radius, and pressure differential at the endpoints. To understand the dependence on these factors, we will do some computer experiments with a piece of virtual garden hose.

The Pressure Builds

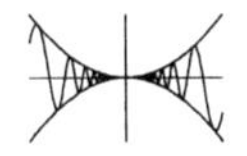

In the lawn and garden section of our local hardware store, there is a variety of garden hoses, with three choices of interior diameter ($\frac{1}{2}$ inch, $\frac{5}{8}$ inch, $\frac{3}{4}$ inch) and three choices of length (50 feet, 75 feet, 100 feet). Assuming fixed viscosity η and pressure P, create a table that indicates the relative number of gallons per minute of water that we can expect from these hoses. Normalize the table by dividing each entry by the smallest entry. Discuss how this table relates to the flow of blood, restricted arteries, and blood pressure.

Eddie's Solution

Assume the velocity of water in a hose is given by

$$v(r) = \frac{P}{4\eta\ell}\left(R^2 - r^2\right)$$

where P denotes the pressure differential at the two ends of the hose, ℓ is the length of hose, η is viscosity, R is the radius of the hose, and r is the distance of a particular water molecule from the center of the hose. All the molecules at the same distance r from the center move at the same velocity, so it is convenient to integrate with respect to r. The following graph depicts a cross section of the hose and a thin annulus of width dr:

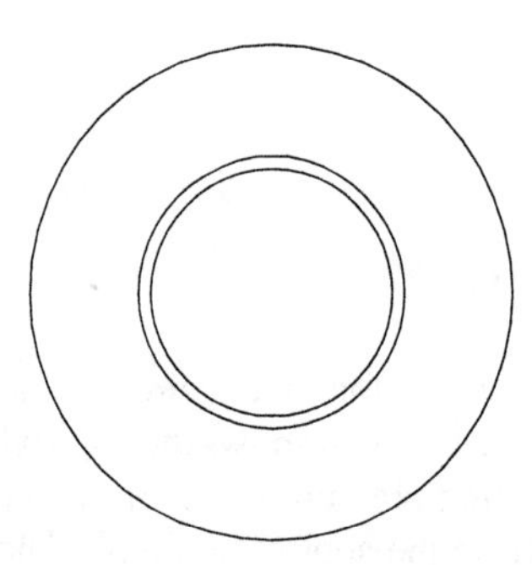

The molecules that lie in this thin annulus of width dr have a cross-sectional area of roughly $2\pi r\,dr$, and each of the molecules is moving with a velocity of roughly $v(r)$. Hence the total flow is given by

$$\int_0^R 2\pi r v(r)\,dr = \int_0^R 2\pi r \frac{P}{4\eta\ell}\left(R^2 - r^2\right) dr = \frac{\pi P R^4}{8\eta\ell}$$

With the insertion point in the equation $v(r) = \frac{P}{4\eta\ell}\left(R^2 - r^2\right)$, choose Define + New Definition. With the insertion point in the integral $\int_0^R 2\pi r \frac{P}{4\eta\ell}\left(R^2 - r^2\right) dr$, choose Evaluate. To make the plot of the cross section, with the insertion point to the right of 1, choose Plot 2D + Polar. Select 1.1 and drag it to the plot. Select 2 and drag it to the plot. Choose Edit + Properties, Axes page. Click Equal Scaling on Each Axis, and change Axis Type to None. Choose OK.

We summarize this formula in the following table. The table has been normalized by dividing each entry by the smallest entry. To get the actual flows, each number in the table would need to be multiplied by a constant. It is interesting to note that a 50-foot garden hose with an interior diameter of $\frac{3}{4}$ inch carries roughly 10 times as much water per minute as a 100-foot garden hose with an interior diameter of $\frac{1}{2}$ inch.

Length \ Radius	$\frac{1}{2}$	$\frac{5}{8}$	$\frac{3}{4}$
50	2.0	4.8829	10.125
75	1.3333	3.2552	6.7501
100	1	2.4414	5.0626

Ⓝ Choose Insert + Table (5 Rows, 5 Columns) and type in the first row and column. For the numerical data, for each of the values $R = \frac{1}{2}, \frac{5}{8}, \frac{3}{4}$; $\ell = 50, 75, 100$ in the expression $\frac{R^4}{\ell}$; choose Evaluate Numerically. To add the lines to the table, select the first two rows of the table, choose Edit + Properties, click the Lines tab, and follow instructions in the dialog. Then select the first two columns and repeat the process.

In terms of blood flow through a blood vessel, a small restriction in the diameter of a blood vessel cuts the amount of blood significantly. Another way to look at this is that to maintain the same amount of blood flow, blood pressure must increase significantly.

Activities

1. In a typical human artery we can take $\eta = .027\,\text{dyne-sec}\,/\,\text{cm}^2$, $R = .008\,\text{cm}$, $\ell = 2\,\text{cm}$, and $P = 4000\,\text{dyne/cm}^2$. Calculate the rate of flow in a typical human artery.

2. The price a consumer is willing to pay for a quantity x of a particular commodity is described by the *demand curve* $y = d(x)$ and the price that a producer is willing to charge for a quantity x of that same commodity is described by the *supply curve* $y = s(x)$. The point of intersection $(x_0, y_0)$of these two curves is called the *equilibrium point.* In the case consumers would have been willing to pay more than the market price, they benefit by the *consumer's surplus* $\int_0^{x_0} d(x)dx - x_0y_0$ and in the case producers would have been willing to charge less than the market price, they benefit by the *producer's surplus* $x_0y_0 - \int_0^{x_0} s(x)dx$. Plot the supply and demand curves for the functions $d(x) = 15e^{-x/8}$ and $s(x) = 2e^{x/10}$. Discuss the geometric significance of the consumer's surplus and the producer's surplus.

9 PARAMETRIC EQUATIONS AND POLAR COORDINATES

Some functions are better described by parametric equations or in a coordinate system based on angles as well as distance.

9.1 Curves Defined by Parametric Equations

To give a precise description of the effects of spin on a golf ball is a difficult physics problem. A golf ball is not merely a sphere; it is an airfoil. Spin about a vertical axis causes the ball to hook or slice, but spin about a horizontal axis can provide lift. Golfers will tell you that a ball hit with the right amount of backspin has a trajectory that is initially concave upward, which would indicate that the lift due to spin is initially greater than the force of gravity.

The Spin Is In

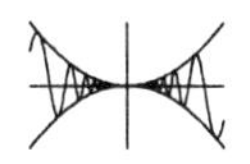

Let v and u denote the horizontal and vertical components of velocity, respectively, of a golf ball in flight, and let (x, y) denote its position. Define the following constants:

$$\begin{aligned} w &= \frac{1.62}{16}, \text{ weight of the ball in pounds} \\ c &= \frac{1}{150}w, \text{ drag term} \\ g &= 32, \text{ acceleration due to gravity in ft/sec}^2 \\ \theta &= \frac{\pi}{8}, \text{ angle of the club head} \\ k &= c, \text{ lift due to backspin} \\ z &= 200, \text{ club head velocity in ft/sec} \end{aligned}$$

Solve the following system of differential equations numerically; then plot (x, y) parametrically to view the flight of a golf ball.

$$\frac{dv}{dt} = -\frac{gcv}{w}$$

$$\begin{aligned}
\frac{du}{dt} &= -\frac{g}{w}(w + cu - kv) \\
\frac{dx}{dt} &= v \\
\frac{dy}{dt} &= u \\
v(0) &= z\cos^2\theta \\
u(0) &= z\cos\theta\sin\theta \\
x(0) &= 0 \\
y(0) &= 0
\end{aligned}$$

Explain why the flight of a golf ball does not follow the parabolic path that is often used to describe the flight of a ball.

Brendan's Solution

To model the motion of a golfball in flight, we define the following constants:

$$\begin{aligned}
w &= \frac{1.62}{16}, \text{ weight of the ball in pounds} \\
c &= \frac{1}{120}w, \text{ drag term} \\
g &= 32, \text{ acceleration due to gravity in ft/sec}^2 \\
\theta &= \frac{\pi}{16}, \text{ angle of the club head} \\
k &= \frac{c}{2}, \text{ lift due to backspin} \\
z &= 200, \text{ horizontal club head velocity in ft/sec}
\end{aligned}$$

N Select each equation separately and choose Define + New Definition: $w = \frac{1.62}{16}$, $c = \frac{1}{120}w$, $g = 32$, $\theta = \frac{\pi}{16}$, $k = \frac{c}{2}$, $z = 200$

A simple model for the flight of a ball is given by

$$\begin{aligned}
\frac{dv}{dt} &= 0 \\
\frac{du}{dt} &= -g \\
\frac{dx}{dt} &= v \\
\frac{dy}{dt} &= u
\end{aligned}$$

where v and u denote the horizontal and vertical components, respectively, of velocity. A solution to this system of ordinary differential equatons is given by

$$\begin{aligned}
x(t) &= C_1 + C_2 t \\
v(t) &= C_2
\end{aligned}$$

$$\begin{aligned} y(t) &= C_3 + C_4 t - \frac{1}{2} g t^2 \\ u(t) &= C_4 - gt \end{aligned}$$

Adding appropriate initial conditions yields the usual parabolic path.

A revised model for the flight of a golf ball is given by the following system of differential equations. Assuming a horizontal drag force proportional to the horizontal component of velocity and a vertical lift that is also proportional to the horizontal component of velocity, we get the revised equations

$$\begin{aligned} \frac{dv}{dt} &= -\frac{gcv}{w} \\ \frac{du}{dt} &= -\frac{g}{w}(w - kv) \\ \frac{dx}{dt} &= v \\ \frac{dy}{dt} &= u \\ v(0) &= z\cos^2\theta \\ u(0) &= z\cos\theta\sin\theta \\ x(0) &= 0 \\ y(0) &= 0 \end{aligned}$$

Enter the system of equations in a display, as shown above, and with the insertion point in the system of equations, choose Solve ODE + Numeric. This will define four functions: v, u, x, y.

The flight of the golf ball is shown in the following graph. Notice that the graph appears to be straight or slightly concave upward initially. Imagine the golf ball traveling very quickly at first, then losing momentum and starting to fall. An angle of $\frac{\pi}{16}$ radians corresponds to $11.25°$, roughly the club head angle for a driver. The indicated distance of over 600 feet corresponds to a drive of over 200 yards. A bounce and roll will add to the total length of the drive.

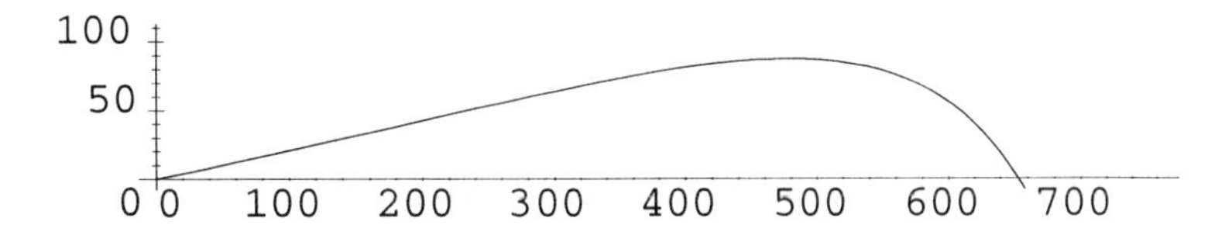

Taking into account the lift and drag components gives a plot that resembles the flight of a golf ball more than a parabolic plot does.

With the insertion point in the pair (x, y), choose Plot 2D + Parametric. Choose Edit + Properties, Plot Components page. Change Domain Interval to $0 < x < 700$. Choose Axes page and check Equal Scaling Along Each Axis.

Activity

Experiment with different values for the constants

$$
\begin{aligned}
c &= \frac{1}{120}w, \text{ drag term} \\
\theta &= \frac{\pi}{16}, \text{ angle of the club head} \\
k &= \frac{c}{2}, \text{ lift due to backspin} \\
z &= 225, \text{ horizontal club head velocity in ft/sec}
\end{aligned}
$$

to see how they affect the golf ball model. Describe some of these changes.

9.2 Tangents and Areas

Bezier curves are widely used in graphics software and for generating shapes such as the letters in a font. The curve segments are determined by the two endpoints and a third point, called an anchor, that controls the shape of the curve segment.

The Bezier Controls the Shape

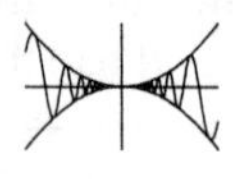

Given three points (a, b), (c, d), and (u, v), find a pair of equations $x(t) = mt^2 + nt + p$ and $y(t) = qt^2 + rt + s$ that satisfy the following six conditions:

$$
\begin{aligned}
x(0) &= a \\
y(0) &= b \\
x(1) &= c \\
y(1) &= d \\
y'(0)\,(a-u) &= x'(0)\,(b-v) \\
y'(1)\,(c-u) &= x'(1)\,(d-v)
\end{aligned}
$$

Discuss how you know that these six conditions yield a pair of parametric equations whose graph goes through the points (a, b) and (c, d), and such that the tangent lines at (a, b) and (c, d) both go through the third point (u, v). Get a general solution. Test your solution using the points

$$
\begin{aligned}
(a,b) &= (0,0) \\
(c,d) &= (0,1) \\
(u,v) &= (2,2)
\end{aligned}
$$

by plotting $(x(t), y(t))$ parametrically, with the domain interval $0 \leq t \leq 1$. Attach the tangent lines by dragging $(2t, 2t)$ and $(2t, 1+t)$ to the frame.

Martin's Solution

Define the parametric equations

$$\begin{aligned} x(t) &= mt^2 + nt + p \\ y(t) &= qt^2 + rt + s \end{aligned}$$

Select with the mouse the equation $x(t) = mt^2 + nt + p$, and choose Define + New Definition. Select with the mouse the equation $y(t) = qt^2 + rt + s$, and choose Define + New Definition.

We need to find coefficients m, n, p and q, r, s so that the graph goes through the points $(x(0), y(0)) = (a, b)$ and $(x(1), y(1)) = (c, d)$. The tangent lines at the endpoints must also go through the point (u, v). Because the slope of the tangent line at the point $(x(t), y(t))$ is given by

$$\frac{y'(t)}{x'(t)}$$

we set the slope of the tangent line equal to the slope of the secant line that joins the point $(x(t), y(t))$ to the point (u, v). This gives the conditions

$$\begin{aligned} \frac{y'(0)}{x'(0)} &= \frac{y(0) - v}{x(0) - u} = \frac{b - v}{a - u} \\ \frac{y'(1)}{x'(1)} &= \frac{y(1) - v}{x(1) - u} = \frac{d - v}{c - u} \end{aligned}$$

The set of six equations has the solution

$$\begin{array}{ll} p = a & s = b \\ n = -2a + 2u & r = -2b + 2v \\ m = a - 2u + c & q = b - 2v + d \end{array}$$

Enter the system of six equations into a display

$$\begin{aligned} x(0) &= a \\ y(0) &= b \\ x(1) &= c \\ y(1) &= d \\ \frac{y'(0)}{x'(0)} &= \frac{b - v}{a - u} \\ \frac{y'(1)}{x'(1)} &= \frac{d - v}{c - u} \end{aligned}$$

and, with the insertion point in the system of equations, choose Solve + Exact (Variables to Solve For: m, n, p, s, r, q). Select with the mouse each of

$$\{p = a, s = b, r = -2b + 2v, n = -2a + 2u, q = b - 2v + d, m = a - 2u + c\}$$

and, for each, choose Define + New Definition.

Using the endpoints $(a, b) = (0, 0)$ and $(c, d) = (0, 1)$ with the anchor point $(u, v) = (2, 2)$, we have

$$x(t) = -4t^2 + 4t \qquad \text{and} \qquad y(t) = -3t^2 + 4t$$

Select each of $a = 0$, $b = 0$, $c = 0$, $d = 1$, $u = 2$, and $v = 2$ with the mouse and, for each, choose Define + New Definition. With the insertion point in the expression $x(t)$, choose Evaluate. With the insertion point in the expression $y(t)$, choose Evaluate.

In the following figure, the graph of the Bezier curve is drawn with a fat pen and the tangent lines are drawn with dotted lines.

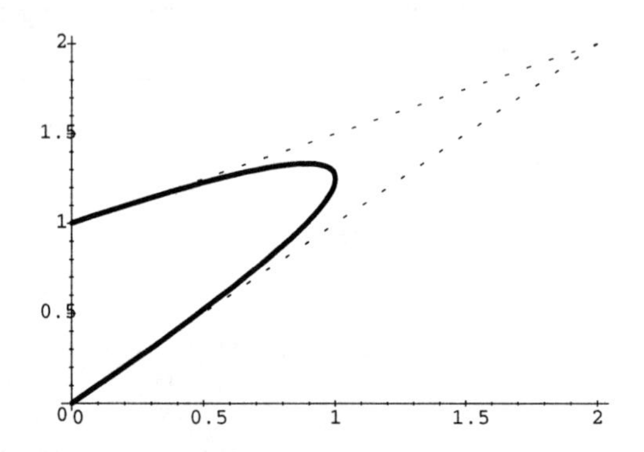

With the insertion point in the list $\left(-4t^2 + 4t, -3t^2 + 4t\right)$, choose Plot 2D + Parametric. Choose Edit + Properties, Plot Components page, and change Domain Interval to $0 \leq t \leq 1$. Select and drag to the frame each of the following: $(2t, 2t)$, $(2t, 1 + t)$.

Assume the endpoints are fixed. Looking at the figure, it appears that different anchor points not on the line containing the endpoints (a, b) and (c, d) will necessarily generate different Bezier curves. However, as long as the tangent lines are not parallel, they will always intersect in a unique anchor point.

Activities

1. In graphics software, Bezier curves are often linked together to create complex designs. Consider the problem of putting two Bezier curves $(x(t), y(t))$ and $(r(t), s(t))$ together so that $(x(1), y(1)) = (r(0), s(0))$, and the slopes of the tangent lines at the common point are equal. Explain why this implies that the two anchor points must be colinear with the common point. Find three Bezier curves that fit together to form a smooth loop based on the three endpoints $(-1, 0)$, $(1, 0)$, and $(0, 1)$ by choosing three appropriate anchor points and solving a system of equations.

2. Create a valentine by creating a pair of Bezier curves that do not fit together smoothly. Take the endpoints of each Bezier curve at $(0, 0)$ and $(0, 1)$ with anchor points $(2, 2)$ and $(-2, 2)$.

9.3 Arc Length and Surface Area

You can find the length of a curve and the area of a surface using integration. Sometimes you can use other techniques to get good estimates of length or area with less computation required.

How Long is the Rope?

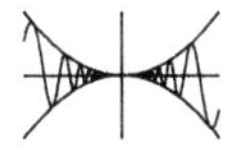

A rope appears neatly coiled on the deck of a ship.

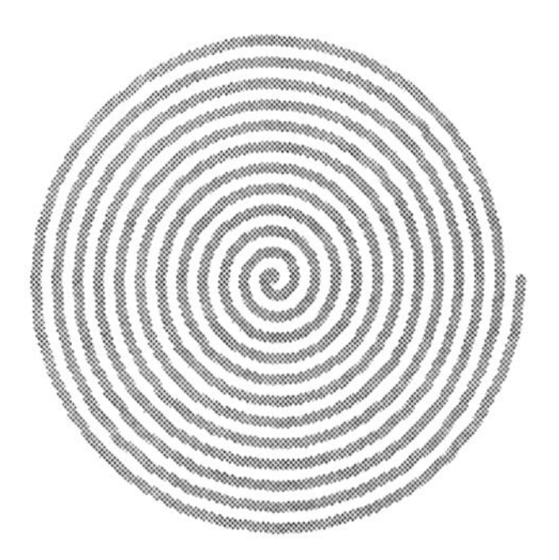

You estimate that the coil is 3 feet in diameter and the rope is $1\frac{1}{2}$ inches in diameter. How long is the rope? Parametric equations for the core (center) of the rope are given by

$$\begin{aligned} x(t) &= \frac{3t}{4\pi}\cos t \\ y(t) &= \frac{3t}{4\pi}\sin t \end{aligned}$$

Estimate the length in two different ways. First, numerically evaluate an integral that represents the arc length. Then calculate the area that the rope covers on the deck. Imagine the rope stretched out along a line. It should cover the same amount of deck area. Equate these two areas and use this to solve for the rope length. Compare and contrast the simplicity and accuracy of the two methods.

Marcia's Solution

Because the diameter of the rope coil is 36 inches, it follows that the radius is 18 inches, and because the rope is $1\frac{1}{2}$ inches in diameter, that means the rope wraps around $\frac{18}{3/2} = 12$ times, so the central angle increases from 0 to $24\pi = 75.398$. The parametric equations for the central core of the rope are

$$x(t) = \frac{3t}{4\pi}\cos t$$

$$y(t) \quad = \quad \frac{3t}{4\pi}\sin t$$

A parametric graph shows the location of the central core of the rope as it wraps around.

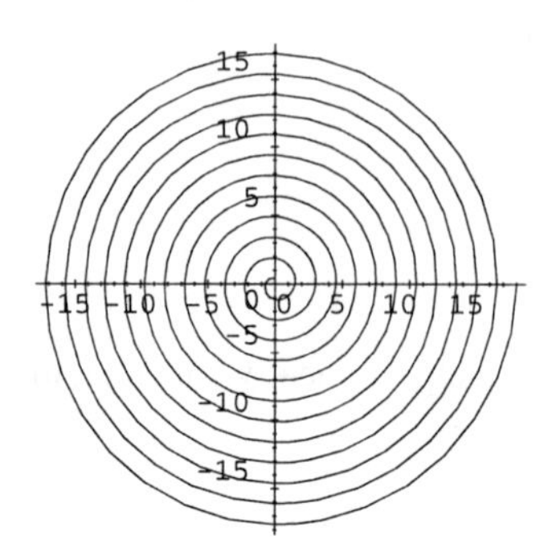

With the insertion point in the pair of expressions $\left(\frac{3t}{4\pi}\cos t, \frac{3t}{4\pi}\sin t\right)$, choose Plot 2D + Parametric. Choose Edit + Properties, Plot Components page, and change Domain Interval to $0 \leq t \leq 75.398$ $(= 24\pi)$.

The length of the rope can be calculated using the following integral for arc length. Numerical integration yields

$$\int_0^{24\pi} \sqrt{(x'(t))^2 + (y'(t))^2}\, dt = 679.24$$

as the length of the rope in inches.

With the insertion point in the integral $\int_0^{24\pi} \sqrt{(x'(t))^2 + (y'(t))^2}\, dt$, choose Evaluate Numerically.

A second approach is to calculate the amount of deck surface area covered by the rope in two different ways and to solve for rope length. The area covered by the coiled rope is

$$A = \pi 18^2$$

If the rope were stretched out and laid down to form a straight line of length ℓ, then the deck area covered by the rope would satisfy the equation

$$A = \frac{3}{2}\ell$$

This gives a rope length of $\ell = 678.58$ inches, very close indeed to the estimate given by the integral.

With the insertion point in the equation $A = \pi 18^2$, choose Define + New Definition. With the insertion point in the equation $A = \frac{3}{2}\ell$, choose Solve + Numeric.

The difference between these estimates is insignificant compared to the probable errors made in the original estimates concerning the diameter of the rope and coil. The computations involved in the second approach have an obvious advantage if you don't have a computer handy.

Activity

Use a piece of string to experiment with the method used to estimate the length of rope in a coil. Carefully wrap the string into a coil and measure the diameter. Count the number of coils and use this information to estimate the diameter of the string. Estimate the string length by equating the area covered by the coil to the length times the diameter of the string. Compare this with direct measurement of the string length. Was your estimate close to the measured value? If not, try to explain the discrepancy.

9.4 Polar Coordinates

You specify a point in polar coordinates by its distance from the origin and by the angle that the line connecting the point to the origin makes with the positive x-axis. Normally, this angle is measured in radians.

You Too Can Be an Artist

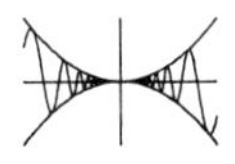

Experiment with the wonderful shapes generated by the following polar equations

$$\begin{aligned} r &= \sqrt{1 - .75\sin^2\theta} \\ r &= 1 + 3\sin\frac{\theta}{3} \\ r &= \sin\frac{7\theta}{4} \end{aligned}$$

and similar equations you get by varying the constants. Pick out your favorite and explain why it should be in an art exhibit. Make sure that you set the Domain Interval large enough to include a complete graph. What clues does the polar equation provide to help you determine the smallest Domain Interval that yields a complete graph?

Tomas's Solution

Here is the graph of $r = \sqrt{1 - .75\sin^2\theta}$ (drawn with a fat pen); with the graphs of $r = \sqrt{1 - .85\sin^2\theta}$ and $r = \sqrt{1 - .95\sin^2\theta}$ (drawn with a regular pen); and the graphs of $r = \sqrt{1 - .65\sin^2\theta}$ and $r = \sqrt{1 - .55\sin^2\theta}$ (drawn with dotted lines). No adjustment of the Domain Interval was necessary.

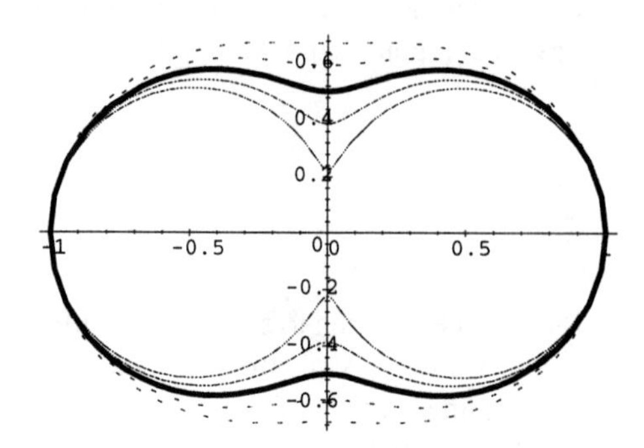

With the insertion point in the equation $r = \sqrt{1 - .75\sin^2\theta}$, choose Plot 2D + Polar. Select and drag to the plot each of $r = \sqrt{1 - .75\sin^2\theta}$, $r = \sqrt{1 - .75\sin^2\theta}$, $r = \sqrt{1 - .75\sin^2\theta}$, and $r = \sqrt{1 - .75\sin^2\theta}$. Choose Edit + Properties, and change Line Style, Line Color, and Thickness as desired. Choose OK.

The following is a graph of $r = 1 + 3\sin\frac{\theta}{3}$ drawn with a fat pen, with the graph of $r = 1+2\sin\frac{\theta}{3}$ drawn with dotted line and $r = 1+4\sin\frac{\theta}{3}$ with regular line. The domain interval was changed to $-3\pi \leq \theta \leq 3\pi$.

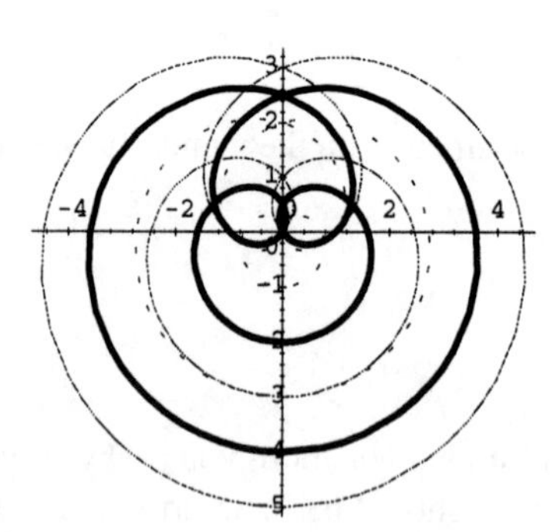

With the insertion point in the equation $r = 1 + 3\sin\frac{\theta}{3}$, choose Plot 2D + Polar. Select and drag to the plot each of $r = 1 + 2\sin\frac{\theta}{3}$ and $r = 1 + 4\sin\frac{\theta}{3}$. Choose Edit + Properties, and change Line Style, Line Color, and Thickness as desired. Change Domain Interval to $-3\pi \approx -9.4248 \leq \theta \leq 9.4248 \approx 3\pi$. (Note that the Domain Interval input boxes require floating-point numbers.) Choose OK.

Here is the graph of $r = \sin\frac{7\theta}{4}$. This is too busy to add the graphs of $r = \sin\frac{5\theta}{4}$ and $r = \sin\frac{9\theta}{4}$. Please try them on your own.

With the insertion point in the equation $r = \sin \frac{7\theta}{4}$, choose Plot 2D + Polar. Choose Edit + Properties and change Line Style, Line Color, and Thickness as desired. Change Domain Interval to $-4\pi \approx -12.6 \leq \theta \leq 12.6 \approx 4\pi$. Choose OK.

To set the domain interval, I found it useful to look at the denominators. If the expression involves sines and cosines of a multiple of θ, then the domain interval $-\pi \leq \theta \leq \pi$ is sufficient. (In fact, a smaller interval may be sufficient.) However, if the expression involves sines and cosines where θ is divided by n, then the domain interval $-n\pi \leq \theta \leq n\pi$ may be required.

I haven't found my favorite yet. I'm still looking.

Activities

1. Experiment with graphs of polar equations of the form $r = a + b \sin \theta$ and $r = a + b \cos \theta$. In what fundamental way is the case $|a| < |b|$ different from $|a| > |b|$?

2. Experiment with graphs of polar equations of the form $r = a \sin b\theta$ and $r = a \cos b\theta$ for integers a and b. What are the significant differences between the graphs for b even and b odd?

3. Solve the equation $r(\sin \theta + r \cos^2 \theta) = 1$ for r, and create a polar plot. Rewrite $r(\sin \theta + r \cos^2 \theta) = 1$ as an equation in rectangular coordinates by replacing $r \cos \theta$ by x and $r \sin \theta$ by y, then create a rectangular plot. Choose domain intervals and view rectangles so that the two graphs look similar.

9.5 Areas and Lengths in Polar Coordinates

The techniques for finding the area of a region bounded by a curve given in rectangular coordinates depend on the formula $A = \ell w$ for the area of a rectangle. For regions bounded by a curve given in polar coordinates, you start with the formula $A = \frac{\theta}{2}r^2$ for the area of a sector of a circle.

Looking At the Rope From Another Angle

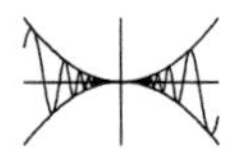

A rope appears neatly coiled on the deck of a ship:

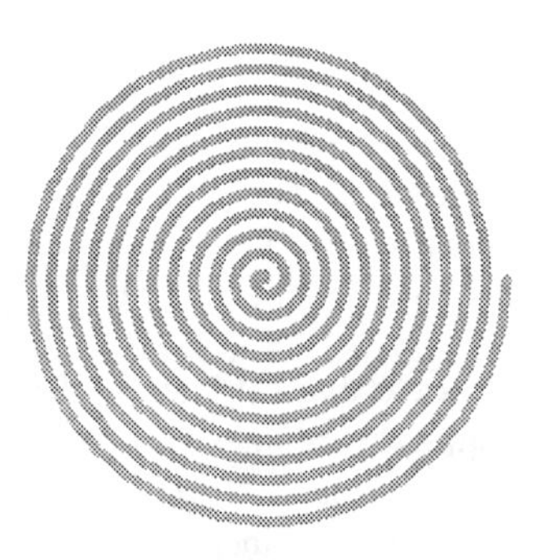

You estimate that the coil is 3 feet in diameter and the rope is $1\frac{1}{2}$ inches in diameter. How long is the rope? You have already seen how to do this using parametric equations. This is also a natural problem to do with polar coordinates. The polar equation for the rope coil is given by

$$r(\theta) = \frac{3}{4\pi}\theta$$

Calculate the length of this polar curve as θ ranges from 0 to 24π (enough to make 12 revolutions). Estimate the length in two different ways:

- First, numerically evaluate an integral that represents the arc length.
- Then calculate the area that the rope covers on the deck. Imagine the rope stretched out along a line. It should cover the same amount of deck area.

Equate these two areas and use this to solve for the rope length. Compare and contrast the simplicity and accuracy of the two methods.

Mary Anne's Solution

Because the diameter of the rope coil is 36 inches, it follows that the radius of the rope coil is 18 inches. The rope is $1\frac{1}{2}$ inches in diameter, so the rope winds around $\frac{18}{3/2} =$ 12 times. This means the central angle increases from 0 to $12 \times 2\pi = 75.398$. A polar graph shows the location of the central core of the rope as it winds around.

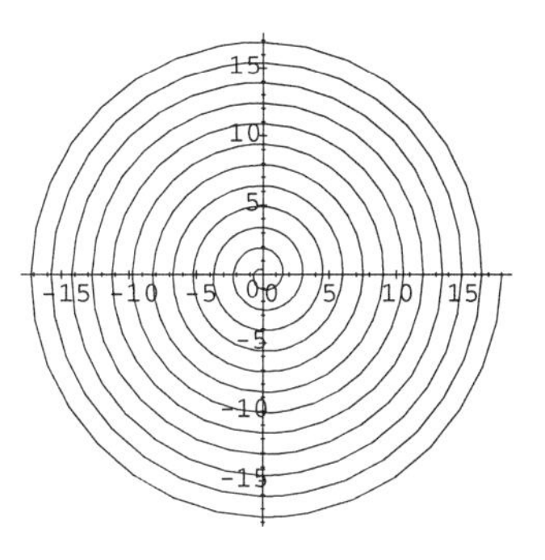

$r = \frac{3}{4\pi}\theta$ with $0 \le \theta \le 24\pi$

With the insertion point in the expressions $\frac{3t}{4\pi}$, choose Plot 2D + Polar. Choose Edit + Properties, Plot Components page, and change Domain Interval to $0 \le \theta \le 75.398$ ($= 24\pi$). Choose Axes page and check Equal Scaling Along Each Axis. Choose OK.

The length of the rope can be calculated using the following integral for arc length. Numerical integration yields

$$\int_0^{24\pi} \sqrt{(r(\theta))^2 + \left(\frac{d}{d\theta} r(\theta)\right)^2}\, d\theta = 679.24$$

as the length of the rope in inches.

Define $r(\theta) = \frac{3}{4\pi}\theta$. With the insertion point in $\int_0^{24\pi} \sqrt{(r(\theta))^2 + \left(\frac{d}{d\theta} r(\theta)\right)^2}\, d\theta$, choose Evaluate Numerically.

A second approach is to calculate the amount of deck surface area covered by the rope in two different ways, and to solve for rope length. The area covered by the coiled rope is

$$A = \pi 18^2$$

If the rope were stretched out and laid down to form a straight line of length ℓ, then the deck area covered by the rope would satisfy the equation

$$A = \frac{3}{2}\ell$$

This gives a rope length of $\ell = 678.58$ inches, very close indeed to the estimate given by the integral.

With the insertion point in the equation $A = \pi 18^2$, choose Define + New Definition. With the insertion point in the equation $A = \frac{3}{2}\ell$, choose Solve + Numeric.

The difference between these estimates does not seem significant compared with the probable errors made in the original estimates concerning the diameter of the rope and coil. In fact, both estimates round to 679 inches. The computations involved in the second approach have an obvious advantage if you are doing the estimate by hand.

Activities

1. What is the minimum length of ribbon needed to make a bow in the shape of $r^2 = \cos 2\theta$, where r is measured in feet? (Use Evaluate Numerically to evaluate an integral.) Draw a polar graph and explain why your estimated length is reasonable.

2. What is the area enclosed by one leaf of $r^2 = \cos 2\theta$? Explain why your evaluated area is reasonable.

3. Find the area enclosed by the inner loop of $r = 3 + 4\sin\theta$. Sketch the graph and explain why your estimate is reasonable.

9.6 Conic Sections

The curves determined by quadratic equations of the form $Ax^2 + Bxy + Cy^2 + Dx + Ey + F = 0$ are often referred to as *conic sections*, as each can be described as the intersection of a plane with a cone. The *discriminant* of the expression

$$Ax^2 + Bxy + Cy^2 + Dx + Ey + F$$

is the number

$$\triangle = B^2 - 4AC$$

The type of curve determined by the quadratic equation depends on the *sign* of the discriminant. In nondegenerate cases, the quadratic equation gives

an ellipse if $\triangle < 0$ a parabola if $\triangle = 0$ a hyperbola if $\triangle > 0$

where a circle is considered as a special case of an ellipse. The degenerate cases give a pair of lines, a line, or a point, or they have imaginary solutions.

A rotation of one or both axes in the plane leads to a change of variables, such as the change

$$\begin{aligned} u &= ax + by \\ v &= cx + dy \end{aligned}$$

from (x, y) to (u, v). The sign of the discriminant (and thus the basic shape of the curve) is invariant under such changes for a quadratic equation $Ax^2 + Bxy + Cy^2 + Dx + Ey + F = 0$ if $ad \neq bc$.

Generating Random Conic Sections

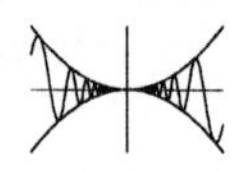

Because the general conic has the form

$$Ax^2 + Bxy + Cy^2 + Dx + Ey + F = 0$$

it appears that a conic is determined by six points. It actually takes only five conditions because multiplication on both sides of the equation by a constant does not change the graph, and one of the coefficients can always be taken to be 1. Define $f(x, y) = Ax^2 + Bxy + Cy^2 + Dx + Ey + 1$. Generate a random conic by starting with five random points and finding an equation of a conic that goes through those five points. Determine what type of conic you have by evaluating the discriminant $\triangle = B^2 - 4AC$ and by plotting the conic in the View Rectangle $-100 \leq x \leq 100$ by $-100 \leq y \leq 100$. (This is enough to view the five data points; you may need to zoom out farther to determine the type of conic section.)

Denise's Solution

Let $f(x, y) = Ax^2 + Bxy + Cy^2 + Dx + Ey + 1$. The random matrix

$$\begin{bmatrix} -85 & -55 \\ -37 & -35 \\ 97 & 50 \\ 79 & 56 \\ 49 & 63 \end{bmatrix}$$

is used to define the system of equations

$$\begin{aligned} f(-85, -55) &= 0 \\ f(-37, -35) &= 0 \\ f(97, 50) &= 0 \\ f(79, 56) &= 0 \\ f(49, 63) &= 0 \end{aligned}$$

whose solution is

$$\begin{aligned} E &= \frac{41118\,83676}{10\,83993\,69155} \\ D &= -\frac{76346256}{30971\,24833} \\ C &= -\frac{15515712}{2\,16798\,73831} \\ B &= \frac{19404996}{10\,83993\,69155} \\ A &= \frac{136751}{30971\,24833} \end{aligned}$$

Choose Matrices + Fill Matrix (Dimensions: 5 Rows by 2 Columns; Fill with: Random). This generates five data points where each coordinate is in the interval $-100 < a < 100$. Choose Insert + Display, and type $f(a, b) = 0$, where $\begin{bmatrix} a & b \end{bmatrix}$ is the first row of your matrix. Press ENTER and type $f(c, d) = 0$, where $\begin{bmatrix} c & d \end{bmatrix}$ is the second row of your matrix. Repeat until you have a system of five equations. Choose Solve + Exact. Select each equation in your solution, one at a time, and choose Define + New Definition.

The discriminant is computed to be

$$\triangle = 1.5845 \times 10^{-7}$$

which is positive, and thus the conic is a hyperbola. To verify this visually, here is the graph.

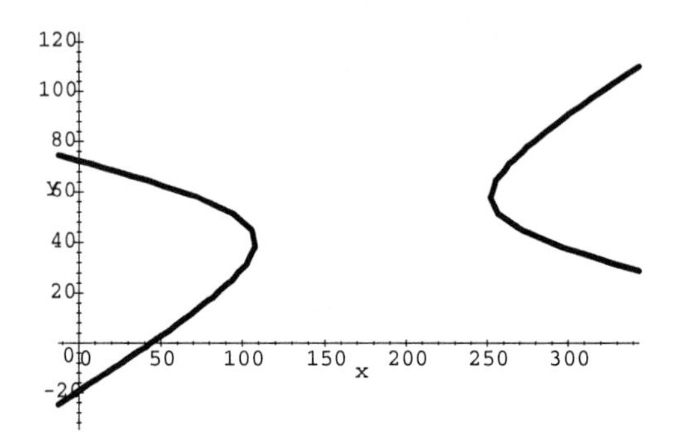

Select each equation in the solution, one at a time, and choose Define + New Definition. Evaluate $\triangle = B^2 - 4AC$ numerically. With the insertion point within $f(x, y) = 0$, choose Plot 2D + Implicit. Zoom and pan as appropriate to view the conic.

Activities

1. Generate random conics until you have at least one hyperbola and one ellipse. Did you get any parabolas? Any circles? Was this a surprise? Speculate about the relative probabilities of obtaining the various conics by this method.

2. A conic section $Ax^2+Bxy+Cy^2+Dx+Ey+F=0$ can be put into standard form by using a rotation of axes. To rotate the axes through an angle θ, use the substitutions
$$x = X\cos\theta - Y\sin\theta$$
$$y = X\sin\theta + Y\cos\theta$$
Evaluate the expression
$$\left[Ax^2 + Bxy + Cy^2 + Dx + Ey + F\right]_{x=X\cos\theta-Y\sin\theta,y=X\sin\theta+Y\cos\theta}$$
and note that the coefficient of XY is given by
$$2(C-A)\cos\theta\sin\theta + B\left(\cos^2\theta - \sin^2\theta\right)$$
Use the identities $2\cos\theta\sin\theta = \sin 2\theta$ and $\cos^2\theta - \sin^2\theta = \cos 2\theta$ to reduce the displayed expression to $(C-A)\sin 2\theta + B\cos 2\theta$. The goal is to replace the original equation by a new equation for which the coefficient of XY is zero. Dividing by $\sin 2\theta$, this condition may be stated as $\cot 2\theta = \frac{A-C}{B}$. Use these facts to rewrite the

equation $x^2 + 12xy + 6y^2 - 2x + 5y + 1 = 0$ in terms of the new coordinates X and Y in such a way that the coefficient of XY is zero.

9.7 Conic Sections in Polar Coordinates

Many physical phenomena are modeled with parabolic models—for example, the flight of a ball thrown into the air in the absence of air friction. Perhaps, however, the ball is actually in suborbital flight, and a better model might be an ellipse. (Because the lines of force all meet at the center of the Earth, they cannot be parallel.) The following activity is designed to show just how delicate the equation of a parabola really is.

A polar equation of the form

$$r = \frac{ed}{1 \pm e\cos\theta} \qquad \text{or} \qquad r = \frac{ed}{1 \pm e\cos\theta}$$

represents a conic section with eccentricity e. The conic is an ellipse if $e < 1$, a parabola if $e = 1$, or a hyperbola if $e > 1$.

Do Parabolas Exist in Nature?

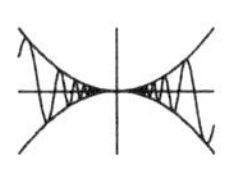

Plot $r = \frac{1}{1+e\cos\theta}$ for several values of e near 1 and at 1. Note how closely ellipses ($e < 1$) and hyperbolas ($e > 1$) fit a parabola. Discuss whether or not you could really distinguish a small piece of a parabola from a small piece of an ellipse or a small piece of a hyperbola.

Larry's Solution

The initial polar plot of $\frac{1}{1+\cos\theta}$ (for $e = 1$) looks like a pair of straight lines. This is caused by the graph being dominated by points determined by θ very close to $-\pi$ and π, where the expression $\frac{1}{1+\cos\theta}$ is faced with division by zero. In order to get a decent graph, I restricted the domain to $-3.1 \leq \theta \leq 3.1$. The following shows the graph of the parabola $r = \frac{1}{1+\cos\theta}$ (drawn with a fat pen) together with the ellipses

$$\begin{aligned} r &- \frac{1}{1+.9\cos\theta} \\ r &= \frac{1}{1+.99\cos\theta} \\ r &= \frac{1}{1+.999\cos\theta} \\ r &= \frac{1}{1+.9999\cos\theta} \end{aligned}$$

(drawn with a regular pen). The gap in the graphs of the ellipses is caused by the fact that $3.1 < \pi$. The graph of $r = \frac{1}{1+.9999\cos\theta}$ is barely distinguishable from the graph of $r = \frac{1}{1+\cos\theta}$.

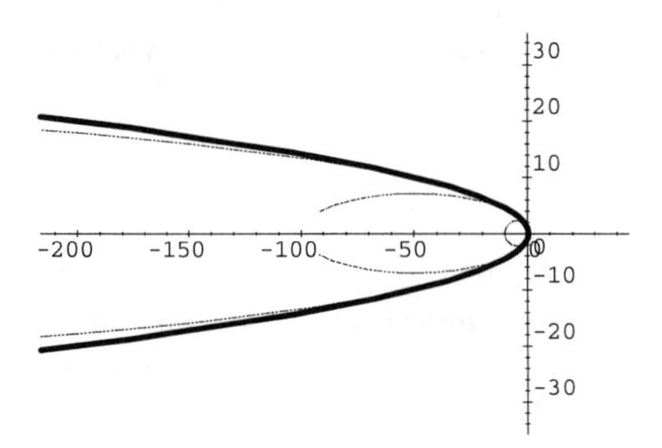

With the insertion point in the expression $\frac{1}{1+\cos\theta}$, choose Plot 2D + Polar. Set Domain Interval to $-3.1 \leq \theta \leq 3.1$ and change Thickness to Medium. Select and drag to the frame each of the following: $\frac{1}{1+.9\cos\theta}$, $\frac{1}{1+.99\cos\theta}$, $\frac{1}{1+.999\cos\theta}$, $\frac{1}{1+.9999\cos\theta}$, and $\frac{1}{1+.99999\cos\theta}$.

In order to plot hyperbolas, I restricted the domain interval to $-3 \leq \theta \leq 3$. The ellipses seemed to line the inside portion of the graph of the parabola. The hyperbolas $r = \frac{1}{1+1.001\cos\theta}$ and $r = \frac{1}{1+1.0001\cos\theta}$ (drawn with a thin pen) seem to line the outside portion of the graph of the parabola. It would be very difficult, given such a picture, to decide whether the graph represents a parabola or a hyperbola (or even an ellipse).

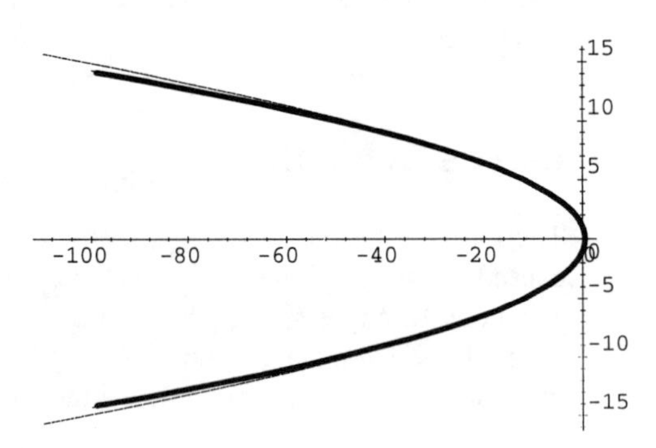

With the insertion point in the expression $\frac{1}{1+\cos\theta}$, choose Plot 2D + Polar. Change the Domain Interval to $-3 \leq \theta \leq 3$ and change Thickness to Medium. Select and drag each of the expressions $\frac{1}{1+1.001\cos\theta}$ and $\frac{1}{1+1.0001\cos\theta}$ to the frame.

Orbits of planets and comets are modeled by ellipses. Hyperbolas are used to model the path of light as it is deflected by a large mass. After looking at these experiments, it is not obvious to me that a parabola would ever occur as an orbit, or as the flight path of a thrown object.

Activities

1. Study the effects of rotation in polar coordinates by replacing θ with $\theta+\alpha$. In particular, draw a graph of each of the following polar equations in one figure:
$$r = \frac{1}{1-\frac{1}{2}\cos\theta}$$
$$r = \frac{1}{1-\frac{1}{2}\cos\left(\theta+\frac{\pi}{6}\right)}$$
$$r = \frac{1}{1-\frac{1}{2}\cos\left(\theta+\frac{\pi}{4}\right)}$$
$$r = \frac{1}{1-\frac{1}{2}\cos\left(\theta+\frac{\pi}{3}\right)}$$
To get a correct geometric perspective, choose Edit + Properties and in the Axes page turn on Equal Scaling Along Each Axis. Discuss how replacing θ by $\theta+\alpha$ affects the graph of $r = \frac{ed}{1\pm e\cos\theta}$ or $r = \frac{ed}{1\pm e\sin\theta}$.

2. Study the effect of e in the equation $r = \frac{ed}{1-e\cos\theta}$ by creating graphs with $e = 0.1$, 0.5, 0.9, 1.0, 1.2, 2.0, and 10.0. Turn on Equal Scaling Along Each Axis. Describe the visual change on the graph as e varies.

10 INFINITE SEQUENCES AND SERIES

In everyday language, the words *sequence* and *series* are used almost interchangeably to mean a number of objects or events arranged or coming one after the other in succession. In mathematics, however, these words are assigned different technical meanings. A *sequence* is a sequence of numbers in more or less the usual sense. However, the word *series* in mathematics refers to the sum of a sequentially ordered finite or infinite set of terms.

10.1 Sequences

Most common examples of sequences are given by a rule or formula for the nth term. For example, $a_n = 1/n$ and $a_n = 2^n$ determine the sequences

$$1, \frac{1}{2}, \frac{1}{3}, \frac{1}{4}, \frac{1}{5}, \ldots$$
$$2, 4, 8, 16, 32, \ldots$$

Sequences $\{a_n\}_{n=1}^{\infty}$ can be visualized by plotting a few points

$$\{(1, a_1), (2, a_2), (3, a_3), \ldots, (n, a_n)\}$$

Visualizing Sequences

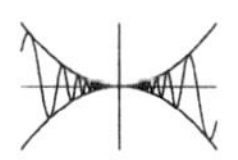

Let $a_n = \left(1 + \frac{1}{n}\right)^n$. Visualize the sequence $\{a_n\}_{n=1}^{\infty}$ by plotting

$$\{(1, a_1), (2, a_2), (3, a_3), \ldots, (10, a_{10})\}$$

View the sequence numerically by creating a table of values for $a_1, a_2, a_3, \ldots, a_{10}$. Find a_{100} and a_{1000} and discuss the limit

$$\lim_{n\to\infty} a_n$$

Discuss the relationship between the limit and the plotted points. Attach a horizontal line to the figure that indicates the limit.

Kitty's Solution

Set

$$a_n = \left(1 + \frac{1}{n}\right)^n$$

The following figure shows the first 10 points of the form (n, a_n):

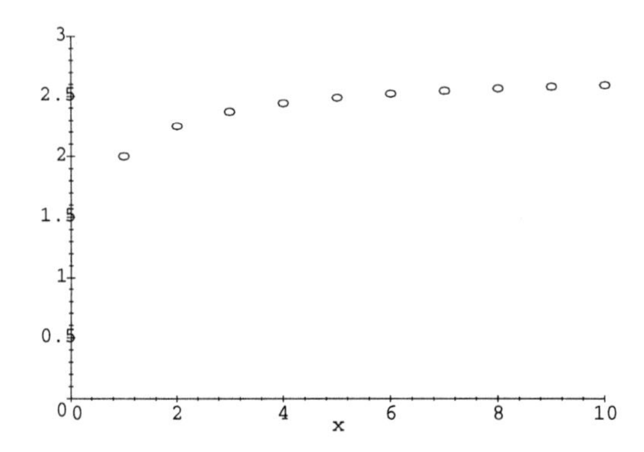

With the insertion point in the equation $a_n = \left(1 + \frac{1}{n}\right)^n$, choose Define + New Definition, and check Interpret Subscript as A Function Argument. Choose OK. With the insertion point in the list of points

$$(1, a_1, 2, a_2, 3, a_3, 4, a_4, 5, a_5, 6, a_6, 7, a_7, 8, a_8, 9, a_9, 10, a_{10})$$

choose Plot 2D + Rectangular. Choose Edit + Properties, Plot Components page. Set Plot Style to Point, Point Style to Circle, and Domain Interval to $1 < x < 10$. Choose View page, and set View Intervals to $0 < x < 10$ and $0 < y < 3$. Choose OK.

The first 10 terms can also be displayed numerically.

n	1	2	3	4	5	6	7	8	9	10
a_n	2.0	2.25	2.237	2.44	2.49	2.52	2.55	2.57	2.59	2.59

To get a better idea about where the sequence is headed, we compute a_{100} and a_{1000} to get

$$a_{100} = 2.7048$$
$$a_{1000} = 2.7169$$

 With the insertion point in the list of points

$$(1, a_1, 2, a_2, 3, a_3, 4, a_4, 5, a_5, 6, a_6, 7, a_7, 8, a_8, 9, a_9, 10, a_{10})$$

choose Evaluate Numerically. Choose Insert + Table, set Columns to 11 and rows to 2, and fill entries. With the insertion point in the expression a_{100}, choose Evaluate Numerically. With the insertion point in the expression a_{1000}, choose Evaluate Numerically.

These numbers are starting to look like $e = 2.7183$. Attaching the line $y = e$ to the figure, we get the following revised figure:

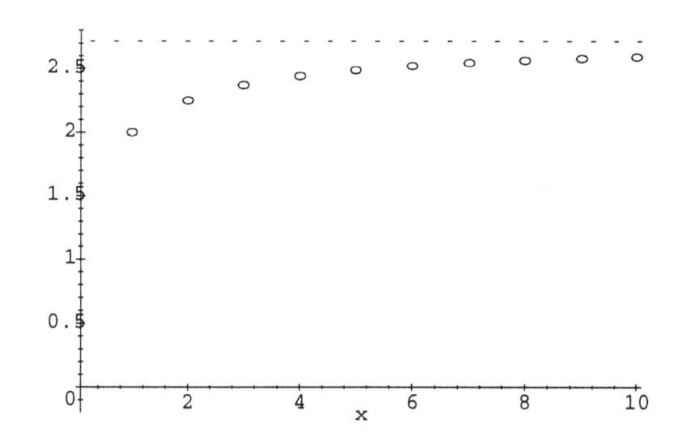

Select and drag to the frame 2.71828. Choose Edit + Properties, Plot Components page. For Item Number 2, change Line Style to Dots. Choose OK.

If we were to extend the figure to the right, we would see the points getting closer and closer to the dotted line, as in the following figure.

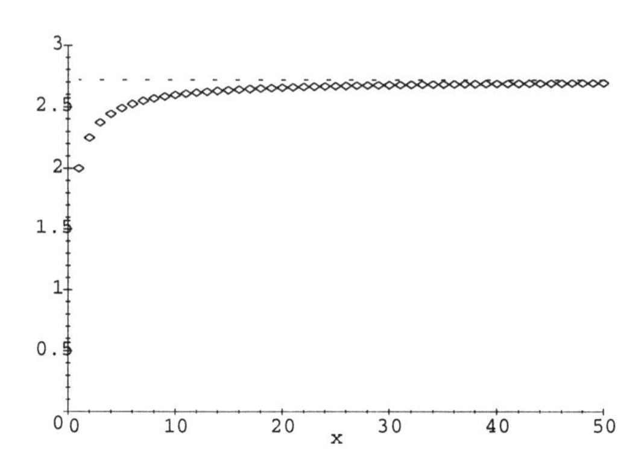

Checking, we see that the symbolic computation system gives

$$\lim_{n\to\infty}\left(1+\frac{1}{n}\right)^n = e$$

Select and drag the expression $\left(1+\frac{1}{n}\right)^n$ to the plot. Choose Edit + Properties, Plot Components page. For the Item Number corresponding to $\left(1+\frac{1}{n}\right)^n$, change Plot Style to Point and Point Symbol to Circle. Change Domain Interval to $0 < x < 50$. Choose OK. With the insertion point in the expression $\lim_{n\to\infty}\left(1+\frac{1}{n}\right)^n$, choose Evaluate.

Finding a Better Way to Approximate e

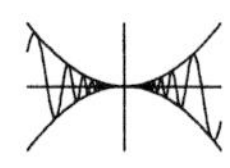

The sequence $a_n = \left(1+\frac{1}{n}\right)^n$ seems to converge to e rather slowly. Continued fractions provide an alternative method for approximating e. A *simple continued fraction* is an expression of the form

$$a_0 + \cfrac{1}{a_1 + \cfrac{1}{a_2 + \cfrac{1}{a_3 + \cdots}}}$$

where the a_i represent positive integers. A sequence $\{c_n\}$ of *convergents* is defined by

$$\begin{aligned} c_0 &= a_0 \\ c_1 &= a_0 + \frac{1}{a_1} = \frac{a_0 a_1 + 1}{a_1} \\ c_2 &= a_0 + \frac{1}{a_1 + \frac{1}{a_2}} = \frac{a_0 a_1 a_2 + a_0 + a_2}{a_1 a_2 + 1} \end{aligned}$$

Experiment with the sequence of convergents determined by the continued fraction expansion of e.

Note With *Scientific Notebook*, you can access any Maple routine. In this activity, you will see how to access the Maple function **cfrac(x,n)** that produces continued fractions. In the Maple function **cfrac(x,n)**, the symbol **x** represents a number and **n** represents the depth of the expansion.

Robert's Solution

We construct the following table of convergents:

$c(e,1)$	$= 2 + \frac{1}{1}$	$= 3$	$= 3.00000\,0000$
$c(e,2)$	$= 2 + \cfrac{1}{1+\frac{1}{2}}$	$= \frac{8}{3}$	$= 2.66666\,6667$
$c(e,3)$	$= 2 + \cfrac{1}{1+\cfrac{1}{2+\frac{1}{1}}}$	$= \frac{11}{4}$	$= 2.75000\,0000$
$c(e,4)$	$= 2 + \cfrac{1}{1+\cfrac{1}{2+\cfrac{1}{1+\frac{1}{1}}}}$	$= \frac{19}{7}$	$= 2.71428\,5714$
$c(e,5)$	$= 2 + \cfrac{1}{1+\cfrac{1}{2+\cfrac{1}{1+\cfrac{1}{1+\frac{1}{4}}}}}$	$= \frac{87}{32}$	$= 2.71875\,0000$
$c(e,6)$	$= 2 + \cfrac{1}{1+\cfrac{1}{2+\cfrac{1}{1+\cfrac{1}{1+\cfrac{1}{4+\frac{1}{1}}}}}}$	$= \frac{106}{39}$	$= 2.71794\,8718$
$c(e,7)$	$= 2 + \cfrac{1}{1+\cfrac{1}{2+\cfrac{1}{1+\cfrac{1}{1+\cfrac{1}{4+\cfrac{1}{1+\frac{1}{1}}}}}}}$	$= \frac{193}{71}$	$= 2.71830\,9859$

Choose Define + Define Maple Name. Fill in Maple Name: cfrac(x,n), Scientific Notebook Name: $c(x,n)$, and Maple Packages Needed: numtheory. With the insertion point in each of $c(e,1)$ to $c(e,7)$, choose Evaluate. For each evaluation, select and delete the plus sign and ellipses $(+\ldots)$ that appear at the bottom. With the insertion point in each expression such as $2+\cfrac{1}{1+\cfrac{1}{2+\cfrac{1}{1+\cfrac{1}{1+\frac{1}{4}}}}}$, choose Simplify, and then choose Evaluate Numerically. You can put the results into a display or copy the entries into a table as shown here.

One interesting observation is that the convergents appear to alternate about the value $e = 2.71828\,1828$, as illustrated in the following figure:

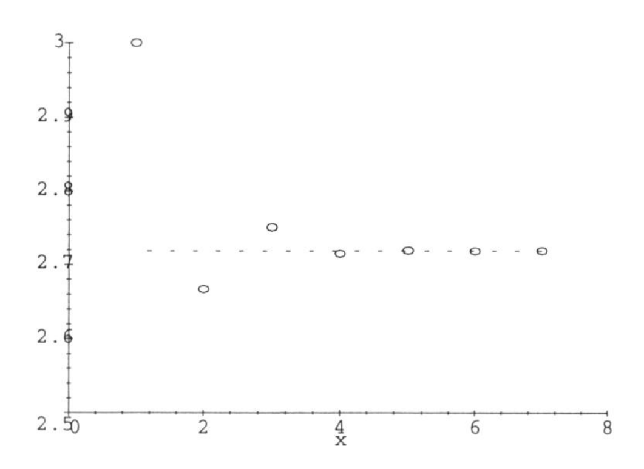

With the insertion point in the list of points

$$(1,3,2,8/3,3,11/4,4,19/7,5,87/32,6,106/39,7,193/71)$$

choose Plot 2D + Rectangular. Select e and drag it to the plot. Choose Edit + Properties, Plot Components page. For Item Number 1, change Plot Style to Point, and Point Symbol to Circle. For Item Number 2, change Line Style to Dots. Change Domain Interval to $0 < x < 8$. Choose OK.

Chaotic Sequences

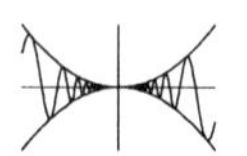

Sequences can be defined by function iteration. For example, let $f(x) = 2.8x(1-x)$ and define a sequence $\{a_n\}$ by $a_1 = \frac{1}{2}$ and $a_{n+1} = f(a_n)$ for $n \geq 1$. Calculate the first 10 terms and plot the graph. What seems to be happening? Repeat with the functions $g(x) = 3.4x(1-x)$ and $h(x) = 3.8x(1-x)$. Do these sequences converge or diverge? Discuss their behavior.

Nancy's Solution

The first 10 iterations on the function $f(x) = 2.8x(1-x)$, starting with $a_0 = \frac{1}{2}$, produce

$$\begin{bmatrix} a_0 \\ a_1 \\ a_2 \\ a_3 \\ a_4 \\ a_5 \\ a_6 \\ a_7 \\ a_8 \\ a_9 \\ a_{10} \end{bmatrix} = \begin{bmatrix} .5 \\ .7 \\ .588 \\ .67832 \\ .61097 \\ .66552 \\ .62329 \\ .65744 \\ .6306 \\ .65225 \\ .6351 \end{bmatrix}$$

With the insertion point in the equation $f(x) = 2.8x(1-x)$, choose Define + New Definition. Choose Calculus + Iterate (Iteration Function: f, Starting Value: .5, Number of Iterations: 10). Choose OK.

Here is a graph of the first 50 terms of $\{a_n\}$.

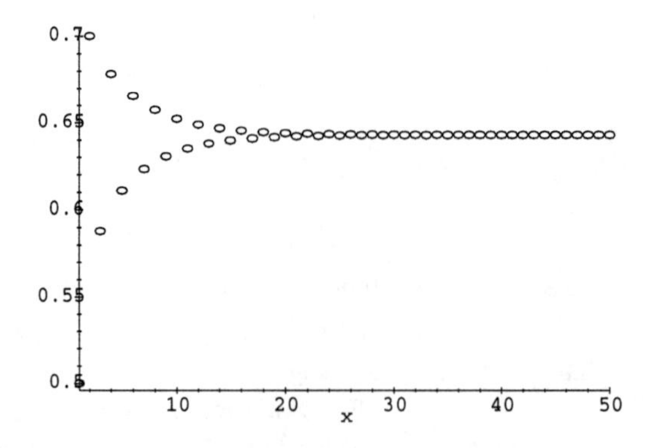

Choose Calculus + Iterate (Iteration Function f, Starting Value .5, Number of Iterations 49). Choose OK. With the insertion point in the expression $R(i, j) = j$, choose Define + New Definition. With the insertion point to the left of the 50-row, 1-column matrix of values, choose Matrix + Fill Matrix. Choose Fill with Defined by Function, 50 rows, 1 column, and type R in the box under Enter Function Name. Choose OK. (You should have a 50-row, 1-column matrix with entries 1 through 50.) Choose Matrix + Concatenate. Choose Plot 2D + Rectangular. Choose Edit + Properties, Plot Components page. Change Plot Style to Point, Point Symbol to Circle, and Domain Interval to $0 < x < 50$. Choose OK.

Notice that the first few terms seem to oscillate back and forth, but then settle down rather quickly. The sequence appears to converge to some number around .64. It is also interesting to look at the line plot spanned by these points.

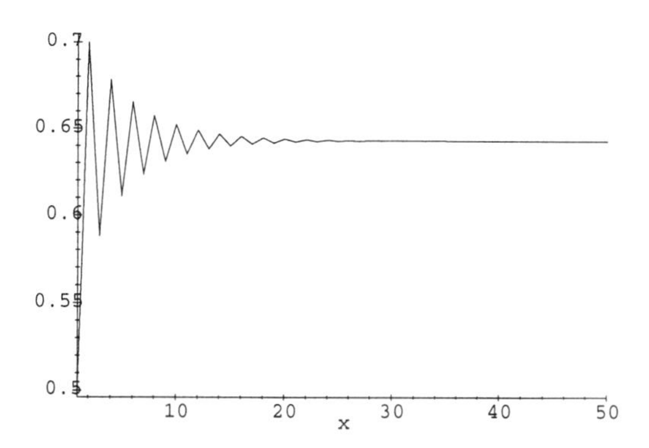

Choose Edit + Properties, Plot Components page. Change Plot Style to Line.

Iteration on the function $g(x) = 3.4x(1-x)$ produces a more surprising result. Here are the first 10 iterations.

$$\begin{bmatrix} b_0 \\ b_1 \\ b_2 \\ b_3 \\ b_4 \\ b_5 \\ b_6 \\ b_7 \\ b_8 \\ b_9 \\ b_{10} \end{bmatrix} = \begin{bmatrix} .5 \\ .85 \\ .4335 \\ .83496 \\ .46852 \\ .84663 \\ .44148 \\ .83836 \\ .46075 \\ .84476 \\ .44587 \end{bmatrix}$$

The following shows the first 50 terms. The values oscillate between about 4.5 and 8.4. It appears that the sequence $\{b_n\}$ does not converge. It looks as though there are two limit points. A sequence must have a unique limit point in order for it to converge.

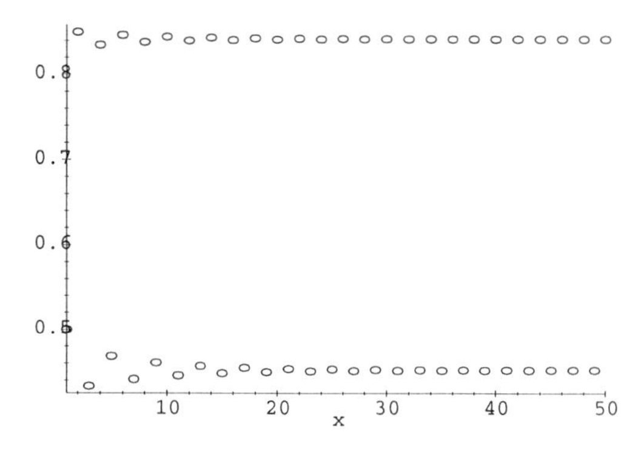

Follow a procedure similar to that described for the first plot.

Define c_n by $c_0 = .5$ and $c_{n+1} = h(c_n)$, where $f(x) = 3.8x(1-x)$. The sequence $\{c_n\}$ looks really wild. The first 10 iterations yield

$$\begin{bmatrix} c_0 \\ c_1 \\ c_2 \\ c_3 \\ c_4 \\ c_5 \\ c_6 \\ c_7 \\ c_8 \\ c_9 \\ c_{10} \end{bmatrix} = \begin{bmatrix} .5 \\ .95 \\ .1805 \\ .5621 \\ .93535 \\ .22979 \\ .67256 \\ .83685 \\ .51882 \\ .94865 \\ .1851 \end{bmatrix}$$

The first 50 terms fail to establish any definite pattern. The points are all over the place. Certainly the sequence $\{c_n\}$ does not converge.

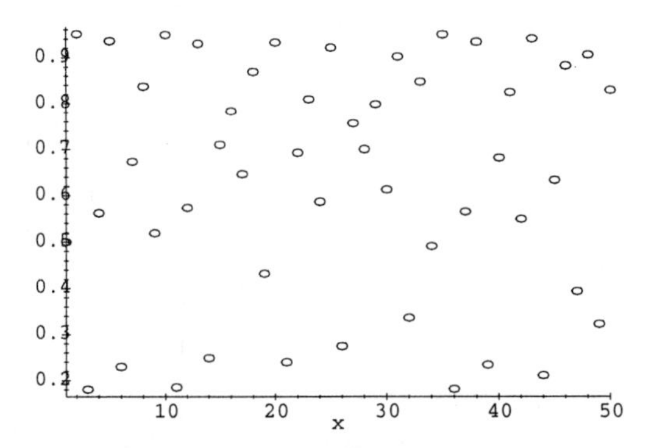

Follow a procedure similar to that described for the first plot.

Remark This is an example of chaos. What this implies is roughly the following: If we start with a slightly different value, $d_0 = .49$, then we get the result

$$\begin{bmatrix} d_0 \\ d_1 \\ d_2 \\ d_3 \\ d_4 \\ d_5 \\ d_6 \\ d_7 \\ d_8 \\ d_9 \\ d_{10} \end{bmatrix} = \begin{bmatrix} .49 \\ .94962 \\ .1818 \\ .56524 \\ .93382 \\ .23483 \\ .68279 \\ .82303 \\ .55348 \\ .93913 \\ .21722 \end{bmatrix}$$

where the terms quickly start to look quite different from the sequence $\{c_n\}$, although the terms start out very close to the terms of that sequence.

Activities

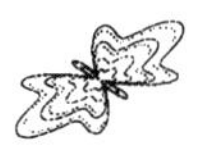

1. Experiment with the sequence $\left\{\sqrt{3}, \sqrt{3\sqrt{3}}, \sqrt{3\sqrt{3\sqrt{3}}}, \ldots\right\}$. Define f by $f(x) = \sqrt{3x}$ and calculate the first 10 terms of the sequence by applying Calculus + Iterate to the function f with starting value 1. Guess a value for a limit a. Verify that $f(a) = a$.

2. Experiment with the sequence of convergents determined by the continued fraction expansion of π. Verify that the first few convergents alternate above and below the value of π. Note that the popular approximations $22/7$ and $355/113$ appear among the convergents.

3. Experiment with iteration on the function $f(x) = ax(1 - x)$ for various values of a. Where is the dividing line between convergence and divergence? Can you find a number a that generates four distinct limit points? What other numbers of distinct limit points are possible?

10.2 Series

The widespread use of infinite series coincided with the early development of the integral calculus. Series can be used to solve a wide variety of problems. Here is an example that involves the idea of expectation.

The Cereal Box Problem

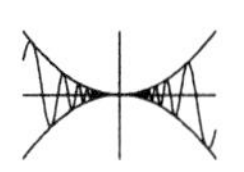

Two different prizes are put inside boxes of Gooey Delight breakfast cereal. If you are very lucky, you may get both prizes by purchasing only two boxes of cereal. But if the first two boxes each contain the same prize, then a third box must be purchased in order to get a full set of prizes. Assuming the prizes are randomly distributed in the cereal boxes, how many boxes do you expect to purchase before you get a complete set of prizes? The expected number can be thought of as the average you would get if you repeated an experiment many times. Let $P(n)$ denote the probability that n boxes must be purchased in order to get a full set of prizes. Show that the expected number of boxes is

$$\sum_{n=2}^{\infty} nP(n)$$

Find an expression for $P(n)$ and use this to find the expected number of cereal boxes. Simulate the collection of prizes by flipping a coin. Heads represents prize A and tails represents prize B. Keep track of how many coin flips are required before you get both prizes. Simulate the collection of prizes by generating 100 random numbers. Discuss whether or not your solution is reasonable based on your experiments.

Sharon's Solution

Evidently $P(1) = 0$, because both prizes are assumed never to appear in the same box. Note that $P(2) = \frac{1}{2}$ because the patterns AB, BA, AA, BB should all be equally likely, so AB or BA should each appear one-fourth of the time. Three boxes will be required exactly in the cases AAB and BBA, and the eight possible strings AAA, AAB, ABA, BAA, ABB, BAB, BBA, BBB should be equally likely, so $P(3) = \frac{1}{4}$. Similarly, $P(4) = \frac{1}{8}$, $P(5) = \frac{1}{16}$, and, in general, $P(i) = \frac{1}{2^{i-1}}$. The expected number of boxes is given by

$$\sum_{i=2}^{\infty} iP(i) = 3$$

With the insertion point in the equation $P(i) = \frac{1}{2^{i-1}}$, choose Define + New Definition. With the insertion point in the expression $\sum_{i=2}^{\infty} iP(i)$, choose Evaluate.

To verify this using coin flips, I repeated an experiment 10 times, getting the following set of outcomes:

$$TH, HT, TTTTH, HT, HT, HT, HT, HT, TTH, HHHHT$$

The average number of coins tossed during each trial was

$$\frac{2+2+5+2+2+2+2+2+3+5}{10} = \frac{27}{10} = 2.7$$

As a second experiment, I generated 100 random numbers between 0 and 2. Numbers between 0 and 1 were interpreted as prize A, and numbers between 1 and 2 as prize B. Here is a summary of the experimental results:

Number of Boxes	2	3	4	5	6	7
Number of Occurrences	22	9	2	0	1	2

The average number of boxes required was thus

$$\frac{2\times 22+3\times 9+4\times 2+5\times 0+6\times 1+7\times 2}{22+9+2+0+1+2} = \frac{11}{4} = 2.75$$

Thus, 3 looks like a reasonable answer.

Choose Statistics + Random Numbers (How Many? 100, Uniform Distribution, Lower End of Range 0, Upper End of Range 2). Choose OK.

Activity

Crunchy Munchies offers a total of three prizes, randomly distributed with one prize in each box. How many boxes do you expect to purchase to collect a complete set of

Crunchy Munchies prizes? Think of a way to simulate this using coins, dice, or playing cards. Conduct a second experiment using 100 random numbers.

10.3 The Integral Test and Estimates of Sums

There is a close relationship between series and definite integrals. If f is a continuous, positive, decreasing function on the interval $[1, \infty)$, and $a_n = f(n)$ for positive integers n, then the series $\sum_{n=1}^{\infty} a_n$ converges if and only if the improper integral $\int_1^{\infty} f(x)dx$ converges. This is known as the *Integral Test.*

Connecting Improper Integrals With Series

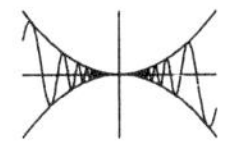

Consider the series

$$\sum_{n=2}^{\infty} \frac{1}{n \ln n}$$

Show that

$$\sum_{n=2}^{10} \frac{1}{n \ln n} > \int_2^{10} \frac{dx}{x \ln x}$$

by plotting a left Riemann sum for the integral using 8 subintervals. Show that

$$\sum_{n=2}^{20} \frac{1}{n \ln n} > \int_2^{20} \frac{dx}{x \ln x}$$

by plotting a Riemann sum for the integral using 18 subintervals. Evaluate the improper integral

$$\int_2^{\infty} \frac{dx}{x \ln x}$$

and explain why you believe that

$$\sum_{n=2}^{\infty} \frac{1}{n \ln n}$$

diverges.

Ron's Solution

The left Riemann sum

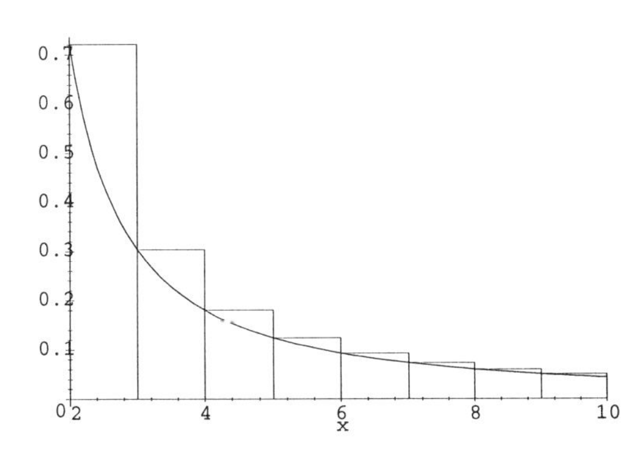

indicates that

$$\sum_{n=2}^{10} \frac{1}{n \ln n} > \int_2^{10} \frac{dx}{x \ln x}$$

and, in fact,

$$\sum_{n=2}^{10} \frac{1}{n \ln n} = 1.6499$$

$$\int_2^{10} \frac{dx}{x \ln x} = 1.2005$$

With the insertion point in the expression $\frac{1}{x \ln x}$, choose Calculus + Plot Approx. Integral (Left Boxes, Number of Boxes 8, Ranges 2 to 10). With the insertion point in the summation $\sum_{n=2}^{10} \frac{1}{n \ln n}$, choose Evaluate Numerically. With the insertion point in the integral $\int_2^{10} \frac{dx}{x \ln x}$, choose Evaluate Numerically.

The left Riemann sum

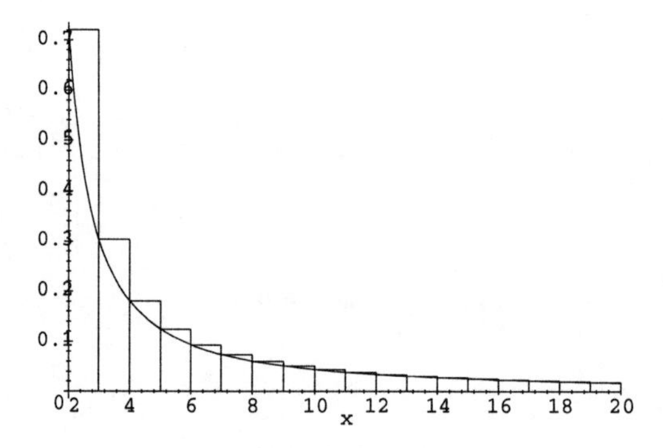

indicates that

$$\sum_{n=2}^{10} \frac{1}{n \ln n} > \int_2^{10} \frac{dx}{x \ln x}$$

and, in fact,

$$\sum_{n=2}^{20} \frac{1}{n \ln n} = 1.9001$$

$$\int_2^{20} \frac{dx}{x \ln x} = 1.4637$$

With the insertion point in the expression $\frac{1}{x \ln x}$, choose Calculus + Plot Approx. Integral (Left Boxes, Number of Boxes 18, Ranges 2 to 20). With the insertion point in the summation $\sum_{n=2}^{20} \frac{1}{n \ln n}$, choose Evaluate Numerically With the insertion point in the integral $\int_2^{20} \frac{dx}{x \ln x}$, choose Evaluate Numerically.

The function $\frac{1}{x \ln x}$ is monotone decreasing. Because

$$\int_2^\infty \frac{dx}{x \ln x} = \int_{\ln 2}^\infty \frac{1}{u}\, du = \infty$$

it follows that

$$\sum_{n=2}^\infty \frac{1}{n \ln n} = \infty$$

With the insertion point in the integral $\int_2^\infty \frac{dx}{x \ln x}$, choose Calculus + Change Variable (Substitution: $u = \ln x$). With the insertion point in the integral $\int_{\ln 2}^\infty \frac{1}{u}\, du$, choose Evaluate.

Activity

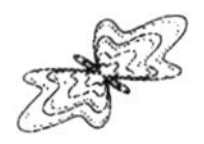

Use the Integral Test to decide whether or not the series

$$\sum_{n=1}^\infty \frac{1}{n^2 + 4}$$

converges. Generate a graph with left boxes or right boxes that shows the relevant comparison with an improper integral.

10.4 The Comparison Tests

The convergence or divergence of many series can be determined by comparing them against a few prototype series with known convergence properties. If $\sum_{n=1}^\infty a_n$ is a series for which $\lim_{n\to\infty} \left|\frac{a_{n+1}}{a_n}\right| = L$ exists, the *Ratio Test* states that the series is absolutely convergent (and therefore convergent) if $L < 1$, and the series is divergent if $L > 1$ (including the case $L = \infty$).

Prototype Series

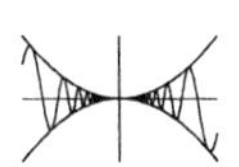

Explore the convergence properties of p-series $\sum_{n=1}^\infty \frac{1}{n^p}$ and geometric series $\sum_{n=0}^\infty ar^n$. Use the Integral Test to determine the values of p for which $\sum_{n=1}^\infty \frac{1}{n^p}$ converges and the values for which it diverges. Use the Ratio Test to determine the values of r for which the geometric series $\sum_{n=0}^\infty ar^n$ converges and for which values of r it diverges.

May's Solution

We look first at the p-series for $p = 1$. Note that

$$\int_1^\infty \frac{dx}{x} = \infty$$

With the insertion point in the integral $\int_1^\infty \frac{dx}{x}$, choose Evaluate.

To show that the series $\sum_{n=1}^{\infty} \frac{1}{n}$ diverges, we apply the Integral Test. The following figure

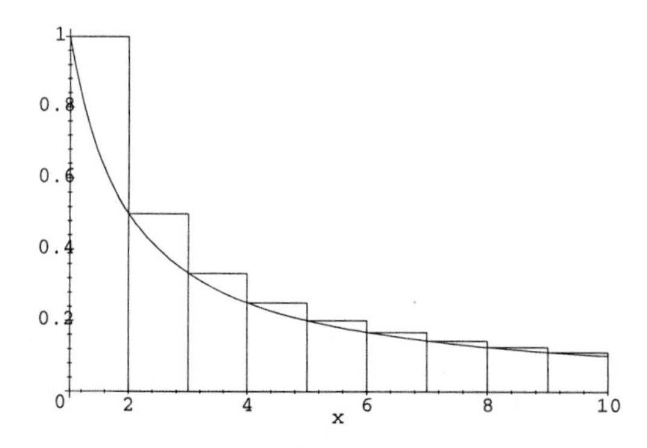

$y = \frac{1}{x}$ and $\sum_{n=1}^{10} \frac{1}{n}$

indicates that

$$\sum_{n=1}^{k} \frac{1}{n} > \int_1^k \frac{dx}{x}$$

and hence that $\sum_{n=1}^{\infty} \frac{1}{n}$ diverges.

With the insertion point to the right of the expression $\frac{1}{x}$, choose Calculus + Plot Approx. Integral (Left Boxes, Number of Boxes 9, Ranges 1 to 10). Choose OK.

For $p > 1$ we have

$$\begin{aligned} \int_1^\infty \frac{dx}{x^p} &= -\frac{1}{p-1} \lim_{x\to\infty} \left(x^{-p+1} - 1\right) \\ &= -\frac{1}{p-1} \lim_{x\to\infty} \left(\frac{1}{x^{p-1}} - 1\right) \\ &= \frac{1}{p-1} \end{aligned}$$

and hence the integral $\int_1^\infty \frac{dx}{x^p}$ converges.

With the insertion point in the integral $\int_1^\infty \frac{dx}{x^p}$, choose Evaluate. Fill in the remaining steps.

The following figure (drawn for $p = 2$)

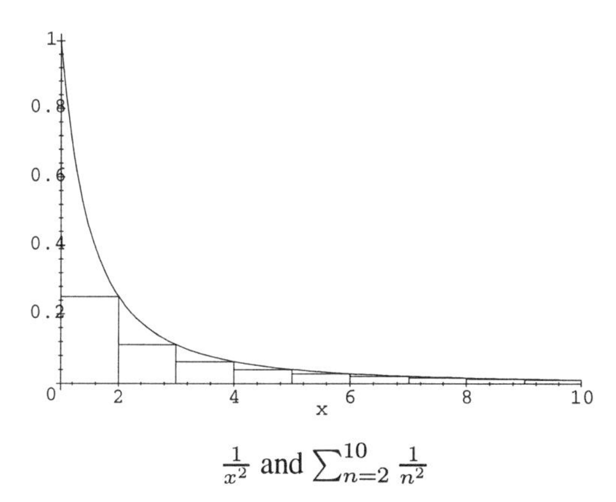

$\frac{1}{x^2}$ and $\sum_{n=2}^{10} \frac{1}{n^2}$

indicates that $\sum_{n=2}^{k} \frac{1}{n^p} < \int_1^k \frac{dx}{x^p}$ and hence the series $\sum_{n=1}^{\infty} \frac{1}{n^p}$ converges.

With the insertion point in the expression $\frac{1}{x^2}$, choose Calculus + Plot Approx. Integral (Right Boxes, Number of Boxes: 9, Ranges: 1 to 10).

To determine the convergence properties of a geometric series $\sum_{n=0}^{\infty} ar^n$, we define $b_n = ar^n$ and calculate the ratios $\frac{b_{n+1}}{b_n}$ of successive terms:

$$\frac{b_{n+1}}{b_n} = \frac{ar^{n+1}}{ar^n} = r$$

Thus

$$\lim_{n \to \infty} \frac{b_{n+1}}{b_n} = r$$

According to the Ratio Test, the series converges if $0 \leq r < 1$ and diverges if $r > 1$.

The Ratio Test does not tell us what happens when $r = 1$. However, for $r = 1$, we can see by inspection that the series $1 + 1 + 1 + \cdots$ and $1 - 1 + 1 - 1 + \cdots$ both diverge.

Activities

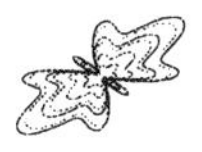

Apply the Comparison Test or the Limit Comparison Test to determine the convergence or divergence of each of the following series:

1. $\sum_{n=1}^{\infty} \frac{n}{n^3+1}$

2. $\sum_{n=1}^{\infty} \frac{1}{2^{n+1}-1}$

3. $\sum_{n=1}^{\infty} \frac{n}{\sqrt{n^5+1}}$

4. $\sum_{n=2}^{\infty} \frac{1}{\ln n}$

10.5 Alternating Series

An alternating series is one whose terms alternate between positive and negative values.

An Improper Alternative

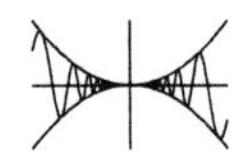

Interpret the improper integral $\int_0^\infty \sin x^2\, dx$ as an alternating series by taking the terms of the series to represent areas of loops of the curve $y = \sin x^2$ above (for the positive terms) or below (for the negative terms) the x-axis. Give an argument that shows that the alternating series converges.

Ann's Solution

Here is a graph of $y = \sin x^2$.

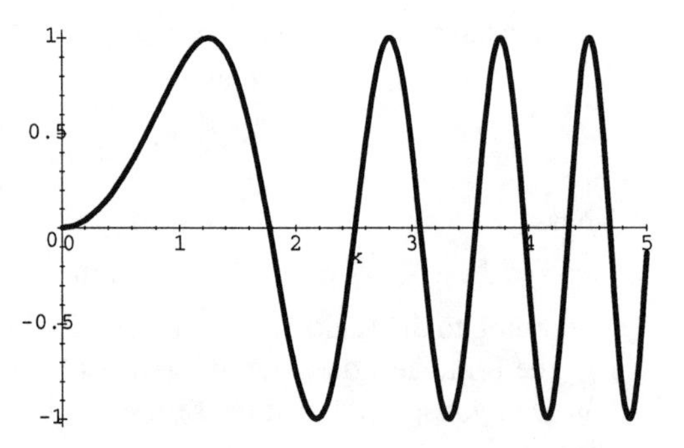

 With the insertion point in the expression $\sin x^2$, choose Plot 2D + Rectangular.

Evidently, the x-intercepts are at $0, \sqrt{\pi}, \sqrt{2\pi}, \sqrt{3\pi}, \sqrt{4\pi}$, and so forth. The improper integral $\int_0^\infty \sin x^2\, dx$ can be interpreted as the sum

$$\sum_{n=0}^{\infty}\left(\int_{\sqrt{n\pi}}^{\sqrt{(n+1)\pi}} \sin x^2\, dx\right)$$

of integrals of the form $\int_{\sqrt{n\pi}}^{\sqrt{(n+1)\pi}} \sin x^2\, dx$ as n takes on integer values from 0 to ∞. The first few terms are given by

$$\int_0^{\sqrt{\pi}} \sin x^2\, dx = .89483$$

$$\int_{\sqrt{\pi}}^{\sqrt{2\pi}} \sin x^2\, dx = -.46442$$

$$\int_{\sqrt{2\pi}}^{\sqrt{3\pi}} \sin x^2 \, dx = .35785$$
$$\int_{\sqrt{3\pi}}^{\sqrt{4\pi}} \sin x^2 \, dx = -.30201$$
$$\int_{\sqrt{4\pi}}^{\sqrt{5\pi}} \sin x^2 \, dx = .2662$$
$$\int_{\sqrt{5\pi}}^{\sqrt{6\pi}} \sin x^2 \, dx = -.24071$$

 With the insertion point inside each of the integrals, choose Evaluate Numerically.

The first few partial sums are given by

$$\begin{aligned}
.89483 &= .89483 \\
.89483 - .46442 &= .43041 \\
.89483 - .46442 + .35785 &= .78826 \\
.89483 - .46442 + .35785 - .30201 &= .48625 \\
.89483 - .46442 + .35785 - .30201 + .2662 &= .75245 \\
.89483 - .46442 + .35785 - .30201 + .2662 - .24071 &= .51174
\end{aligned}$$

 Place the insertion point inside each sum and choose Evaluate.

The infinite series is an alternating series, the sequence of absolute values of the terms of the series is monotone decreasing, and

$$\lim_{n\to\infty} \int_{\sqrt{n\pi}}^{\sqrt{(n+1)\pi}} \sin x^2 \, dx = 0$$

because $-1 \le \sin x^2 \le 1$ and

$$\lim_{n\to\infty} \left(\sqrt{(n+1)\pi} - \sqrt{n\pi} \right) = 0$$

It follows that the improper integral $\int_0^\infty \sin x^2 dx$ must converge. In fact,

$$\int_0^\infty \sin x^2 dx = \frac{1}{4}\sqrt{2}\sqrt{\pi} \approx .62666$$

is compatible with the first few partial sums.

With the insertion point inside the integral $\int_0^\infty \sin x^2 dx$, choose Evaluate; then choose Evaluate Numerically.

Activity

Experiment with the improper integral $\int_0^\infty \sin x^{1+\varepsilon}\, dx$ for $\varepsilon > 0$, and argue that this improper integral can be interpreted as a convergent alternating series. Explain why the improper integral $\int_0^\infty \sin x\, dx$ does not exist. Consider the one-sided limit

$$\lim_{\varepsilon \to 0+} \left(\int_0^\infty \sin\left(x^{1+\varepsilon}\right) dx \right)$$

Generate some evidence that the limit exists by evaluating the following improper integrals:

$$\int_0^\infty \sin\left(x^{4/3}\right) dx$$

$$\int_0^\infty \sin\left(x^{9/8}\right) dx$$

$$\int_0^\infty \sin\left(x^{16/15}\right) dx$$

$$\int_0^\infty \sin\left(x^{101/100}\right) dx$$

Evaluate

$$\lim_{\varepsilon \to 0+} \left(\int_0^\infty \sin\left(x^{1+\varepsilon}\right) dx \right)$$

and discuss whether or not what you get is reasonable. Discuss how this limit relates to the divergence of the improper integral

$$\int_0^\infty \sin x\, dx$$

10.6 Absolute Convergence and the Ratio and Root Tests

A series $\sum_{n=1}^\infty a_n$ is called *absolutely convergent* if the series $\sum_{n=1}^\infty |a_n|$ converges. A convergent series that is not absolutely convergent is called *conditionally convergent*.

You Control the Sum

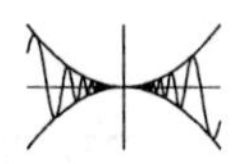

Consider the conditionally convergent series

$$\sum_{n=1}^{\infty} (-1)^{n+1} \frac{1}{n}$$

Use the terms of this series as building blocks to construct the first 10 terms of a conditionally convergent series that converges to $\frac{1}{e} \approx .36788$.

Joe's Solution

We have the positive building blocks $1 + \frac{1}{3} + \frac{1}{5} + \frac{1}{7} + \frac{1}{9} + \frac{1}{11} + \cdots$ and the negative building blocks $-\frac{1}{2} - \frac{1}{4} - \frac{1}{6} - \frac{1}{8} - \frac{1}{10} - \cdots$ with which to work. Becausee $\frac{1}{e}$ is positive, we start first with a positive building block and form a series with the following partial sums:

$$
\begin{aligned}
1 &> .36788 \\
1 - \frac{1}{2} &= .5 > .36788 \\
1 - \frac{1}{2} - \frac{1}{4} &= .25 < .36788 \\
1 - \frac{1}{2} - \frac{1}{4} + \frac{1}{3} &= .58333 > .36788 \\
1 - \frac{1}{2} - \frac{1}{4} + \frac{1}{3} - \frac{1}{6} &= .41667 > .36788 \\
1 - \frac{1}{2} - \frac{1}{4} + \frac{1}{3} - \frac{1}{6} - \frac{1}{8} &= .29167 < .36788 \\
1 - \frac{1}{2} - \frac{1}{4} + \frac{1}{3} - \frac{1}{6} - \frac{1}{8} + \frac{1}{5} &= .49167 > .36788 \\
1 - \frac{1}{2} - \frac{1}{4} + \frac{1}{3} - \frac{1}{6} - \frac{1}{8} + \frac{1}{5} - \frac{1}{10} &= .39167 > .36788 \\
1 - \frac{1}{2} - \frac{1}{4} + \frac{1}{3} - \frac{1}{6} - \frac{1}{8} + \frac{1}{5} - \frac{1}{10} - \frac{1}{12} &= .30833 < .36788 \\
1 - \frac{1}{2} - \frac{1}{4} + \frac{1}{3} - \frac{1}{6} - \frac{1}{8} + \frac{1}{5} - \frac{1}{10} - \frac{1}{12} + \frac{1}{7} &= .45119 > .36788
\end{aligned}
$$

N To generate the expression

$$1 - \frac{1}{2} - \frac{1}{4} + \frac{1}{3} - \frac{1}{6} = .41667 > .36788$$

Type $1 - \frac{1}{2} - \frac{1}{4} + \frac{1}{3} - \frac{1}{6} =$. Select and drag to the right of the equality sign $1 - \frac{1}{2} - \frac{1}{4} + \frac{1}{3} - \frac{1}{6}$. While the selection is still highlighted, press and hold CTRL/COMMAND, and choose Evaluate Numerically. Type $> .36788$ or $< .36788$, as appropriate.

Activities

1. Use the terms of the conditionally convergent series $\sum_{n=1}^{\infty} \frac{1}{n}$ as building blocks to construct the first 10 terms of a conditionally convergent series that converges to 0.

2. Explain how you could use the terms of the conditionally convergent series $\sum_{n=1}^{\infty} \frac{1}{n}$ as building blocks to design a series that diverges to $+\infty$.

Hint Use positive building blocks to create a partial sum that exceeds 10, then use the first negative building block $-\frac{1}{2}$, then use more positive terms until the sum exceeds 20, then use another negative building block, and so on.

10.7 Strategy for Testing Series

How does a computer algebra system fit in with other strategies for testing series?

Activity

Write a summary of all the strategies you have learned for testing the convergence or divergence of series. Why do you think it might be important to understand these strategies even when you have a computer algebra system to help you test series?

10.8 Power Series

When plotting the graph of a power series using a computer algebra system, beware of ghosts. What a computer algebra system often does is to find a simple expression that is represented by the power series on its interval of convergence, then plots the graph of the expression, including regions outside the interval of convergence.

Will the Real Interval of Convergence Please Stand Up?

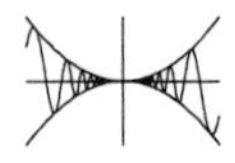

Plot a graph of the power series $f(x) = \sum_{i=0}^{\infty} x^i$. Try to determine the interval of convergence from the graph. Determine the interval of convergence algebraically. Test the interval of convergence by evaluating the power series at sample points inside and outside the interval of convergence. Evaluate the power series to see what function the computer algebra system is graphing. Plot a few partial sums to test the interval of convergence.

Candice's Solution

Looking at the graph of

$$\sum_{i=0}^{\infty} x^i$$

it appears that there is a vertical asymptote at $x = 1$, but otherwise the curve is visible between $x = -2$ and $x = 2$. It is confusing to try to visualize the interval of convergence by looking at the graph.

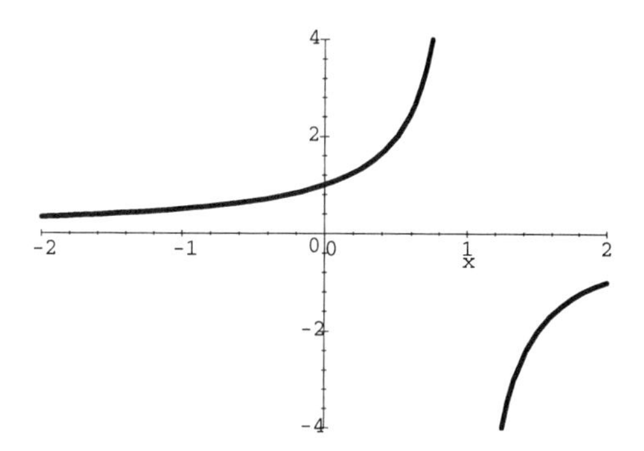

With the insertion point in the series $\sum_{i=0}^{\infty} x^i$, choose Plot 2D + Rectangular (set Domain Interval to $-2 \leq x \leq 2$ and set the View Intervals to $-2 \leq x \leq 2$ by $-4 \leq y \leq 4$).

According to the Ratio Test, the series should converge for $|x| < 1$ and diverge for $|x| > 1$. Defining

$$f(x) = \sum_{i=0}^{\infty} x^i$$

we get

$$\begin{aligned} f(-2) &= \sum_{i=0}^{\infty} (-2)^i = 1 - 2 + 4 - 8 + 16 - 32 + \cdots \\ f\left(-\frac{3}{4}\right) &= \frac{4}{7} \\ f\left(\frac{1}{2}\right) &= 2 \\ f\left(\frac{99}{100}\right) &= 100 \\ f\left(\frac{100}{99}\right) &= \infty \end{aligned}$$

With the insertion point in the equation $f(x) = \sum_{i=0}^{\infty} x^i$, choose Define + New Definition. With the insertion point in each of $f(-2)$, $f\left(-\frac{3}{4}\right)$, $f\left(\frac{1}{2}\right)$, $f\left(\frac{99}{100}\right)$, and $f\left(\frac{100}{99}\right)$, choose Evaluate.

Although evaluation of $f(a)$ for $a < -1$ returns an expression for an infinite series, that series evidently diverges because the terms of the series do not tend to zero. Sample evaluations $f(a)$ for a inside the interval of convergence seem to give reasonable results, and sample evaluations $f(a)$ for $a > 1$ return ∞.

Evaluating the series

$$\sum_{i=0}^{\infty} x^i$$

provides a clue as to what is happening. The graph that appears is actually the graph of

$$y = \frac{1}{1-x}$$

which has a vertical asymptote at $x = 1$ but is defined everywhere else. On the interval $-1 < x < 1$, the series represents the function $y = \frac{1}{1-x}$.

 With the insertion point in the series $\sum_{i=0}^{\infty} x^i$, choose Evaluate.

To get a better view of the limitations imposed by the interval of convergence, we plot the series together with several partial sums. The graphs of the partial sums appear to try to match the graph of the series inside the interval of convergence, and quickly deviate from the graph for $|x| > 1$.

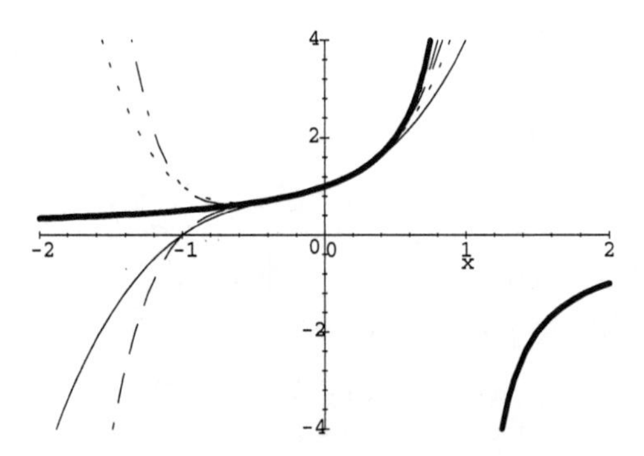

$\sum_{i=0}^{\infty} x^i$, $\sum_{i=0}^{3} x^i$, $\sum_{i=0}^{4} x^i$, $\sum_{i=0}^{5} x^i$, $\sum_{i=0}^{6} x^i$

With the insertion point in the series $\sum_{i=0}^{\infty} x^i$, choose Plot 2D + Rectangular. Select and drag each of the series $\sum_{i=0}^{3} x^i$, $\sum_{i=0}^{4} x^i$, $\sum_{i=0}^{5} x^i$, $\sum_{i=0}^{6} x^i$ to the plot. Choose Edit + Properties, and change Line Style and Line Color as you wish. Set Domain Interval to $-2 \leq x \leq 2$, and set the View Intervals to $-2 \leq x \leq 2$ by $-4 \leq y \leq 4$. Choose OK.

Because the series diverges for $|x| \geq 1$, a better representation of the graph of

$$y = \sum_{i=0}^{\infty} x^i$$

can be made by specifying the Domain Interval to be $-1 < x < 1$ and the View Intervals to be $-2 \leq x \leq 2$ by $-4 \leq y \leq 4$.

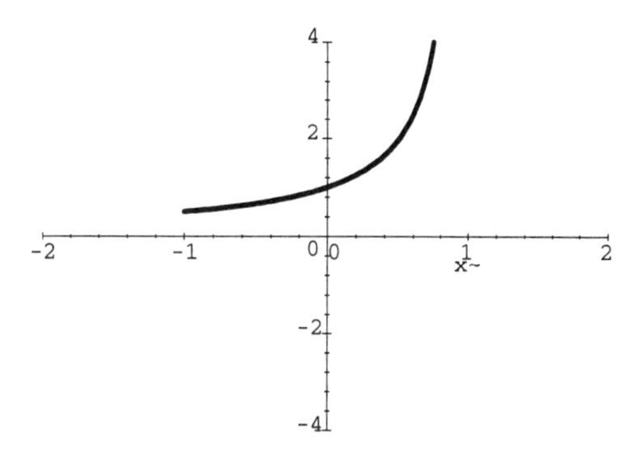

With the insertion point in the series $\sum_{i=0}^{\infty} x^i$, choose Plot 2D + Rectangular. Choose Edit + Properties, and change Thickness to Medium. Set Domain Interval to $-1 \leq x \leq 1$ and set the View Intervals to $-2 \leq x \leq 2$ by $-4 \leq y \leq 4$. Choose OK.

Activities

Experiment with the interval of convergence for each of the following series. If the interval of convergence is finite, then set the Domain Interval to be the interval of convergence and plot the graph of the series using a larger set of View Intervals. Evaluate the series at sample points inside and outside the interval of convergence to verify convergence or divergence at these points.

1. $\sum_{n=1}^{\infty} \frac{x^n}{n2^n}$

2. $\sum_{n=1}^{\infty} \frac{n^2}{1+n^2} x^n$

3. $\sum_{n=2}^{\infty} \frac{x^n}{\ln n}$

4. $\sum_{n=1}^{\infty} \left(\frac{n}{50}\right)^n x^n$

10.9 Representations of Functions as Power Series

Given a function $f(x)$, when can you find a power series $\sum_{n=0}^{\infty} a_n x^n$ such that $f(x) = \sum_{n=0}^{\infty} a_n x^n$ on some interval?

Series Solutions: One Coefficient at a Time

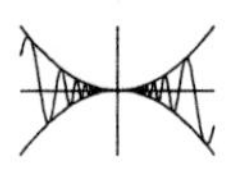

Assume that a power series can be differentiated term by term. Construct a power series

$$p(x) = \sum_{n=0}^{\infty} a_n x^n$$

such that $p''(x) = -p(x)$, $p(0) = 0$, and $p'(0) = 1$. (Pretend not to notice that $p(x) = \sin x$ is a solution.)

David's Solution

We find the required series by using term-by-term differentiation, and then using the equation $p''(x) = -p(x)$ to equate terms. If we assume

$$p(x) = \sum_{n=0}^{\infty} a_n x^n = a_0 + a_1 x + a_2 x^2 + a_3 x^3 + \cdots$$

then

$$\begin{aligned} p'(x) &= \sum_{n=0}^{\infty} a_n x^n \frac{n}{x} \\ &= \sum_{n=0}^{\infty} a_n x^{n-1} n \\ &= a_1 + 2a_2 x + 3a_3 x^2 + 4a_4 x^3 + \cdots \end{aligned}$$

and

$$\begin{aligned} p''(x) &= \sum_{n=0}^{\infty} \left(a_n x^n \frac{n^2}{x^2} - a_n x^n \frac{n}{x^2} \right) \\ &= \sum_{n=0}^{\infty} (n-1)\, n a_n x^{n-2} \\ &= (2 \cdot 1)\, a_2 + (3 \cdot 2)\, a_3 x + (4 \cdot 3)\, a_4 x^2 + (5 \cdot 4)\, a_5 x^3 + \cdots \end{aligned}$$

With the insertion point in the equation $p(x) = \sum_{i=0}^{\infty} a_i x^i$, choose Define + New Definition. With the insertion point in the expression $p'(x)$, choose Evaluate, and choose Simplify. With the insertion point in the expression $p''(x)$, choose Evaluate, Simplify, and Factor (in place).

The condition $p(0) = 0$ implies $a_0 = 0$ and $p'(0) = 1$ implies $a_1 = 1$. Equating coefficients between the series for $p(x)$ and $p''(x)$, we see that $a_2 = 0$, $a_4 = 0$, and so forth. Similarly, $6a_3 = -a_1 = -1$ implies $a_3 = -\frac{1}{6}$ and $20a_5 = -a_3 = -\left(-\frac{1}{6}\right) = \frac{1}{6}$ implies $a_5 = \frac{1}{5!}$. Continuing this approach, we get the series

$$x - \frac{1}{3!}x^3 + \frac{1}{5!}x^5 - \frac{1}{7!}x^7 + \cdots = \sum_{n=0}^{\infty} (-1)^n \frac{1}{(2n-1)!} x^{2n-1}$$

Here is a graph of $y = x - \frac{1}{3!}x^3 + \frac{1}{5!}x^5 - \frac{1}{7!}x^7$ (in regular pen) with the graph of

$y = \sin x$ (drawn with a fat pen). Notice that for x close to zero, the truncated series $x - \frac{1}{3!}x^3 + \frac{1}{5!}x^5 - \frac{1}{7!}x^7$ provides a reasonable approximation to $\sin x$.

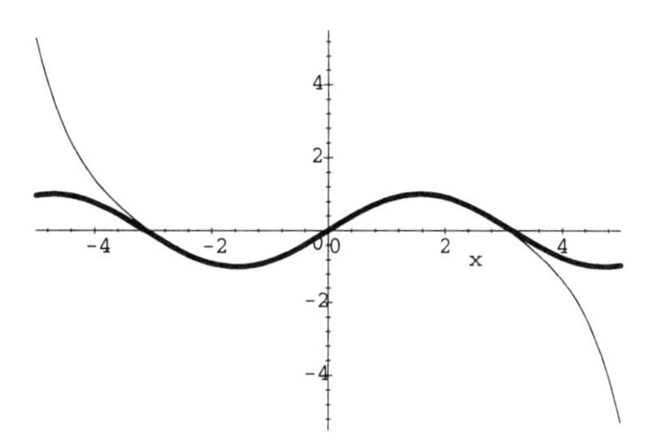

$y = x - \frac{1}{3!}x^3 + \frac{1}{5!}x^5 - \frac{1}{7!}x^7$ and $y = \sin x$

With the insertion point in the expression $\sin x$, choose Plot 2D + Rectangular. Select the expression $x - \frac{1}{3!}x^3 + \frac{1}{5!}x^5 - \frac{1}{7!}x^7$ and drag it to the plot. Change Thickness to Medium for $\sin x$.

Activities

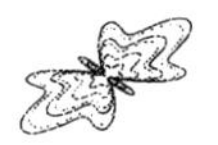

1. Find a series solution to the initial-value problem $y' = y$, $y(0) = 1$. Compare the graph of a truncated series solution with the graph of $y = e^x$.

2. Find a series solution to the initial-value problem $f'(x) = \frac{1}{x^2+1}$, $f(0) = 0$. Compare a graph of a truncated series solution with the graph of $y = \tan x$.

10.10 Taylor and Maclaurin Series

Taylor series are determined by looking closely at a function f at a particular point a. Let

$$p(x) = \sum_{i=0}^{\infty} c_i (x-a)^i$$

A *Taylor series of the function f at a* is a series $p(x)$ such that $p(a) = f(a)$, $p'(a) = f'(a)$, $p''(a) = f''(a)$, and so forth. A *Maclaurin series* is a Taylor series where $a = 0$.

Taylor Made Polynomials

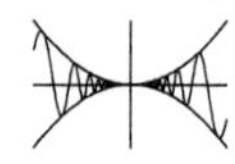

Let $f(x)$ be a generic function and define

$$p(x) = \sum_{i=0}^{5} c_i(x-a)^i$$

Evaluate p and its first five derivatives at a, then solve the following system of equations to determine the coefficients c_i. Use this to find a Taylor polynomial for $f(x) = \sin x$ at 0. Plot a graph of f together with this Taylor polynomial.

Keith's Solution

Let $f(x)$ be a generic function, and define p by the equation

$$p(x) = \sum_{i=0}^{5} c_i(x-a)^i$$

Evaluating p and its first five derivatives at a, we get

$$\begin{aligned} p(a) &= c_0 \\ p'(a) &= c_1 \\ p''(a) &= 2c_2 \\ p'''(a) &= 6c_3 \\ p^{(4)}(a) &= 24c_4 \\ p^{(5)}(a) &= 120c_5 \end{aligned}$$

N With the insertion point in the equation $p(x) = \sum_{i=0}^{5} c_i(x-a)^i$, choose Define + New Definition. Enter the system of expressions as shown, and with the insertion point in the system, choose Evaluate. Select and drag the values into the display.

Solving the system of equations

$$\begin{aligned} c_0 &= f(a) \\ c_1 &= f'(a) \\ 2c_2 &= f''(a) \\ 6c_3 &= f^{(3)}(a) \\ 24c_4 &= f^{(4)}(a) \\ 120c_5 &= f^{(5)}(a) \end{aligned}$$

we get,

$$\begin{aligned} c_0 &= f(a) \\ c_1 &= f'(a) \\ c_2 &= \frac{1}{2}f''(a) \end{aligned}$$

$$
\begin{aligned}
c_3 &= \frac{1}{6} f'''(a) \\
c_4 &= \frac{1}{24} f^4(a) \\
c_5 &= \frac{1}{120} f^5(a)
\end{aligned}
$$

With the insertion point in the expression $f(x)$, choose Define + New Definition. Enter the system of equations as shown, and with the insertion point in the system, choose Solve + Exact.

In particular, if $f(x) = \sin x$ then

$$
\begin{aligned}
f(0) &= 0 \\
f'(0) &= 1 \\
f''(0) &= 0 \\
f'''(0) &= -1 \\
f^{(4)}(0) &= 0 \\
f^{(5)}(0) &= 1
\end{aligned}
$$

and hence $p(x) = x - \frac{1}{6}x^3 + \frac{1}{120}x^5$.

With the insertion point in the equation $f(x) = \sin x$, choose Define + New Definition. Enter the system of equations as shown, and with the insertion point in the system, choose Evaluate.

The graphs of f and p are shown in the following figure:

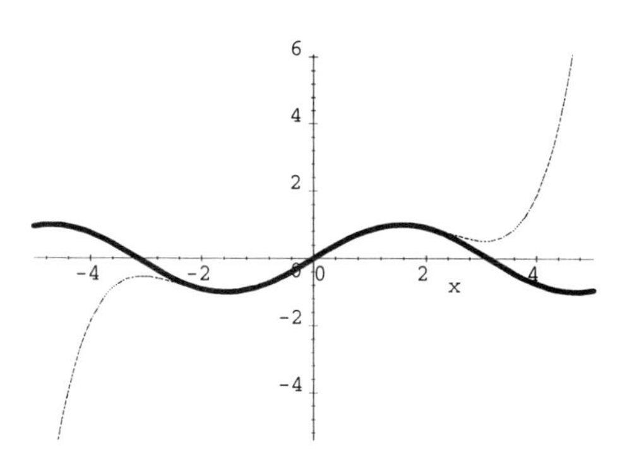

With the insertion point in the list of expressions $\sin x, x - \frac{1}{6}x^3 + \frac{1}{120}x^5$, choose Plot 2D + Rectangular.

Visualizing Taylor Polynomials

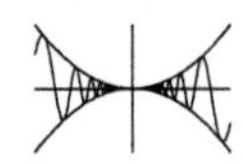

Find Taylor polynomials of the function $f(x) = \tan x$ at 0 of degrees 1, 3, 5, and 7, and compare their graphs with the graph of $y = \tan x$. Plot graphs of the error functions $p_n(x) - f(x)$ and use these graphs to explain whether or not you think enough terms would generate a function that looks like $\tan x$ on the entire interval $-\pi \leq x \leq \pi$.

Jeannine's Solution

When requested to return two terms of the Taylor series of $\tan x$ at 0, the symbolic computation system produced $x + O\left(x^2\right)$. The "two terms" are actually 0 and x. The expression $O\left(x^2\right)$ indicates that all the remaining terms in the Taylor series have a factor of at least x^2. Four terms look like $x + \frac{1}{3}x^3 + O\left(x^4\right)$. The Taylor polynomial is actually $0 + x + 0x^2 + \frac{1}{3}x^3$, and $O\left(x^4\right)$ indicates that all the other terms have a factor of at least x^4. In a similar manner, we get the Taylor polynomials $x + \frac{1}{3}x^3 + \frac{2}{15}x^5$ and $x + \frac{1}{3}x^3 + \frac{2}{15}x^5 + \frac{17}{315}x^7$.

With the insertion point in the expression $\tan x$, choose Power Series (Number of Terms 2, Expand in Powers of x). Repeat with Number of Terms 4, Number of Terms 6, and Number of Terms 8.

Graphs of $y = \tan x$ and the Taylor polynomials are given in the following figure. Notice how the graphs of the polynomials of higher degree try harder and harder to look like the graph of $y = \tan x$, but only on the interval $-\frac{\pi}{2} \leq x \leq \frac{\pi}{2}$.

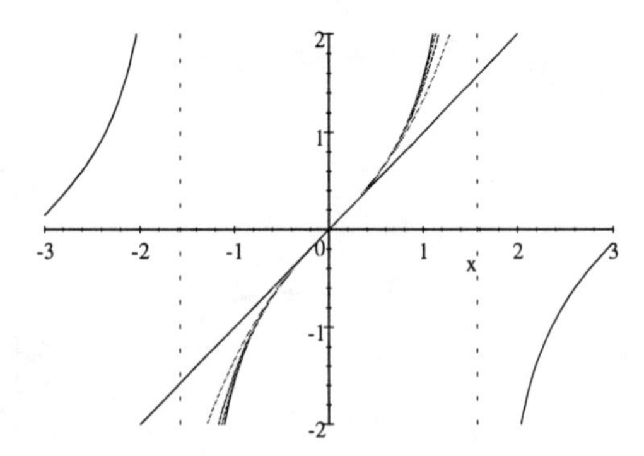

With the insertion point in the expression $\tan x$, choose Plot 2D + Rectangular. Select and drag to the frame each of x, $x+\frac{1}{3}x^3$, $x+\frac{1}{3}x^3+\frac{2}{15}x^5$, $x+\frac{1}{3}x^3+\frac{2}{15}x^5+\frac{17}{315}x^7$. Change Domain Interval to $-2 < x < 2$ and View Intervals to $-2 < x < 2$ and $-3.2 < y < 3.2$.

The following figure includes the error functions $x - \tan x$, $x + \frac{x^3}{3} - \tan x$, $x +$

$\frac{x^3}{3} + \frac{2x^5}{15} - \tan x$, and $x + \frac{x^3}{3} + \frac{2x^5}{15} + \frac{17x^5}{315} - \tan x$:

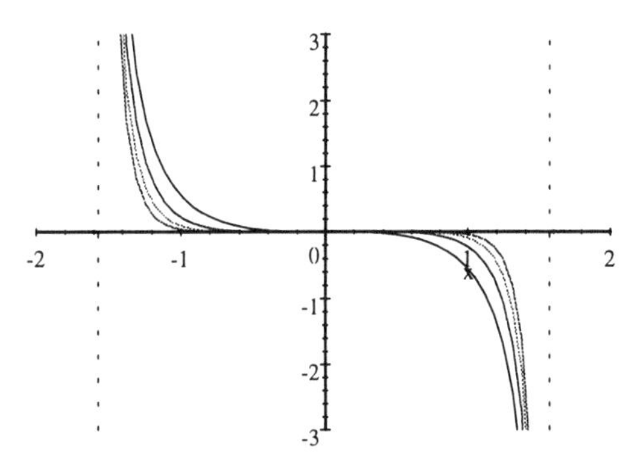

Notice how these error functions seem attached to more and more of the x-axis, but there appears to be a natural boundary at $x = \pm\frac{\pi}{2}$. It appears doubtful that any approximating polynomial could reasonably represent $\tan x$ on the entire interval $-\pi \leq x \leq \pi$.

Activities

1. Find Taylor polynomials of $\sin x$ at $\frac{\pi}{4}$ (expand in powers of $x - \frac{\pi}{4}$) of degrees 1, 2, 3, 4, and 5. Plot the graph of $y = \sin x$ together with these five polynomials, and discuss how well the graphs of these polynomials fit the graph of $y = \sin x$. Use the polynomials to estimate $\sin\frac{\pi}{2}$. Which polynomial provides the best estimate?

2. Find Maclaurin polynomials of $\ln(1 - x)$ and $\frac{1}{x^2+1}$ of degree 5; then expand the product of these two polynomials. Compare the terms of this polynomial with the Maclaurin polynomial of $\frac{\ln(1-x)}{x^2+1}$ of degree 10. Explain any discrepancies that you observe.

3. Find Maclaurin polynomials of the function $g(x) = \cos x$ of degrees 2, 4, and 6, and compare their graphs with the graph of $y = \cos x$. Discuss whether or not you think that enough terms would generate a function that looks like $\cos x$ on the entire interval $-100 \leq x \leq 100$.

10.11 The Binomial Series

The binomial coefficient $\binom{n}{k}$ represents the number of ways that k objects can be selected from among n objects. They can be calculated using the equation

$$\binom{n}{k} = \frac{n!}{k!\,(n-k)!} = \frac{n(n-1)(n-2)\cdots(n-k+1)}{k(k-1)(k-2)\cdots 1}$$

Evidently, this is always a positive integer, with n and k restricted to integers that satisfy $0 \leq k \leq n$.

The Binomial Expansion that Would Not End

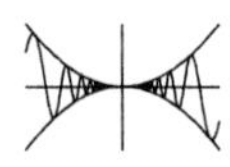

Make sense of the symbol $\binom{r}{n}$ for negative and noninteger values of r by defining $\binom{r}{n}$ by the equation

$$(1+x)^r = \sum_{n=0}^{\infty} \binom{r}{n} x^n$$

where the right side of the equation is the Maclaurin series for $(1+x)^r$. Observe that $\binom{r}{n} = 0$ if r is a positive integer and $n > r$. Investigate the binomial coefficient $\binom{r}{n}$ in the special cases $r = -2$ and $r = \frac{1}{2}$. In particular, construct a table of values in each of these cases for $n = 0, 1, 2, 3, 4, 5$. Test the Pascal identity

$$\binom{r}{n} + \binom{r}{n+1} = \binom{r+1}{n+1}$$

for the special cases $r = -2$ and $r = \frac{1}{2}$.

Harry's Solution

The Maclaurin series of $(1+x)^{-2}$ is

$$(1+x)^{-2} = 1 - 2x + 3x^2 - 4x^3 + 5x^4 - 6x^5 + 7x^6 - 8x^7 + 9x^8 - 10x^9 + O\left(x^{10}\right)$$

which indicates that

$$\binom{-2}{n} = (-1)^n (n+1)$$

With the insertion point in the expression $(1+x)^{-2}$, choose Power Series (Number of Terms: 10, Expand in Powers of: x). To create the expression $\binom{-2}{n}$, choose Insert + Binomial (Line, check None). Choose OK. Enter numbers in input boxes.

Similarly, the Maclaurin series of $(1+x)^{1/2}$ is given by

$$\begin{aligned}(1+x)^{1/2} &= 1 + \frac{1}{2}x - \frac{1}{8}x^2 + \frac{1}{16}x^3 - \frac{5}{128}x^4 + \frac{7}{256}x^5 \\ &\quad - \frac{21}{1024}x^6 + \frac{33}{2048}x^7 - \frac{429}{32768}x^8 + \frac{715}{65536}x^9 + O\left(x^{10}\right)\end{aligned}$$

and here the formula for $\binom{1/2}{n}$ is somewhat more complicated. However, in both cases we have

$$\binom{r}{n} = \frac{r(4-1)(r-2)\cdots(r-n+1)}{n!}$$

With the insertion point in the expression $(1+x)^{1/2}$, choose Power Series (Number of Terms 10, Expand in Powers of x).

Here is a table of values for selected binomial coefficients.

$$\binom{-2}{0} = \frac{1}{0!} = 1 \qquad \binom{1/2}{0} = \frac{1}{0!} = 1$$

$$\binom{-2}{1} = \frac{-2}{1!} = -2 \qquad \binom{1/2}{1} = \frac{1/2}{1!} = \frac{1}{2}$$

$$\binom{-2}{2} = \frac{(-2)(-3)}{2!} = 3 \qquad \binom{1/2}{2} = \frac{(1/2)(-1/2)}{2!} = -\frac{1}{8}$$

$$\binom{-2}{3} = \frac{(-2)(-3)(-4)}{3!} = -4 \qquad \binom{1/2}{3} = \frac{(1/2)(-1/2)(-3/2)}{3!} = \frac{1}{16}$$

$$\binom{-2}{4} = \frac{(-2)(-3)(-4)(-5)}{4!} = 5 \qquad \binom{1/2}{4} = \frac{(1/2)(-1/2)(-3/2)(-5/2)}{4!} = -\frac{5}{128}$$

$$\binom{-2}{5} = \frac{(-2)(-3)(-4)(-5)(-6)}{5!} = -6 \qquad \binom{1/2}{5} = \frac{(1/2)(-1/2)(-3/2)(-5/2)(-7/2)}{5!} = \frac{7}{256}$$

Select each of the binomial coefficients

$$\binom{-2}{0} \quad \binom{-2}{1} \quad \binom{-2}{2} \quad \binom{-2}{3} \quad \binom{-2}{4} \quad \binom{-2}{5}$$
$$\binom{1/2}{0} \quad \binom{1/2}{1} \quad \binom{1/2}{2} \quad \binom{1/2}{3} \quad \binom{1/2}{4} \quad \binom{1/2}{5}$$

and, for each, choose Evaluate.

Notice that for any nonnegative integer n and any real number r we have

$$\begin{aligned}
\binom{r}{n} + \binom{r}{n+1} &= \frac{r(r-1)(r-2)\cdots(r-n+1)}{n!} \\
&\quad + \frac{r(r-1)(r-2)\cdots(r-(n+1)+1)}{(n+1)!} \\
&= \frac{r(r-1)(r-2)\cdots(r-n+1)}{n!}\left(1 + \frac{(r-(n+1)+1)}{n+1}\right) \\
&= \frac{r(r-1)(r-2)\cdots(r-n+1)}{n!}\left(\frac{1+r}{n+1}\right) \\
&= \frac{(r+1)r(r-1)(r-2)\cdots(r-n+1)}{(n+1)!} \\
&= \binom{r+1}{n+1}
\end{aligned}$$

and hence the Pascal identity always holds. For example,

$$\begin{aligned}
\binom{1/2}{3} &= \frac{1}{16} \\
\binom{1/2}{4} &= -\frac{5}{128} \\
\binom{1/2}{3} + \binom{1/2}{4} &= \frac{3}{128} \\
\binom{3/2}{4} &= \frac{3}{128}
\end{aligned}$$

Select each expression and choose Evaluate.

Activities

1. Find a Maclaurin polynomial of $f(x) = \dfrac{1}{\sqrt{1+x^2}}$ of degree 6:

   a. By rewriting $\dfrac{1}{\sqrt{1+x^2}}$ in the form $(1+u)^n$ and calculating appropriate binomial coefficients

   b. By calculating $f(0)$ and derivatives $f'(0)$, $f''(0)$, ..., $f^{(6)}(0)$ and applying the Taylor formula

   c. By direct use of the menu item Power Series...

2. Find a Maclaurin polynomial of $f(x) = \sqrt[3]{1+x}$ of degree 3:

   a. By rewriting $\sqrt[3]{1+x}$ in the form $(1+u)^n$ and calculating appropriate binomial coefficients

   b. By calculating $f(0)$ and derivatives $f'(0)$, $f''(0)$, $f'''(0)$ and applying the Taylor formula

   c. By direct use of the menu item Power Series.

10.12 Applications of Taylor Polynomials

Simply viewing a function together with a Taylor polynomial does not give very precise information about the accuracy of the approximation. In this activity, we view the error function, which is the difference between the function and the polynomial.

Keeping a Lid on the Error

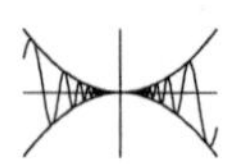

Let $f(x) = \sin x$ and let $T_n(x)$ denote the Taylor polynomial of f at 0 of degree n. Explore the accuracy of $T_n(x)$ by plotting the graph of $y = f(x) - T_n(x)$ for various values of n. Determine the smallest value of n such that $T_n(x)$ provides an error of less than .0001 over the entire interval $-2 \le x \le 2$.

Melba's Solution

Because $\sin x$ is an odd function, only odd-degree terms appear in the Taylor series expansion at 0. Here are a few Taylor polynomials of $\sin x$ at 0:

$$\begin{aligned} T_1(x) &= x \\ T_3(x) &= x - \frac{1}{6}x^3 \end{aligned}$$

$$T_5(x) = x - \frac{1}{6}x^3 + \frac{1}{120}x^5$$
$$T_7(x) = x - \frac{1}{6}x^3 + \frac{1}{120}x^5 - \frac{1}{5040}x^7$$
$$T_9(x) = x - \frac{1}{6}x^3 + \frac{1}{120}x^5 - \frac{1}{5040}x^7 + \frac{1}{362880}x^9$$

With the insertion point in the expression $\sin x$, choose Power Series (Number of Terms 2, Expand in Powers of x). Repeat for Number of Terms 4, 6, 8, and 10. Choose OK.

The following graph includes the error functions $T_7(x) - \sin x$ and $T_9(x) - \sin x$. Evidently $T_7(x)$ is too large for negative x and too small for positive x. However, it appears that $T_9(x)$ is too small for negative x and too large for positive x. The largest errors seem to appear at the endpoints, and indeed $T_7(-2) - \sin(-2) = 1.3609 \times 10^{-3}$ and $T_7(2) = -1.3609 \times 10^{-3}$. These errors exceed the required tolerance, but $T_9(-2) - \sin(-2) = -5.0016 \times 10^{-5}$ and $T_9(2) - \sin 2 = 5.0016 \times 10^{-5}$ indicate that $T_9(x)$ may provide an acceptable error.

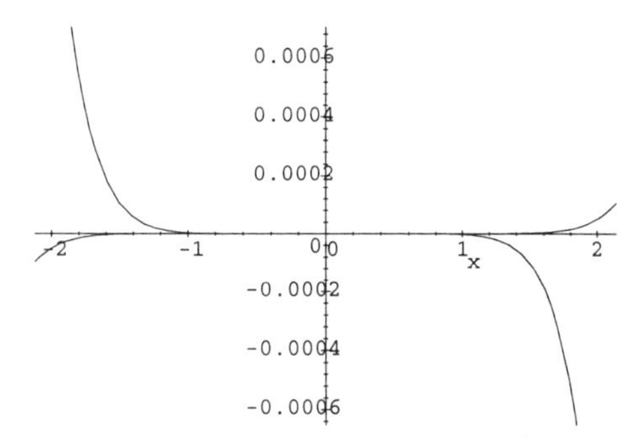

With the insertion point in the expression $\left(x - \frac{1}{6}x^3 + \frac{1}{120}x^5 - \frac{1}{5040}x^7\right) - \sin x$, choose Plot 2D + Rectangular. Select and drag to the plot the expression $x - \frac{1}{6}x^3 + \frac{1}{120}x^5 - \frac{1}{5040}x^7 + \frac{1}{362880}x^9 - \sin x$. Change the Domain Intervals to $-2.2 < x < 2.2$, and change the View Intervals to $-2.2 < x < 2.2$ and $-0.00062 < y < 0.00062$. With the insertion point in the expression $E_7(x) = \left(x - \frac{1}{6}x^3 + \frac{1}{120}x^5 - \frac{1}{5040}x^7\right) - \sin x$, choose Define + New Definition. Repeat for the expression $E_9(x) = x - \frac{1}{6}x^3 + \frac{1}{120}x^5 - \frac{1}{5040}x^7 + \frac{1}{362880}x^9 - \sin x$. With the insertion point in the expression $E_7(2)$, choose Evaluate Numerically. Repeat for the expression $E_9(2)$.

Approximations by Rational Functions

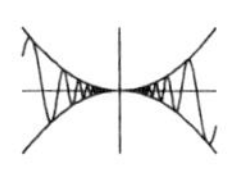

A rational function is a function $f(x) = \frac{p(x)}{q(x)}$, where $p(x)$ and $q(x)$ are polynomials. Experiment with the use of rational functions as approximations to $\tan x$.

Angela's Solution

The continued fraction expansion of $\tan x$ at 0 is given by

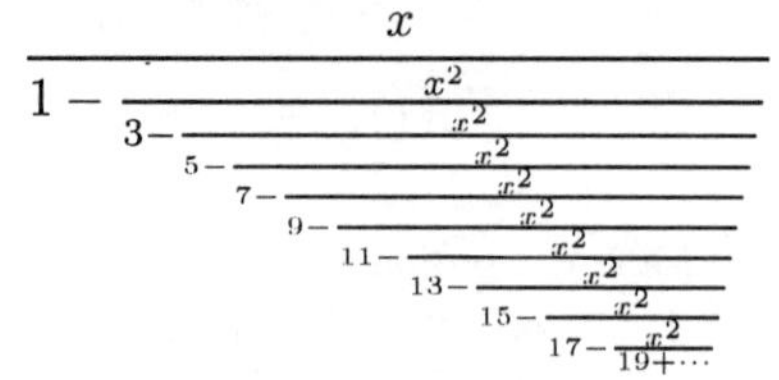

$$\cfrac{x}{1-\cfrac{x^2}{3-\cfrac{x^2}{5-\cfrac{x^2}{7-\cfrac{x^2}{9-\cfrac{x^2}{11-\cfrac{x^2}{13-\cfrac{x^2}{15-\cfrac{x^2}{17-\cfrac{x^2}{19+\cdots}}}}}}}}}}$$

where the "$+\cdots$" indicates that the pattern continues.

Choose Define + Define Maple Name (Maple Name: cfrac(x), Scientific Notebook Name: $c(x)$, File (leave blank), Maple Packages Needed: numtheory). Choose OK. With the insertion point in the expression $c(\tan x)$, choose Evaluate.

The expansion can be truncated and simplified to yield the rational function

$$-55x\frac{11904165 - 1670760x^2 + 51597x^4 - 468x^6 + x^8}{-654729075 + 310134825x^2 - 18918900x^4 + 315315x^6 - 1485x^8 + x^{10}}$$

Select and delete "$+\ \cdots$". With the insertion point in the resulting expression, choose Simplify.

The graph of this rational function is rather remarkable. It extends well beyond the interval of convergence that would be expected from the corresponding Taylor series. It is necessary to zoom out significantly to view the difference between the graph of this rational function and the graph of $y = \tan x$. Notice in particular how well the rational function locates the vertical asymptotes and x-intercepts of $y = \tan x$.

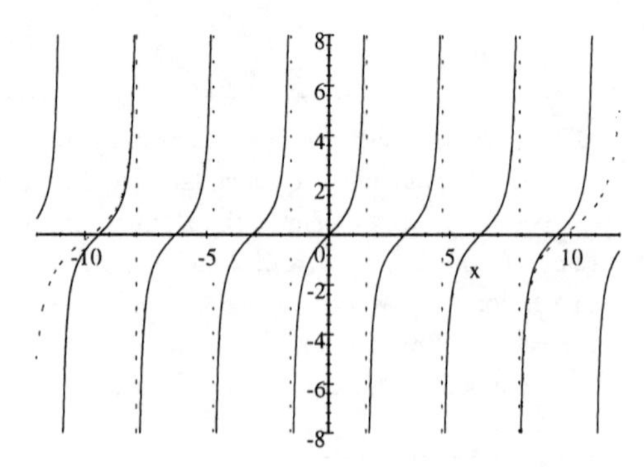

$$y = -55x\frac{11904165-1670760x^2+51597x^4-468x^6+x^8}{-654729075+310134825x^2-18918900x^4+315315x^6-1485x^8+x^{10}}$$

With the insertion point in the expression

$$-55x\frac{11904165-1670760x^2+51597x^4-468x^6+x^8}{-654729075+310134825x^2-18918900x^4+315315x^6-1485x^8+x^{10}}$$

choose Plot 2D + Rectangular. Select and drag $\tan x$ to the plot. For Item Number 1, change Line Style to Dots. Change Domain Interval to $-12 < x < 12$ and View Intervals to $-12 < x < 12$ and $-3 < y < 3$.

To study the differences more easily, consider the following graph of the error function:

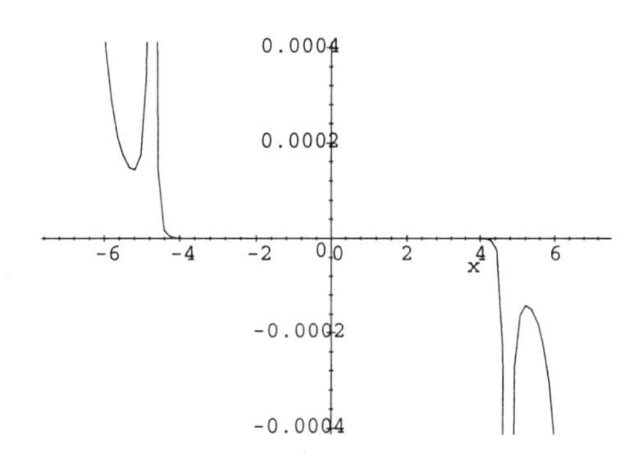

With the insertion point in the difference

$$-55x\frac{11904165-1670760x^2+51597x^4-468x^6+x^8}{-654729075+310134825x^2-18918900x^4+315315x^6-1485x^8+x^{10}} - \tan x$$

choose Plot 2D + Rectangular. Change Domain Interval to $-7 < x < 7$ and change View Intervals to $-7 < x < 7$ and $-0.00045 < y < 0.00045$.

Activities

1. Evidently Taylor polynomials of a function f at a provide very good accuracy near a, but the accuracy gets worse the farther x gets from a. Here is another strategy for approximating a function with a polynomial. Select a few sampling points a_0, a_1, ... , a_n and find a polynomial $p(x)$ such that the following equations are satisfied for each i:

$$f(a_i) = p(a_i), f'(a_i) = p'(a_i), \ldots, f^{(k)}(a_i) = p^{(k)}(a_i)$$

Find a polynomial $p(x) = c_0 + c_1x + c_2x^2 + c_3x^3 + c_4x^4 + c_5x^5$ of degree at most 5 that satisfies the following six conditions for $f(x) = \sin x$:

$$f(-2) = p(-2)$$
$$f'(-2) = p'(-2)$$
$$f(0) = p(0)$$
$$f'(0) = p'(0)$$
$$f(2) = p(2)$$

$$f'(2) = p'(2)$$

Visually compare the error function with the error $x - \frac{1}{6}x^3 + \frac{1}{120}x^5 - \sin x$ given by a Taylor polynomial of degree 5.

2. Develop your own strategy for generating polynomials that provides good approximations over intervals. What criteria might you use to decide whether one approximating polynomial is "better" than a second approximating polynomial? Include at least one concrete example in your discussion, and include graphs of error functions, with error functions corresponding to the use of Taylor polynomials.

Index

Absolute convergence, 252
Alternating series, 250
Antiderivative, 141
Arc length, 203, 221, 225
Area, 148
Area between curves, 163
Automatic substitution, 32
Average value of a function, 173

Bezier curves, 218
Binomial coefficient, 263
Binomial series, 263

Catenary, 105, 203
Center of mass, 207
Chain rule, 66
Change of variable, 179
Check, 4
Choose, 4
Circle, 19
Click, 4
Common symbols toolbar, 2
Compound interest, 139
Compute toolbar, 2
Concave downward, 118
Concave upward, 118
Conditional convergence, 252
Conic sections, 228
Constant of integration, 191
Continued fraction, 238
Continuous function, 41
Cubic spline, 73
Cylindrical shells, 168

Define
- Clear definitions, 10
- New definition, 9
- Show definitions, 10

Definite integral, 145
Derivative, 53
- Higher derivatives, 72
- Logarithmic functions, 93
- Trigonometric functions, 64

Differential equations, 141, 201
Differentials, 79
Differentiation formulas, 59
Direction field, 144
Discriminant, 228
Disks, 166
Divide and average, 86
Double-click, 4
Drag, 4
Drag and drop, 4

Elementary functions, 66
Entering mathematics
- Polynomials, 6
- Radicals, 6
- Rational expressions, 6

Entry area, 1
Error function, 269
Evaluate, 10
Evaluation, 79
Expectation, 243
Exponential functions, 87
Exponential growth, 96
Extended precision, 25

Field toolbar, 2
Fitting curves to data, 134
Floating-point arithmetic, 25
Floor function, 140
Fragments toolbar, 2
Fundamental theorem of calculus, 153

Generic function, 10
Graphing
- General guidelines, 124
- Polynomial functions, 125
- Rational functions, 126

Greatest integer function, 140

Hyperbolic functions, 101

Implicit differentiation, 68
Implicit plots, 68
Improper integrals, 197
In-place replacement, 5
Indeterminate forms, 39, 106
Insertion point, 1
Integral test, 245, 247
Integration by parts, 177
Interval of convergence, 254
Inverse function, 89
Inverse relation, 89
Inverse trigonometric functions, 99
Irreducible polynomial, 184
Iteration, 239

l'Hospital's rule, 106
Limit, 32
- Definition, 38

Linear approximation, 79
Linearization of a function, 79
Links toolbar, 3
Logarithmic differentiation, 96
Logarithmic functions, 92, 160

Maclaurin series, 259
Main window, 1
Maple functions, 238
Math name, 79
Math toolbar, 2
Mathematical models, 46, 47
- Fitting curves to data, 134

Matrix, 10
Maximum and minimum, 111, 132
Mean value theorem, 113
Menu bar, 1
Middle boxes, 148
Midpoint rule, 148, 151
Moments and centers of mass, 207
Monotonic functions, 116

Navigate toolbar, 2
Newton iteration function, 83
Newton's method, 82
Normal distribution, 195
Numerical integration, 192

Parabolic reflector, 205
Parametric equations, 215
Parametric plots, 89, 217
Partial fractions, 183

Piecewise-defined function, 32
Plot properties dialog, 8
Plotting
  Defined functions, 11
  Expressions, 6
  Multiple expressions, 8
Plotting tools, 7
Point of inflection, 118
Point plot, 18
Polar coordinates
  2D Plots, 223
  Areas and lengths, 225
  Conic sections, 231
Polynomial functions, 15
Power series, 254, 257
Pressure and force, 209

Quadratic approximation, 81
Quadratic equation, 228

Rates of change, 60
Ratio test, 247
Rational function, 183
Rational function approximations, 267
Rationalizing substitutions, 187
Related rates, 76
Riemann sum, 158
Rotation of axes, 230

Sandwich theorem, 36
Select, 4
Sequences, 235
Series, 235, 243
Shifting and scaling graphs, 21
Sigma notation for sums, 145
Simpson's rule, 193
Slope, 29
Solving equations
  Exact solutions, 12
  Numerical solutions, 13
Spline, 73
Squeeze theorem, 36
Substitution, 179
Substitution rule, 156
Surface of revolution, 205
Symbol toolbar, 2
Systems of equations, 12

Table of values, 10, 34
Tag toolbar, 2
Tangent line, 30, 53
Tangent line approximation, 79
Taylor polynomials, 266
Taylor series, 259
Toggle Math/Text button, 3
Toolbar
  Common symbols, 2
  Compute, 2
  Field, 2
  Fragment, 2
  Fragments, 2
  Links, 3
  Math, 2
  Symbol, 2
  Tag, 2
Tooltip, 1
Trigonometric substitution, 182

Volume of revolution, 165

Work, 171

Zoom tools, 7